METAL IONS IN BIOLOGICAL SYSTEMS

VOLUME 19

Antibiotics and Their Complexes

METAL IONS IN BIOLOGICAL SYSTEMS

Edited by

Helmut Sigel
Institute of Inorganic Chemistry
University of Basel
Basel, Switzerland

with the assistance of Astrid Sigel

VOLUME 19
Antibiotics and Their Complexes

MARCEL DEKKER, INC.   New York and Basel

Library of Congress Catalog Number: 79-640972

MARCEL DEKKER, INC.
270 Madison Avenue, New York, New York 10016

ISSN: 0161-5149
ISBN: 0-8247-7425-6

Current printing (last digit):
10 9 8 7 6 5 4 3 2 1

PRINTED IN THE UNITED STATES OF AMERICA

PREFACE TO THE SERIES

Recently, the importance of metal ions to the vital functions
of living organisms, hence their health and well-being, has become
increasingly apparent. As a result, the long-neglected field of
"bioinorganic chemistry" is now developing at a rapid pace. The
research centers on the synthesis, stability, formation, structure,
and reactivity of biological metal ion-containing compounds of low
and high molecular weight. The metabolism and transport of metal
ions and their complexes is being studied, and new models for complicated natural structures and processes are being devised and
tested. The focal point of our attention is the connection between
the chemistry of metal ions and their role for life.

No doubt, we are only at the brink of this process. Thus,
it is with the intention of linking coordination chemistry and
biochemistry in their widest sense that the series METAL IONS IN
BIOLOGICAL SYSTEMS reflects the growing field of "bioinorganic
chemistry." We hope, also, that this series will help to break
down the barriers between the historically separate spheres of
chemistry, biochemistry, biology, medicine, and physics, with the
expectation that a good deal of the future outstanding discoveries
will be made in the interdisciplinary areas of science.

Should this series prove a stimulus for new activities in
this fascinating "field," it would well serve its purpose and would
be a satisfactory result for the efforts spent by the authors.

Fall 1973   Helmut Sigel

PREFACE TO VOLUME 19

The story of ionophores commences with earlier studies of bioenergetics. These substances have moderate molecular weights
(roughly between 200 and 2000) and are produced by various microorganisms, from which they may be extracted. They all have the
unusual property of catalyzing the translocation of metal cations
from one side of a lipid bilayer membrane to the other; some function as carriers and others as pore-formers or ion channels. The
antibiotic action of these molecules is linked to these transport
processes through cell membranes, as they may discharge the concentrations created by the metabolism of a cell. Several of these
compounds, also termed antibiotics, are used against a variety of
human infections due to bacteria, while others are employed in the
treatment of cancers. This volume is an attempt to improve our
understanding of the coordination chemistry and action of these
biologically important compounds and to stimulate further research
in this area.

The volume opens with an historical account of the discovery
of ionophores, and then describes in the following nine chapters
the properties of tetracyclines, daunorubicin, streptonigrin,
bleomycin, valinomycin, beauvericin, and other enniatins, gramicidins, nactins, lasalocid, monensin, calcimycin, D-cycloserine,
and some related antibiotics; many derivatives are also covered.
The volume closes with a contribution on iron-containing antibiotics and a consideration of the factors governing the selective
cation-ionophore interactions.

Helmut Sigel

CONTENTS

Chapter 6

BEAUVERICIN AND THE OTHER ENNIATINS   139

Larry K. Steinrauf

Chapter 7

COMPLEXING PROPERTIES OF GRAMICIDINS   173

James F. Hinton and Roger E. Koeppe II

Chapter 8

NACTINS: THEIR COMPLEXES AND BIOLOGICAL PROPERTIES   207

Yoshiharu Nawata, Kunio Ando, and Yoichi Iitaka

CONTRIBUTORS

Numbers in parentheses indicate the pages on which the authors' contributions begin.

Kunio Ando Research Laboratories, Chugai Pharmaceutical Co. Ltd., Takada, Toshima, Tokyo, Japan (207)

K. R. K. Easwaran Molecular Biophysics Unit, Indian Institute of Science, Bangalore, India (109)

Nohad Gresh Institut de Biologie Physico-Chimique, Laboratoire de Biochimie Théorique associé au CNRS, Paris, France (335)

Joseph Hajdu[*] Department of Chemistry, Boston College, Chestnut Hill, Massachusetts (53)

James F. Hinton Department of Chemistry, University of Arkansas, Fayetteville, Arkansas (173)

Yoichi Iitaka Faculty of Pharmaceutical Sciences, University of Tokyo, Hongo, Bunkyo, Tokyo, Japan (207)

Roger E. Koeppe II Department of Chemistry, University of Arkansas, Fayetteville, Arkansas (173)

R. Bruce Martin Chemistry Department, University of Virginia, Charlottesville, Virginia (19)

Yoshiharu Nawata Research Laboratories, Chugai Pharmaceutical Co. Ltd., Takada, Toshima, Tokyo, Japan (207)

J. B. Neilands Department of Biochemistry, University of California, Berkeley, California (313)

Paul O'Brien Department of Chemistry, Queen Mary College, London, United Kingdom (295)

George R. Painter Wellcome Research Laboratories, Research Triangle Park, North Carolina (229)

---

[*]Present affiliation: Department of Chemistry, California State University, Northridge, California

Berton C. Pressman Department of Pharmacology, University of Miami
    Medical School, Miami, Florida (1, 229)

Alberte Pullman Institut de Biologie Physico-Chimique, Laboratoire
    de Biochimie Théorique associé au CNRS, Paris, France (335)

Larry K. Steinrauf Department of Biochemistry, Indiana University
    School of Medicine, Indianapolis, Indiana (139)

Yukio Sugiura Faculty of Pharmaceutical Sciences, Kyoto University,
    Kyoto, Japan (81)

Tomohisa Takita Institute of Microbial Chemistry, Tokyo, Japan
    (81)

Hamao Umezawa Institute of Microbial Chemistry, Tokyo, Japan (81)

J. R. Valenta Smith Kline and French Laboratories, Philadelphia,
    Pennsylvania (313)

CONTENTS OF OTHER VOLUMES

_____

*Out of print

---

*Out of print

Other volumes are in preparation.

Comments and suggestions with regard to contents, topics, and the
like for future volumes of the series would be greatly welcome.

METAL IONS IN BIOLOGICAL SYSTEMS

VOLUME 19

Antibiotics and Their Complexes

Chapter 1

THE DISCOVERY OF IONOPHORES:
AN HISTORICAL ACCOUNT

Berton C. Pressman
Department of Pharmacology
University of Miami Medical School
Miami, Florida

I have always held a mystic reverence for the creative act, producing
something from nothing, whether it takes place in art or in science.
These days we are rewarded with research grants for our skill in
casting our work into the strict formats designed by pedantic admin-
istrators. This type of organization presents less difficulty to
investigators whose primary skill lies in rearranging known facts
into new permutations and testing the resultant conclusions with conventional techniques. I for one have repeatedly agonized over making
the required distinctions between research plan, objectives, specific
aims, and so on. I doubt if such formulations, however eloquent,
could have persuaded a reviewing body of the likelihood of anyone
stringing together numerous diverse bits of information into the discovery of the ionophores. The chain of logic leading to discovery
does not often fit a conventional mold and I hope that recounting
the various personal experiences underlying my contribution to the
recognition of the remarkable properties of ionophores will interest
those who are intrigued by the history and methodology of science. I
have been encouraged in the value of this endeavor by the prior publication of a brief article by Webb, *The Story Behind the Story: Ionophores* [1]. The present account will be more personal and anecdotal.

The story of ionophores commences with earlier studies of bioenergetics, i.e., how mitochondria convert metabolic energy into ATP, and

how this process is inhibited by certain guanidine derivatives. As
a postdoctoral student in Stockholm (1954-1955) I had reviewed the
thesis of Gunner Hollunger on the inhibition by guanidine of mitochondrial respiration and phosphorylation and how one particular
guanidine derivative, synthalin, used briefly in the 1920s to reduce
blood sugar in diabetics, produces a similar effect at considerably
lower concentrations [2]. I deduced that the difference in potency
of these inhibiting agents was due to synthalin being more lipophylic
and more capable of interacting with mitochondrial membranes; accordingly, I synthesized a series of alkylguanidines with progressively
longer alkyl chains and found that for each carbon the alkyl chain
was lengthened, the potency for the alkylguanidines to inhibit mitochondria increased by a factor of approximately 2.5 [3]. Thus no
attribute of synthalin other than its lipophylicity is necessary to
explain its potency and furthermore the factor of 2.5 corresponds to
the increased partition of alkylguanidines from water into lipid for
each carbon the alkyl chain is lengthened.

Characteristically, guanidine derivatives produce a slow inhibition of the respiration of mitochondria actively synthesizing ATP
from ADP and inorganic phosphate (Pi) by means of energy derived from
NAD-linked substrates, e.g., malate, hydroxybutyrate, etc. The inhibition of respiration is reversed by compounds termed *uncoupling*
agents which stimulate maximal respiratory rates but prevent conservation of metabolic energy for ATP synthesis. Furthermore the respiration of succinate, which enters the respiratory chain below the level
of NAD, is not inhibited [2,3]. DBI, a biguanide related to guanidine (Fig. 1), has also been used as an oral hypoglycemic agent and
was also reported to inhibit mitochondria [4]. I found its inhibition
of mitochondrial respiration to resemble that of the guanidines in its
slow onset and reversibility by uncoupling agents; however, it differs
in extending to succinate as well as to NAD-linked substrates [3].

Other differences between the effects of phenethylbiguanide
(DBI) and octylguanidine (adopted as the prototype alkylguanidine)
emerged in the effect of uncoupling agents on relieving their inhibition of mitochondrial respiration. Inhibition by octylguanidine was

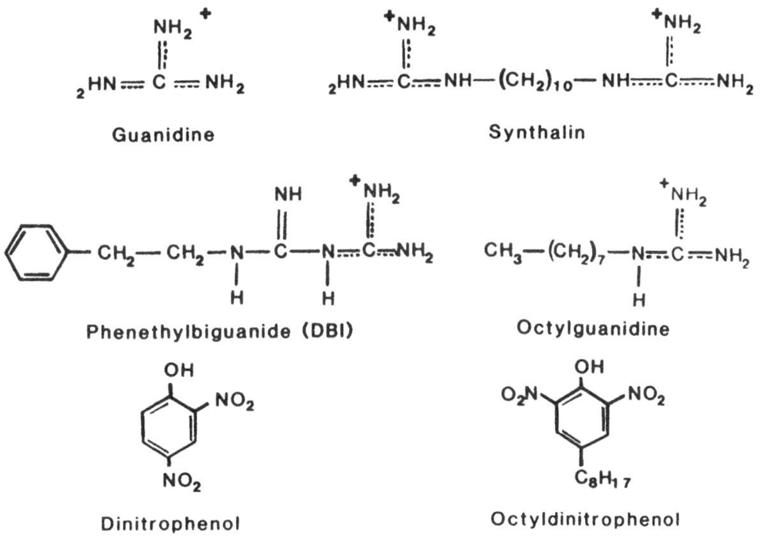

FIG. 1. Structures of mitochondrial inhibitors.

relieved 36% by the classical uncoupler 2,4-dinitrophenol (DNP) and
54% by pentachlorophenol; both uncouplers, however, fully released
DBI-inhibited respiration. A second subclass of uncoupling agents
including octyl-DNP and dicumarol appeared virtually incapable of
releasing octylguanidine inhibition but was very effective in releasing DBI inhibition [5]. In retrospect, I would presume the
difference to be somehow related to the second uncoupler subclass
being more lipophylic than the first.

On the basis of these data, it appeared that we had an experimental method for dividing all uncoupling agents into two distinct
subclasses and we set about surveying the ones available to us,
particularly those of diverse structure. I was particularly intrigued by the report of McMurray and Begg that the cyclic depsipeptide valinomycin acted as an uncoupling agent at concentrations
as low as 10^{-8} M, which meant it was the most powerful one known at
the time [6]. Surprisingly, by the criteria of our test system, it
appeared distinct from the other two subclasses in that it was only
slightly active in reversing the inhibition of mitochondria by octylguanidine and produced a slowly increasing partial release of DBI

inhibition not previously observed. The existence of a third uncoupler subclass was confirmed by finding another substance which exhibited the same behavior as valinomycin, namely, the linear polypeptide
gramicidin [5].

By permitting the chemical bond energy of ATP to be dissipated
as heat, uncoupling agents stimulate the hydrolysis of ATP by enabling
the mitochrondrial ATP synthetase to run backward. This stimulation
of the latent ATPase of mitochondria provided a simple system for
studying the interaction of uncoupling agents and mitochondria. The
conventional way of measuring ATPase at the time was assaying for the
release of Pi colorometrically. Having just arrived at the biophysics
department of the Johnson Foundation, I was reluctant to reveal my
dependence on mundane wet chemical procedures. A more sophisticated
technique prevalent in the department was to track ATPase, which at
pH 7 liberates approximately 0.7 H^+/ATP hydrolysed, continuously with
a recording pH meter [7]:

$$\text{Adenosine-O-}\overset{\overset{O}{\|}}{\underset{\underset{O^-}{|}}{P}}\text{-O-}\overset{\overset{O}{\|}}{\underset{\underset{O^-}{|}}{P}}\text{-O-}\overset{\overset{O}{\|}}{\underset{\underset{O^-}{|}}{P}}\text{-O}^- + H_2O \longrightarrow \text{adenosine-O-}\overset{\overset{O}{\|}}{\underset{\underset{O^-}{|}}{P}}\text{-O-}\overset{\overset{O}{\|}}{\underset{\underset{O^-}{|}}{P}}\text{-O}^- + \overset{\overset{O}{\|}}{\underset{\underset{OH}{/}\underset{O^-}{\backslash}}{P}}\text{-O}^{-0.7} + 0.7H^+$$

My initial encounter with this instrumental procedure was frustrating.
While DNP-treated mitochondria produced a drop in pH which was linear
with time, valinomycin produced an acute drop in pH which in less than
a minute halted abruptly, reversed, and then resumed at a slower rate
(Fig. 2). The pattern was entirely reproducible, which dispelled my
initial suspicion that the apparatus was malfunctioning and indirect
biophysical techniques were not to be trusted. Because of its pattern
this tracing has been referred to as the *drainpipe* phenomenon [5].

The origin of the drainpipe became clearer when samples were
periodically removed for more conventional Pi analysis [8]. From
the known relationship of H^+ produced/ATP hydrolysed and the experimentally determined buffer capacity of the medium, the true effect of
ATP hydrolysis on pH could be determined. When this function was
subtracted from the experimental curve, a derived curve resulted
which I concluded represented a migration of protons across the mito-

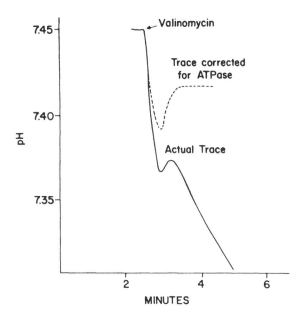

FIG. 2. Drainpipe phenomenon. (Solid tracing) Actual pH tracing of
valinomycin-induced acid production by mitochondria in presence of
ATP. (Dashed tracing) Calculated residual acid production when
experimental trace is corrected for acid production attributable to
hydrolysis of ATP. (Reproduced by permission from Ref. 5.)

chondrial membrane supported by ATP hydrolysis. This was confirmed
by adding valinomycin to mitochondria energized by substrate oxidation
which per se produces negligible metabolic acid. In this case the
tracing indicated the true course of energy-dependent proton trans-
migration without distortion. The conclusion that the pH changes
sensed by the electrode were due to a proton translocation rather
than metabolic acid production was verified by the addition of a
detergent to destroy the phase bounding mitochondrial membrane; this
quickly restored the pH of the system to its initial value. The
energy-dependent nature of the valinomycin-induced proton movement
was established by the fact that it could be prevented by various
metabolic inhibitors or classical uncoupling agents, such as DNP [5].

    The energy-dependent proton translocation induced by valinomycin
resembled that which was known to attend the energy-dependent uptake

of Ca^{2+} [9] or Mn^{2+} [10] by mitochondria; however, in this case a
simple stoichiometry obtains between cation uptake and proton ejec-
tion. In our system thousands of protons were ejected for each
equivalent of valinomycin added. The protons ejected had to be
electrostatically balanced by some sort of charge movement. I
therefore concluded that valinomycin was inducing the mitochondria
to take up some cation in the reaction system in exchange for the
protons which were expelled.

My Ph.D. thesis work concerned the K^+ requirement of metabolizing
mitochondria, particularly as potentiated by fatty acids [11,12]. This
suggested to me that K^+ was a likely candidate for the counterion to
the ejected protons. At this point the problem was assigned to a post
doctoral student, Cyril Moore, whose arrival was delayed several months
unexpectedly in order for him to complete his Ph.D. thesis. I awaited
anxiously. When he finally did start work in our laboratory his first
experiment was to add valinomycin to a K^+-free mitochondrial system.
Most significantly, nothing happened, confirming that K^+ was indeed
the ion that valinomycin was inducing mitochondria to accumulate [13]
exactly the way others had observed Ca^{2+} [9] and Mn^{2+} [10] to be
accumulated spontaneously. I realized at that moment that the phenomenon we had uncovered would undoubtedly determine the course of
my work for many years to come.

Parenthetically, we now know that guanidine derivatives do not
affect conversion of energy to ATP but rather inhibit mitochondrial
electron transport directly. Their involvement in energy transduction
suggested by release of their inhibition of respiration by uncouplers
is due to the fact that they are spontaneously accumulated by energized mitochondria exactly like Ca^{2+} or K^+ in the presence of valinomycin and that their inhibition of electron transport is related to
their intramitochondrial concentration [14]. Many interesting aspects
of the interaction of guanidine derivatives with mitochondria remain
to be resolved; however, the transfer of interest demanded by the
ionophores preempted continuation of the guanidine studies.

We were fortunate that at just this time Beckman began marketing
a glass cation-selective electrode which could sense changes in K^+

activity potentiometrically the same way the more established glass
pH electrode senses changes in H^+ activity. We applied it to our
reaction medium and, just as we had hoped, the valinomycin-induced
increase in H^+ activity sensed by the pH electrode was echoed by a
decrease in K^+ activity as this cation left the medium to enter the
mitochondria [13,15]. No such changes were sensed by another Beckman
glass electrode highly selective for Na^+ over K^+. Thus the evidence
appeared solid that valinomycin induced the energy-dependent mitochondrial uptake of K^+ in exchange for H^+!

Our conclusion about the effects of valinomycin on mitochondria
has withstood the test of time. Our initial hypothesis for the molecular basis of its action was somewhat less durable. We obtained a
series of valinomycin (Fig. 3) analogs from Prof. M. M. Shemyakin and

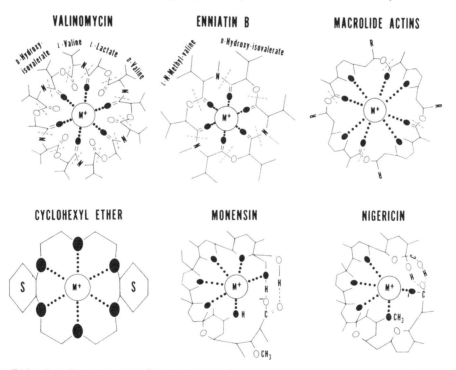

FIG. 3. Structures of complexes of representative ionophores. Those
oxygen atoms liganding the cations as indicated by x-ray crystallography are filled in black. (Reproduced by permission from *Ann. N.Y.
Acad. Sci., 264,* 373, 1978.)

found that inversion of a single D-valine destroyed 87% of mitochondrial activity while inversion of a single L-lactate destroyed
98% of activity. Enlargement or contraction of the ring by a single
repeating four-residue sequence also destroyed activity [15]. These
structure-activity relationships and the fact that the response of
mitochondria to as little as 10^{-11} M valinomycin could be detected
led us to believe that it was interacting as a hormone with a pur-
poseful mitochondrial receptor site which controlled monovalent
transport [13,15].

An alternative hypothesis suggested by Chappell was that
valinomycin introduced pores in otherwise unpermeable membranes
through which ions could flow in response to electrical and concentration gradients [15,16]. Thus according to molecular models the
presumed cyclic octadepsipeptide valinomycin would provide an opening
in a membrane which would permit passage of K^+ (hydrated ionic radius
2.32 Å) but not Na^+ (hydrated ionic radius 2.76 Å) while the larger
cyclic decapeptide gramicidin would permit the larger hydrated Na^+ to
pass through as well as Li^+ (3.40 Å). Ironically, his suggestion was
based on errors of the structure assumed for the two agents. Originally, valinomycin was thought to consist of only D- and L-hydroxy-
isovalerate and D- and L-valine (amidomycin) [17]. Brockmann correctly identified the residues as D- and L-valine, L-lactate, and D-
hydroxyisovalarate but thought that the sequence repeated only twice
as an octadepsipeptide [18]. By actual synthesis Shemyakin proved
that the biologically active depsipeptide repeated its sequence thrice
to form a dodecadepsipeptide [19] with an annulus larger than that
assumed by Chappell; the analagous octadepsipeptide (two repeats) as
well as the hexadecadepsipeptide (four repeats) were inactive in
Shemyakin's biological system (inhibition of growth of *Mycobacterium
phlei* [20] as they were in our mitochondrial test system. Thus
Chappell's mechanism for molecular discrimination of K^+ and Na^+
could not be correct.

The structure assumed for gramicidin, a cyclic decapeptide,
was even more in error since the transport active gramicidin, the
Dubos gramicidin [21], actually consists of three pairs of analagous
pentadecapeptides designated as A, B, and C [22]. By coincidence, a

cyclic decapeptide discovered by the Russians [23] was designated as
gramicidin C (for Soviet in Cyrillic) and usually, but not always,
transliterated as either S or C in the Roman alphabet. Thus the
designation gramicidin C is ambiguous, as is the term gramicidin,
which is often used without further qualification. Both terms can
refer to either the linear, transport-inducing Dubos pentadecapeptides or the cyclic Russian decapeptide which alters membrane permeability by its nonspecific surface activity. Chappell had assigned
the structure of the Russian gramicidin to the Dubos gramicidin we
had used and hence his conclusions again were invalid. We now know
that the Dubos gramicidins alter membrane permeability by inserting
themselves into membranes as dimeric helices which form cation conducting channels [24]. Since at any position within the channel the
cation is surrounded by an ionophorelike configuration of oxygen
atoms with which they form ion-dipole interactions, we have referred
to such agents as quasi-ionophores [25]. In our laboratory we also
found that gramicidin S does not induce mitochondrial ion uptake and
the octadepsipeptide enniatins, only half the size of valinomycin
(18 ring atoms vs. 36 ring atoms), induce mitochondrial ion uptake
with an ion selectivity similar to that of valinomycin. This shows
that the ring size per se could not be determinative for the ion
selectivity of ionophore-induced transport.

Information from a number of sources combined to clarify the
mechanisms of action of ionophores. Rudin and Mueller [26] and
Harris in our laboratory found that ionophores produce alterations
in the permeability of artificial lipid bilayers. This work was
independently corroborated by Lev and Buzhinsky [27]. It was obvious
that if membranes of pure lipids could respond to ionophores no biological receptors were obliged to mediate the effects. Shemyakin and
his group synthesized the enantiomer of enniatin B which they appropriately called enantio-enniatin [28]. This proved fully active in
their microbiological growth assay as did the samples they provided
us in our mitochondria systems. This also strongly mitigated against
a biological receptor mediating the effects of ionophores on ion
transport and suggested that the remarkable properties of ionophores

resides intrinsically in some molecular attribute for which precise
conformation is all determinative. Thus the conformation of the
enniatin enantiomers would be identical mirror images of each other
but would be different in diastereoisomers in which the configura-
tions of some, but not all, of the residues were inverted. The third
clue was the detection by means of vapor pressure osmometry of complex
formation between the macrolide nactins and alkali cation salts [29].

The Lardy group had been studying the strange inhibitory effects
of nigericin toward mitochondria for a number of years. After valinomycin and the ion-specific electrodes became established tools for
studying mitochondria, they found that this agent caused a reversal
of the valinomycin-induced cation transmigrations [30]. In keeping
with classical biological concepts they postulated that nigericin forms
a K^+ complex which competes with K^+ at a hypothetical mitochondrial
K^+ pump and the inhibited pump permits K^+ to be released from mitochondria down its concentration gradient [31]. In our laboratory we
had observed that nigericin induced a K^+-for-H^+ exchange in erythrocytes; hence, like valinomycin, its transport properties were inherent
to the molecule and did not depend on any mitochondrial component
[32]. One other property of nigericin which was known was the unusual
high solubility of its Na^+ salt in organic solvents such as hydro-
carbons [33].

We were now in a position to fit together the various pieces
of the jigsaw puzzle. The lipid solubility of the alkali salts of
nigericin suggested that it might transport ions across lipid barriers,
such as the interior of biological membranes as lipid soluble complexes. In its carboxyl protonated form it could likewise transport
protons. Accordingly, the key to the biological mechanism of action
of the ionophores could be their complexing properties.

The system we chose to study the complexation behavior of
ionophores was a water:toluene/butanol mixture our laboratory was
using to extract phosphomolybdic acid from water in order to separate
Pi from adenine nucleotides [34]. As expected, alkali cations distributed strongly in favor of water. Ionophores, being lipophylic,
partitioned into the organic phase. When the phases were shaken,

alkali cations migrated from the water into the organic phase signi-
fying the formation of a lipophylic complex. This was conveniently
monitored as the migration of radioactive $^{86}Rb^+$ from the water phase
into the butanol/toluene. With nigericin and a related compound
dianemycin, complexation increased sharply as the aqueous pH rose
indicating competition between Rb^+ and H^+ for association with these
ionophores.

In marked contrast the complexation of Rb^+ by valinomycin was
relatively insensitive to pH. On the other hand the ability of
valinomycin to complex Rb^+ and transfer it to the organic phase was
aided by the inclusion of a lipid-compatible anion such as laurate.
This was consistent with the Rb^+ complex of the unionized valinomycin acquiring the charge of the complexed cation. In order for
an appreciable amount of Rb^+ to transmigrate, it must be accompanied
with a lipid-compatible anion in order to prevent prohibitively large
charges from building up between the phases. In the case of nigericin,
the complex formed is an electrically neutral zwitterion in which the
charge of the cation is offset by the negative charge on the carboxyl
group. Complexation in this case is essentially a migration of cation
into the organic phase in exchange for a proton which countermigrates
into the aqueous phase. This process is promoted by a low proton concentration, i.e., high pH, in the aqueous phase but is not affected
by the availability of a lipophylic anion [35].

The ionophore-induced migration of cations in two phase systems
was used to ascertain the ion selectivity of various ionophores.
Initial qualitative determinations based on competition of test
cations with $^{86}Rb^+$ as the radioactive indicator ion gave the series
$K^+ > Rb^+ > Na^+ > Cs^+$ (pH 10.4) for nigericin. For valinomycin, in
the presence of 0.01 M laurate, we obtained the series $Rb^+ > K^+ >$
$Cs^+ \gg Na^+$ which correlated with its affects on the permeability of
the artificial lipid bilayer [26,27] as well as its interaction with
mitochondria [35]. However, the mitochondrial test system is not a
precise indicator of true ionophore complexation affinity because it
superimposes the intrinsic ion selectivity of mitochondrial metabolism
on that of the ionophore.

These studies established that there are two groups of substances that can alter the permeability of biological membranes by
carrying ions across lipid barriers as lipid-soluble complexes: the
valinomycin group, which transports cations as lipid-soluble, charged
complexes, and the nigericin group, which contains a charged carboxyl
group and transports cations as lipid-soluble, electrically neutral
zwitterions. Accordingly, we proposed that such substances be classified generically as *ionophores* (Greek *ion bearer*) or *ionophorous*
agents (not *ionophoretic* or *ionophoric* since *-ic* is a Latin suffix)
[35]. Subsequently, we termed the valinomycin subclass *neutral*
ionophores and the nigericin subclass *carboxylic* ionophores [25].

The difference in charge of complexes of the two subclasses of
ionophores is entirely sufficient to account for their very different
behavior toward mitochondria. The neutral ionophores catalyze the
net transport of charged ions, i.e., *ionophoretic* transport. The
thermodynamic force driving this mode of transport is a function of
both concentration gradient and membrane potential; expressed as
free energy this would be

$$1320 \cdot \left(\log \frac{[M^+]_i}{[M^+]_o} + \frac{E_m}{59 \text{ mV}} \right)$$

where $[M^+]_i$ is the concentration of M^+ inside the mitochondria; $[M^+]_o$
the concentration of M^+ outside the mitochondria; and E_m the mitochondrial potential membrane in millivolts.

The carboxylic ionophores transport cations as electrically
neutral complexes, hence the membrane potential would not affect
transport. These ionophores catalyze what has been classically called
the *exchange diffusion* mode of transport whereby one ion species is
exchange for N^+, e.g., the expulsion of H^+. The prevailing K^+
would be

$$1320 \cdot \log \left(\frac{[M^+]_i}{[M^+]_o} \cdot \frac{[N^+]_o}{[N^+]_i} \right)$$

In the case of mitochondria, species M^+ could be K^+ entering in
exchange for N^+, i.e., the expulsion of H^+. The prevailing K^+

gradient in a representative mitochondrial system would be about
20:1 out to in. Valinomycin would then promote the uptake of K^+
provided that the membrane potential is sufficiently large (>77 mV,
inside negative) to overcome the concentration gradient. On the
other hand, nigericin would promote an egress of K^+ down its concentration gradient in exchange for H^+ unless the H^+ gradient were
greater in the opposite direction. If these relationships are
operative, in the presence of a membrane potential > 77 mV valinomycin, which catalyzes electrophoretic cation transport, would promote the uptake of K^+ while under the same circumstances nigericin,
which catalyzes exchange diffusion transport, would promote the
release of K^+ in exchange for the uptake of H^+. The electrically
silent nature of carboxylic ionophore-induced transport is confirmed
by its inability to induce electrometrically detectable changes in
the artificial lipid bilayer [35].

More detailed studies of the complexation behavior of ionophores with the two-phase system indicated that at alkaline pH (10.4)
the saturation of carboxylic ionophores by cations follows an ideal
Langmuir saturation isotherm, i.e., the plot of the reciprocal of
complex formation is linear with the reciprocal of the concentration
of the complexing cation in the water phase (Figure 4). The slope
of the line indicates the complexation affinity while the intercept
at infinite cation concentration indicates a complex with a 1:1
ratio of cation to ionophore [36]. Ideal saturation isotherms could
also be obtained with neutral ionophores if laurate was replaced by
SCN^- as a lipid-compatible anion because the fatty acid tended to
form troublesome emulsions. In this fashion the extent of variability
of ion selectivity of ionophores became evident. While nigericin prefers K^+ over Na^+ by a factor of 50, valinomycin prefers K^+ over Na^+
by a factor of 10^4. Lasalocid (X-537) yields the lyotropic series
($Cs^+ > Rb^+ > K^+ > Na^+ > Li^+$) while monensin prefers Na^+ over K^+ by
a factor of 10 [36]. Although a great many new ionophores have been
discovered since these early studies, very few have been characterized
quantitatively for ion selectivity by this technique.

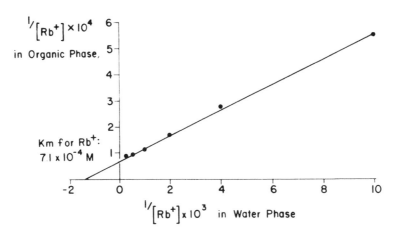

FIG. 4. Ideal linear double-reciprocal Langmuir saturation isotherm of aqueous $^{86}Rb^+$ saturating nigericin in 70% toluene:30% n-butanol. From the slope the K_D expressed as $([nigericin^-]_{org} \cdot [Rb^+]_{H_2O})/([nigericin^- \cdot Rb^+]_{org})$ can be estimated while the intercept with the ordinate indicates a 1:1 nigericin-Rb^+ complex ([nigericin] = ~ 1.5×10^{-4} M). (Reproduced by permission from Ref. 36.)

The true molecular basis of the ion selectivity of ionophores became clearer when the Zürich group established the three-dimensional structure of the K^+ complex of nonactin by x-ray crystallography. Its 40-atom backbone folded in three dimensions like the "seam of a tennis ball" deploying four ether oxygens and four carboxyl oxygens at the apices of a cube [37]. A desolvated K^+ fits snugly in the cavity, while Na^+ fits more loosely accounting for the preference of nonactin for complexing K^+ over Na^+. The three-dimensional structures of valinomycin (by both x-ray crystallography [38] and NMR [39]) and monensin [40] soon followed. The general pattern of the ionophore complexes now became evident. Short-chain ionophores like enniatin and the synthetic crown ether dicyclohexyl-18-crown-6 (both 18-ring atoms) [41] can just manage to wrap themselves about a cation in a more or less planar configuration. Larger ionophores such as the nactins and valinomycin convolute themselves in space to define a three-dimensional cavity which forms an even more critical size-dependent fit with cations. Carboxylic ionophores, which in general are not held in rings by true covalent bonds, fold into rings because of the chirality of their heterocyclic rings and other steric factors.

Head-to-tail hydrogen bonding also helps to stabilize the rings.
All of these structures deploy oxygen atoms strategically about a
defined cavity so that they can abut a cation within. The actual
complexation bond is the induced ion-dipole interaction between the
critical oxygen atoms and the desolvated cations. Another way of
viewing it is that the ionophore replaces the shell of solvent
molecules which is normally held about the cations by polar forces.
In carboxylic ionophores an ionic bond with the carboxyl may
(nigericin) or may not (monensin) supplement the ion-dipole bond
energy. Complexation then consists of a concerted stripping away
of solvent from cations as the ionophore engulphs it. While such
complexes are stable within bulk organic solvents and the low-
polarity interior of biological membranes, they are easily broken
at the aqueous interface of membranes by interaction with water,
which resolvates the complexed cation and also stabilizes the broken
hydrogen bonds of the extended ionophore conformation. Some conformational fluctuations are required to permit cations to enter and leave
some of the more highly defined three-dimensional cavities such as
that of valinomycin.

  We have now completed the account of how the unique biological
properties of the ionophores ultimately led to a firm understanding
of the general molecular basis by which these substances transport
cations across membranes as lipophylic complexes. The detailed
dynamics of ionophore complexation continue to attract the attention
of many investigators who have investigated the process by a variety
of sophisticated biophysical techniques, such as x-ray crystallography, NMR, circular dichroism measurements, and computer modeling
[42]. Such studies may serve as a prototype for analogous studies
of the dynamics of operation of more complicated biological molecules
such as enzymes. Our original hope that ionophores would turn out to
be models for biological carriers has not been borne out but they have
provided us with clues as to the architecture of conducting channels
in biological membranes and the mechanism of biological ion selectivity.

  Ionophores have a number of practical applications, such as
ion-selective electrodes and feed supplements for farm animals [44].

They are widely employed as tools for the rational perturbation of biological systems, particularly the divalent ionophore A-23187 [43] because Ca^{2+} is such an important ion in biological control. Their powerful pharmacological effects arising from their ability to perturb transcellular ion gradients which control metabolism may turn out to have important therapeutic applications [25,44]. New ionophores are continuing to be discovered from natural sources and synthesized in the laboratory. It is intriguing to speculate what new applications will be found for these fascinating compounds.

ACKNOWLEDGMENT

I should like to acknowledge my gratitude for the contributions of my collaborators C. Moore, E. J. Harris, M. Hoffer, R. Cockrell, and D. H. Haynes; to P. Mueller, L. K. Steinrauf, Yu. A. Ovchinnikov, M. M. Shemyakin, W. Simon, and J. D. Dunitz, who made important information available to us prior to publication; and for the long-term support and encouragement of B. Chance, Director of the Johnson Foundation, University of Pennsylvania, where the early work on the ionophores and their mechanism of action took place.

ABBREVIATIONS

ADP	adenosine-5'-diphosphate
ATP	adenosine-5'-triphosphate
ATPase	adenosine triphosphatase
DBI	phenethylbiguanide
DNP	2,4-dinitrophenol
NAD	nicotinamide adenine dinucleotide
Pi	inorganic phosphate

REFERENCES

1. J. Webb, *J. Chem. Educ.*, *56*, 502 (1979).

2. G. Hollunger, *Acta Pharmacol. Toxicol. II*, Suppl. 1 (1955).

3. B. C. Pressman, *J. Biol. Chem.*, *238*, 401 (1962).

4. A. B. Falcone, R. L. Mao, and E. Shrago, *J. Biol. Chem.*, *237*, 904 (1962).

5. B. C. Pressman, in *Energy-linked Functions of Mitochondria* (B. Chance, ed.), Academic Press, New York, 1963, pp. 181-203.

6. W. McMurray and R. W. Begg, *Arch. Biochem. Biophys.*, *84*, 546 (1959).

7. M. Nishimura, *Biochim. Biophys. Acta*, *59*, 183 (1962).

8. O. H. Lowry and J. A. Lopez, *J. Biol. Chem.*, *162*, 421 (1946).

9. N. E. Saris, *Soc. Sci. Fennica, Commentationes Phys. Math.*, *28*, 11 (1963).

10. J. B. Chappell, M. Cohn, and G. D. Greville, in *Energy-linked Functions of Mitochondria* (B. Chance, ed.), Academic Press, New York, 1963, pp. 219-231.

11. B. C. Pressman and H. A. Lardy, *J. Biol. Chem.*, *197*, 547 (1952).

12. B. C. Pressman and H. A. Lardy, *Biochem. Biophys. Acta*, *21*, 458 (1956).

13. C. Moore and B. C. Pressman, *Biochem. Biophys. Res. Commun.*, *15*, 562 (1964).

14. J. Z. Fields, Ph.D. thesis: Guanidines, Sedatives and Mitochondria: Inhibitory Mechanisms, Univ. of Miami, 1976.

15. B. C. Pressman, *Proc. Natl. Acad. Sci. USA*, *53*, 1076 (1965).

16. J. B. Chappell. Presented at Symposium on Structure-Function Relationships in Mitochondria, Sixth International Congress of Biochemistry, New York, July 1964.

17. L. C. Vining and W. A. Tabor, *Can. J. of Microbiol.*, *3*, 953 (1957).

18. H. Brockmann and H. Geeren, *Ann. Chem.*, *603*, 217 (1957).

19. M. M. Shemyakin, E. I. Vinogradova, M. Yu Feigund, and N. A. Aldanova, *Tetrahedron Lett.*, 1921 (1963).

20. M. M. Shemyakin, Yu. A. Ovchinnikov, V. T. Ivanov, A. A. Kiryushkin, G. L. Zhdanov, and I. D. Ryabova, *Experientia*, *19*, 566 (1963).

21. R. J. Dubos and R. D. Hotchkiss, *J. Exp. Med.*, *73*, 629 (1941).

22. R. Sarges and B. Witkop, *Biochemistry*, *4*, 2491 (1965).

23. R. L. Synge, *Biochem. J.*, *39*, 363 (1945).

24. D. W. Urry, M. C. Goodall, J. D. Glickson, and D. F. Mayers, *Proc. Natl. Acad. Sci. USA*, *63*, 1907 (1971).

25. B. C. Pressman, *Ann. Rev. Biochem.*, *45*, 501 (1976).

26. P. Mueller and D. O. Rudin, *Biochem. Biophys. Res. Commun.*, *26*, 398 (1967).

27. A. A. Lev and E. P. Buzhinsky, *Cytologia USSR*, *9*, 102 (1967).

28. M. M. Shemyakin, Yu. A. Ovchinnikov, V. T. Ivanov, and A. V. Evstratov, *Nature*, *213*, 412 (1967).

29. L. A. R. Pioda, H. A. Wacter, R. E. Dohner, and W. Simon, *Helv. Chim. Acta*, *50*, 1373 (1967).

30. S. N. Graven, S. Estrada-O, and H. A. Lardy, *Proc. Natl. Acad. Sci. USA*, *53*, 1076 (1965).

31. H. A. Lardy, S. N. Graven, and S. Estrada-O, *Fed. Proc.*, *26*, 1355 (1967).

32. E. J. Harris and B. C. Pressman, *Nature*, *216*, 918 (1967).

33. R. L. Harned, P. H. Harter, C. J. Corum, and K. L. Jones, *Antibiot. Chemiother.*, *1*, 592 (1951).

34. L. Ernster, R. Zetterstrom, and O. Lindberg, *Acta Chem. Scand.*, *4*, 942 (1950).

35. B. C. Pressman, E. J. Harris, W. S. Jagger, and J. H. Johnson, *Proc. Natl. Acad. Sci. USA*, *58*, 1949 (1967).

36. B. C. Pressman, *Fed. Proc.*, *27*, 1282 (1968).

37. B. T. Kilbourn, J. D. Dunitz, L. A. R. Pioda, and W. Simon, *J. Mol. Biol.*, *30*, 559 (1967).

38. M. Pinkerton, L. K. Steinrauf, and P. Dawkins, *Biochem. Biophys. Res. Commun.*, *35*, 512 (1969).

39. V. T. Ivanov, I. A. Laine, N. D. Abdullaev, L. B. Senyavina, E. M. Popov, Yu. A. Ovchinnikov, and M. M. Shemyakin, *Biochem. Biophys. Res. Commun.*, *34*, 803 (1969).

40. L. K. Steinrauf, M. Pinkerton, and J. W. Chamberlin, *Biochem. Biophys. Res. Commun.*, *33*, 29 (1970).

41. C. J. Pedersen, *J. Am. Chem. Soc.*, *89*, 7017 (1967).

42. G. W. Painter, R. Pollack, and B. C. Pressman, *Biochemistry*, *231*, 5613 (1982).

43. P. W. Reed and H. A. Lardy, *J. Biol. Chem.*, *247*, 6970 (1972).

44. B. C. Pressman and M. Fahim, *Ann. Rev. Pharmacol. Toxicol.*, *22*, 465 (1982).

Chapter 2

TETRACYCLINES AND DAUNORUBICIN

R. Bruce Martin
Chemistry Department
University of Virginia
Charlottesville, Virginia

1. INTRODUCTION

The antibiotics tetracycline and daunorubicin share neither a common chemistry nor common biological functions. However, certain formal similarities in both proton and metal ion binding justify their inclusion in a single chapter. Both antibiotics possess three acidic groups, and in both cases there is controversy concerning associations of proton-binding sites with experimentally determined pK_a values. This chapter summarizes the proton-binding information in each case and

proposes comprehensive assignments of all proton-binding sites. The
proton-binding sites will then be applied to the metal ion-binding
properties of each, antibiotic.

2. TETRACYCLINES

The tetracycline family of broad-spectrum antibiotics has been used
for over 30 years for treatment of a variety of infections due to
bacteria (especially *Rickettsia*) and other agents [1]. They are
thought to act by inhibiting protein synthesis at the ribosome where
a Mg^{2+} complex may be important [2-4]. The drugs are transported as
Ca^{2+} complexes through the plasma, where there is little unbound
drug [5]. Their ability to bind Ca^{2+} probably accounts for inhibition of bone development [6]. Because tetracyclines may cause dis-
coloration of growing teeth, they are not administered to children
or during pregnancy [7]. The not always consistent reports of the
roles of metal ions on the biological effects of tetracyclines have
been summarized [8]. As already noted, this chapter has a different
emphasis.

2.1. Structures

Figure 1 depicts the structure of tetracycline as it occurs in neutral
aqueous solutions. The molecule is a zwitterion with a positive
charge located on the protonated dimethylammonium group. The negative
charge is delocalized over the O1-C1-C2-C3-O3 system in the A ring.
The C1-C2 and C2-C3 bonds exhibit nearly equal lengths as do the C1-O1
and C3-O3 bonds, according to crystal structure determinations [9].
The delocalization extends into the amide group which is nearly planar
and takes up a cis or trans position with respect to the O1 to O3
system. Tetracycline betaine adopts a nearly identical structure [10].
     The tetracycline molecule is not flat as implied by representa-
tions such as that in Fig. 1. If a plane is arranged to pass as well

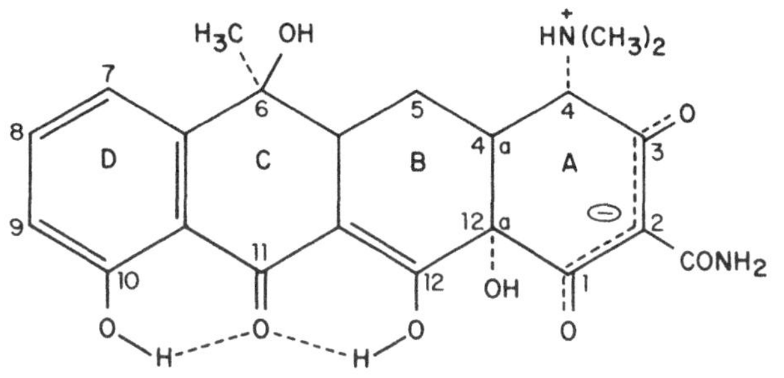

FIG. 1. Zwitterion form of tetracycline as it occurs in neutral to
acidic aqueous solutions. The negative charge is delocalized over
the A ring O1-C1-C2-C3-O3 system, which lies above the plane of the
page that nearly accommodates the B, C, and D rings. In oxytetracycline an -OH group substitutes for the hydrogen below the plane of
the page at C5 in the B ring. In chlorotetracycline a -Cl substitutes for the hydrogen at C7 in the unsaturated D ring.

as possible through the B, C, and D rings, C1, C2, C3, and the amide
carbon will lie above this plane [9]. The dimethylammonium group
projects away from the rest of the molecule in what may be termed the
extended conformation. This conformation is important in metal ion
binding because the juxtaposition of O1, O12, and O11 provides a tridentate chelate site with donors from substituents on the A, B, and
C rings.

When crystallized from nonaqueous media, a neutral tautomer of
the zwitterion shown in Fig. 1 occurs with the proton at N4 transferred to O3 [9,11]. The proton transfer breaks up the delocalization along O1 to O3 and yields an -OH group at C3, a double bond
between C3 and C2, a single bond between C2 and C1, and a ketonic
O1 oxygen. This neutral tautomer is also not flat but adopts an
alternative conformation due to a rotation about the C4a-C12a bond
of the A-ring groups that lie above the paper plane. In the alternative conformation the amino group swings in the direction of O6 in
what may be termed the folded conformation.

The neutral tautomer in the folded conformation is said to be
stabilized by a strong intramolecular hydrogen bond by the proton
at O3 to the amide oxygen [9,11]. The zwitterion in the extended

conformation is stabilized by a weaker hydrogen bond from the protonated ammonium group to O3. Though the two kinds of hydrogen bonds
occur exclusively in each of the two conformations, they both occur
only among A-ring substituents, and it has not been explained why
they have such decisive roles in stabilizing the conformations. The
only intramolecular hydrogen bond between substituents on two different rings that is different in the two conformations is a weak HO12a-O1
hydrogen bond that might help stabilize the neutral tautomer. The
energy difference between the two conformations is probably low. The
conformation with the lower dipole moment may simply be favored in
lower dielectric media.

The solution equilibrium between the extended and folded conformations has been followed by circular dichroism [12]. Assuming
complete conversion to the folded form in 100% ethanol, the two conformations occur to equal extents in about 85% by volume ethanol.
It is commonly assumed that in the extended conformation adopted on
the water side the molecule exists exclusively as a zwitterion and
in the folded conformation on the alcohol side solely as the neutral
tautomer [12]. Though reasonable, this exact one-to-one correspondence does not seem proved. The driving force for the change from
zwitterion to neutral tautomer results from the combined effect of
an increase in ammonium group acidity and O3 basicity as water is
replaced by another solvent. At some stage in the replacement N4,
which is about 10^3 times more basic than O3 in water, becomes less
basic (Sec. 2.2).

Even in solutions of low water content where the neutral tautomer predominates, metal ions appear to complex preferentially with
the zwitterion form (Sec. 2.3). Stronger metal ion binding occurs
with anionic tetracyclines; which anionic species predominates concerns us next.

2.2. Proton Binding

2.2.1. *Macroscopic Constants*

In the pH region accessible to potentiometric titrations, tetracycline displays three acidic groups with acidity constants pK_1 =
3.3, pK_2 = 7.5, and pK_3 = 9.3 in aqueous solutions at about 0.1 M
ionic strength and 25-37°C. Table 1 lists literature values of
tetracycline acidity constants. The values refer to the usual mixed
constants in terms of hydrogen ion activity. Several of the references in Table 1 give pK_a values for other tetracyclines. Chlorotetracycline (aureomycin) and oxotetracycline (terramycin) are only
slightly more acidic than tetracycline. The structures are described
in Fig. 1. Due to the strong similarity in structures and acidities,

TABLE 1

Tetracycline Macroscopic pK_a Values

pK_1	pK_2	pK_3	Solvent, temp., I^a	Ref.
3.35	7.82	9.57	Water, 20°, 0.01	13
3.30	7.68	9.69	Water, 25°, 0.00[b]	14
3.33	7.75	9.61	Water, 25°, 0.01	15
3.33	7.70	9.50	Water, 25°, 0.01	16
3.42	7.52	9.07	Water, 25°, 0.10	17
—	7.50	9.18	Water, 24°, 0.14	18
3.69	7.63	9.24	Water, 30°, 0.01	19
3.52	7.27	8.80	Water, 30°, 0.10	20
3.28	7.39	9.17	Water, 37°, 0.15[c]	8
4.4	7.8	9.4	50% CH_3OH, 26-30°, 0.01[d]	21
4.4	8.1	9.8	50% DMSO, 25°, 0.01	22

[a]I designates ionic strength molarity. A value of I = 0.01 M indicates that the ionic strength is that due to reagents and is uncon-
trolled.
[b]Thermodynamic values by extrapolation to zero ionic strength.
[c]To obtain mixed constants 0.12 log units have been added to concentration constants.
[d]Values not given but calculated from microconstants.

conclusions drawn for tetracycline itself are also applicable to
other family members.

There has been a great deal of discussion and disagreement
concerning the assignment of specific groups on tetracycline to the
macroscopic acidity constants K_1, K_2, and K_3. The original assignment of K_1 to the O3 site [14] is generally accepted. (However,
crystals of α-6-deoxyoxytetracycline hydrohalides prepared from 90%
ethanol show protonation of the amide oxygen with a strong hydrogen
bond to either O3 or O1 [23].) The same authors also argued that K_2
should be assigned to the dimethylammonium group and K_3 to the BCD
ring system, not venturing to distinguish HO10 from HO12 [14]. Part
of their argument depends on a misapprehension of the effects of
added Ca^{2+}. They also stated in a footnote that recent (but undis-
closed) work placed the K_2 and K_3 assignments in serious jeopardy
[14]. Based on comparisons with model compounds, the assignments
were later reversed, identifying K_2 mainly with the BCD ring system
and K_3 with the dimethylammonium group [15]. Two studies using chemical shifts in NMR spectroscopy proposed that K_2 and K_3 contained
contributions from both deprotonations [21,22]. These two studies
are discussed in detail near the end of Sec. 2.2.2.

Assignment of groups contributing to the tetracycline K_2 and
K_3 deprotonations may be made by use of model compounds, either
simple models or those that actually block a specific deprotonation.
For zwitterionic tetracycline the dimethylammonium deprotonation
might be expected to be similar to that of dimethylglycine where the
charge separation is comparable. For dimethylglycine, $pK_a = 9.9$
[24], only slightly greater than the pK_3 values and much greater
than the pK_2 values listed for tetracycline in Table 1. This result
suggests that the K_3 deprotonation in tetracycline is primarily from
the dimethylammonium group. As a model for the deprotonation from
O12 we consider benzoylacetone for which $pK_a = 8.7$ [25]. To apply
to tetracycline the acid-strengthening effect of the hydroxy group
at C12a must be incorporated. For ethanolamine $pK_a = 9.5$ reduced
by the hydroxy group from 10.6 in ethylamine [26]. We apply this
difference of 1.1 log units to benzoylacetone to estimate $pK_a =$

8.7 - 1.1 = 7.6 for HO12 in tetracycline. This estimated value is
within the range of pK_2 values in Table 1 and much less than the pK_3
values. This result suggests that the K_2 deprotonation in tetracycline occurs primarily from the hydroxy group at C12. The conclusions from these two comparisons in aqueous solutions consistently
identify K_2 with ionization at HO12 and K_3 with deprotonation from
the dimethylammonium group.

Most investigators have chosen not to discriminate between the
HO12 and HO10 ionizations in the tetracycline BCD ring system. We
have suggested above that the tetracycline K_2 deprotonation is from
HO12. This view is supported by considering as a model for the HO10
ionization 2-hydroxyacetophenone for which pK_a = 10.3 [27]. This
model is imperfect, however, and it is anticipated that the additional
groups in tetracycline, especially an ionized O12, will increase the
basicity of HO10 to pK_a >12, where it is inaccessible to potentiometric titrations. The ionization of HO12 and not HO10 is supported
by ^{13}C NMR of tetracycline where there is a large chemical shift of
C12 and an insignificant shift of C10 on passing through the basic
pH region [22]. Results of metal ion-binding investigations also
indicate that it is the O12 site that undergoes both ionization and
metalation. The di K^+ salt of oxytetracycline shows, according to
the crystal structure, K^+ coordination to O12 among other sites and
HO10 remains protonated [28]. At pH 8.6 selective broadening in
^{1}H NMR indicates that Gd^{3+} chelates to O12-O11 [29]. Thus three
entirely different lines of evidence, model compound comparisons,
^{13}C NMR of the free ligand, and metal ion-binding sites, all indicate that the only ionization taking place in the BCD ring system
is from O12.

On the basis of potentiometric and spectrophotometric titrations of tetracycline methiodide in 50% methanol a pK_a = 10.67 was
claimed to demonstrate a fourth acidic group in tetracycline [21].
Due to the low dielectric media this value should be about 1 log
unit less when transferred to water. There is, however, no evidence
for such a fourth acid group on tetracycline in water. Furthermore,
we find by potentiometric titration of tetracycline no sign of a

fourth deprotonation in 50% dimethylsulfoxide (DMSO) [30]. In addition, there are no changes in NMR chemical shifts of tetracycline by
[13]C in 50% DMSO [22] or of tetracycline and its methiodide by ^{1}H in
50% methanol [21] that indicate a fourth acidic group. The last
negative result appears in the same paper in which the additional
acidic group was proposed. Probably the claimed fourth acidic group
is an artifact of the solvent system or salt employed and is due to
a reaction of methoxide or hydroxide with tetracycline betaine, which
is expected to be unstable in basic solutions [31].

2.2.2. *Microscopic Constants*

A more quantitative analysis of proton-binding sites and their solvent
dependence is gained by considering the microconstants for each acidic
group, the values of which depend on the charges elsewhere in the molecule. A microconstant scheme for tetracycline deprotonations appears
in Fig. 2. The letter A denotes the A ring system in Fig. 1, the
symbol BH the B-ring system with a proton at O12, and the symbol NH[+]
the dimethylammonium group at C4. With three different acidic groups

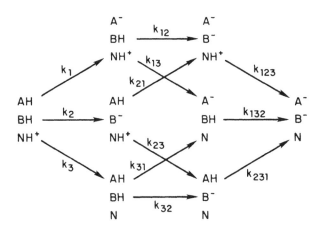

FIG. 2. Tetracycline deprotonation scheme. Subscript 1 refers to
the A-ring ionization, subscript 2 to ionization in the B ring at
O12, and subscript 3 to deprotonation at the dimethylammonium group
substituted at C4. The last group undergoing deprotonation is indicated by the last subscript on the acidity microconstants.

the scheme is formally analogous to that solved for the amino acid tyrosine [32]. Each of the 12 acidity microconstants in Fig. 2 is of the form $k = (H^+)[B]/[HB]$, where B represents the basic and HB the acidic species in each acid-base reaction. Some relations between the macroconstants determined by titration, K_1, K_2, and K_3, and the microconstants defined in Fig. 2 are given by

$$K_1 = k_1 + k_2 + k_3 \tag{1}$$
$$K_1 K_2 K_3 = k_1 k_{12} k_{123} = k_1 k_{13} k_{132} \tag{2}$$
$$K_3^{-1} = k_{123}^{-1} + k_{132}^{-1} + k_{231}^{-1} \tag{3}$$

Many other relations among the macro- and microconstants have been derived [32].

For tetracycline there is general agreement that the first deprotonation occurs from the A ring by the k_1 pathway in Fig. 2. With this conclusion Eqs. (1)-(3) reduce to

$$K_1 = k_1$$
$$K_2 K_3 = k_{12} k_{123} = k_{13} k_{132} \tag{4}$$
$$K_3^{-1} = k_{123}^{-1} + k_{132}^{-1} \tag{5}$$

and we obtain

$$K_2 = k_{12} + k_{13} \tag{6}$$

The uncertainty has concerned which branch predominates from the (A^-, BH, NH^+) microform: via the BCD ring system ionization, k_{12}, or via the dimethylammonium group deprotonation, k_{13}. To resolve this uncertainty it is necessary to supply at least one item of additional information beyond the macroconstants that are obtained from potentiometric titration on tetracycline. The additional information may be gained either by pK_a determinations on model compounds that block a deprotonation or by measuring the extent of one of two or more competing deprotonations by an appropriate spectroscopic method. Results from both approaches are now described.

In order to estimate the k_{12}/k_{13} ratio in tetracycline with model compounds, it is desirable to block one or the other of the two competing sites. A single study performed under the same conditions in water reports that for tetracycline $pK_2 = 7.75$ and for

tetracycline methiodide pk_{12} = 7.80 [15]. Since the k_{13} pathway is
not possible in the trimethylammonium derivative, only the k_{12} deprotonation from HO12 can occur. From Eq. (6) we calculate that pk_{13} =
8.7(1) and hence k_{12}/k_{13} ≃ 8. Unfortunately the calculation depends
on the difference of two large numbers. For 0.02 log unit variation
the resulting ratio of k_{12}/k_{13} varies from 6 to 13. A more accurate
ratio would be available if pk_{13} could be estimated from 12-methoxy-
tetracycline. This result provides strong semiquantitative support
for the conclusions made in Sec. 2.2.1: in water the tetracycline
HO12 hydroxy group is significantly more acidic than the dimethylammonium group.

The preceding results of pk_{12} = 7.80 and pk_{13} = 8.7(1) in
tetracycline may be expanded to include a calculation of pk_{123} and
pk_{132}. In the same study [15] for tetracycline pK_3 = 9.61 and with
pK_2 = 7.75 we calculate from Eqs. (4) and (5) that pk_{123} = 9.56 and
pk_{132} = 8.6(5). From these values the effect of a protonated dimethylammonium group on the HO12 deprotonation and vice versa may
be expressed as $\Delta pk_{23} = pk_{132} - pk_{12} = pk_{123} - pk_{13}$ = 0.8(5). This
value for the interaction between the two sites may be compared with
that for the two amino groups in ornithine (2,5-diaminopentanoate)
where there is also a difference of five bonds between the basic
sites. From a NMR microconstant study [33] the ratio of the two
zwitterionic microforms of ornithine may be calculated to be 7.6 and
when combined with ornithine pK_2 = 8.86 and pK_3 = 10.65 [34] yields
Δpk_{23} = 0.80. This value, determined from results at about 0.1 M
ionic strength, would be greater at lower ionic strength and hence
almost identical to the 0.85 value calculated above for tetracycline
from results at 0.01 M ionic strength. This comparison provides convincing support that the pK_2 and pK_3 deprotonations have been correctly
assigned to HO12 and the dimethylammonium group, respectively. Further
support appears in the next paragraph. If instead of HO12 the ionization occurred from HO10, the interaction with the dimethylammonium
group would be much less, Δpk ~ 0.4 as for daunorubicin in Sec. 3.2.
(More discussion of the use of Δpk values for evaluating interactions
between acidic sites appears in Refs. 32 and 34.)

Already 5 of the 12 microconstants in the scheme of Fig. 2 have been evaluated and the values of pk_1, pk_{12}, pk_{123}, pk_{13}, and pk_{132} are collected in Table 2. With two more items of information it is possible to estimate values for the remaining 7 microconstants. First, for desdimethylaminotetracycline, where only a pair of hydrogens occur at C4 in the structure of Fig. 1, potentiometric titration gives pK_1 = 5.97 and pK_2 = 8.56 [15]. The pK_2 difference with tetracycline from the same study of 8.56 - 7.75 = 0.81 is a measure of the N4-O12 interaction and agrees with the value of Δpk_{23} = 0.85 deduced independently above and with Δpk_{23} = 0.80 from use of ornithine as a model. The pK_1 difference with tetracycline of 5.97 - 3.33 = 2.64 is an approximate measure of the N4-O3 interaction. Due to the nearness of the two groups the difference requires a small adjustment. Deamination of glycine to give acetic acid results in a pK_1 difference of 2.35 - 4.76 = -2.4 log units [26]. In glycine itself the effect of

TABLE 2

Macroscopic and Microscopic Acidity Constants[a]

pK	Tetracycline[b]	Daunorubicin[c]
pK_1	3.33	8.61
pK_2	7.75	10.0
pK_3	9.61	13.7
pk_1	3.33	9.87
pk_2	7.1	9.87
pk_3	6.4	8.66
pk_{12}	7.80	12.95
pk_{13}	8.7(1)	9.11
pk_{21}	4.0	12.95
pk_{23}	7.3	9.11
pk_{31}	5.6	10.32
pk_{32}	8.0	10.32
pk_{123}	9.56	9.56
pk_{132}	8.6(5)	13.4
pk_{231}	6.3	13.4

[a]Microscopic constants defined in scheme of Fig. 2 and text.
[b]In water at 25°C and 0.01 M ionic strength.
[c]In water at 20°C and 0.15 M ionic strength. See Sec. 3.2. From Ref. 54.

a protonated vs. basic amino group on the carboxylic acid deprotonation and vice versa is 2.1 log units [35]. The small difference of
2.4 - 2.1 = 0.3 log units is subtracted from 2.64 to yield Δpk_{13} =
2.3 for the N4-O3 interaction in tetracycline. From the microconstant scheme in Fig. 2, from among several relations [32] we have
Δpk_{13} = 2.3 = pk_{13} - pk_3 = pk_{31} - pk_1 = pk_{123} - pk_{23}, from which
pk_3 = 6.4, pk_{31} = 5.6, and pk_{23} = 7.3. Once pk_3 is estimated, the
value of pk_{31} = 5.6 also follows from the cyclic system result,
$k_3 k_{31} k_{132}$ = $K_1 K_2 K_3$. The identical value of pk_{23} = 7.3 is also available from the N4-O12 interaction from which Δpk_{23} = pk_{123} - pk_{13} =
0.85 = pk_{23} - pk_3. Other relations of this type among the microconstants have been given [32].

In order to estimate the remaining four microconstants a second
item needs to be assumed. One way to approach the missing information
is to estimate the magnitude of the O12-O3 interaction, Δpk_{12}. Since
the through space distance on a molecular model from O12 is slightly
longer to O3 than to N4, we set Δpk_{12} = 0.7 to a slightly smaller
value than Δpk_{23} = 0.85. The relatively long O12-O3 distance makes
the assigned value of Δpk_{12} = 0.7 a good approximation. From one of
several relations [32] we write 0.7 = Δpk_{12} = pk_{12} - pk_2 from which
pk_2 = 7.1. The remaining three microconstants k_{21}, k_{32}, and k_{231} may
now be found as components of cyclic systems. Values for all 12
tetracycline microconstants in water are collected in Table 2.

Two NMR studies attempted to assign tetracycline microconstants
from analysis of chemical shifts. In the first study conducted in 50%
by weight methanol, the dimethylammonium group protons were used to
monitor that group's deprotonation and aromatic D-ring resonances the
BCD ring system ionization [21]. In their analysis the authors make
several invalid assumptions leading to unreasonable conclusions. Even
after the addition of 1 eq of base to the cation they calculate that
19% of the A ring is still protonated despite sharp breaks in the
chemical shift vs. pH curves. They also calculate that after the
addition of the second equivalent of base that the A ring is 33%
protonated, 14% more than after 1 eq. These exceedingly unlikely
conclusions are the result of use of invalid assumptions, only

approximate equations, and use of only a few data points in the
analysis. The microconstant values appearing in this paper are not
reliable and should not be used. More than half the microconstant
values presented cannot be evaluated from the data given by the
methods described. The magnitudes of interactions between pairs of
sites expressed as Δ pk values are invalid.

The second NMR study employed several ^{13}C chemical shifts to
monitor tetracycline deprotonations in 50% dimethyl sulfoxide (DMSO)
[22]. These authors also used incorrect equations in attempting to
evaluate microconstant values. They also appear to have combined
acidity macroconstants determined in an H_2O-DMSO mixture with chemical shifts obtained in a D_2O-DMSO mixture. The effect of D_2O is to
increase appreciably basicities [36] and results from the two systems
should not be combined. The authors conclude that after addition of
1.5 eq base that the dimethylammonium group is slightly more deprotonated than the BCD ring system, but that after addition of 2.0 and
2.5 eq base the BCD ring system is more than twice as deprotonated
as the dimethylammonium group. This unreasonable conclusion is due
to use of incorrect equations and invalid assumptions. The microconstant values appearing in this study are not reliable and should
not be used.

The proper way to evaluate microconstants from NMR chemical
shift (or absorption intensity) vs. pH data is by nonlinear least
squares fitting of the observable such as chemical shift in terms of
pH and five parameters to be determined: the limiting chemical shifts
of the most acidic and basis forms, and three independent microconstants. Points should be included over the entire course of the
titration curve for the two competing deprotonations. The macroconstants may then be calculated from the parameterized microconstants. As an independent check, the macroconstants should be determined from potentiometric titration under the same conditions of
temperature, solvent (e.g., D_2O instead of H_2O) and ionic strength,
which should be controlled at some fixed value. Such an NMR chemical
shift or absorption intensity analysis has never been performed on
tetracycline or related compounds.

The two deficient NMR chemical shift studies furnish the only data for resolving the pK_2 and pK_3 tetracycline deprotonations by spectroscopic means. Leaving aside the faulty interpretations in these papers, what may we learn from the data presented in figures? In the ^{13}C study the authors used only the C8 chemical shifts in their analysis [22]. Yet inspection of all their plots indicates that virtually every ^{13}C shift from all parts of tetracycline exhibits the same midpoint pH = 8.8 in 50% DMSO. It is possible in ^{13}C NMR that the effects of both deprotonations are delocalized efficiently throughout the entire molecule. More plausible, however, is the likelihood that both deprotonations occur to about the same extent and $k_{12} \simeq k_{13}$ in 50% DMSO. Examination of ^1H chemical shifts for the dimethyl group and H9 in 50% DMSO also indicates that the dimethylammonium and BCD ring deprotonations occur to comparable extents in this solvent system [30]. In the ^1H NMR study performed in 50% methanol, inspection of their Fig. 2 reveals that the midpoint of the tetracycline dimethyl group shift occurs at somewhat lower pH than the midpoint of the D-ring shifts [21]. It is difficult to quantify the results but perhaps $k_{13} \simeq 3k_{12}$ in 50% methanol.

The consequences of performing experiments in nonaqueous solvent mixtures has not been explicitly considered. Since it involves creation of two new charges, the tetracycline BCD ring ionization becomes less acidic in lower dielectric constant media. The dimethylammonium group deprotonation involves only proton transfer from an ammonium to water and typically such neutral reactions become more acidic in lower dielectric constant media [26]. For 50% by weight methanol at 25°C the dielectric constant is reduced to 55 from 78.5 in water. For acetic acid in 50% by weight methanol pK_a = 5.7, 0.9 units greater than pK_a = 4.76 in water [26]. A similar difference is also evident from the pK_1 ionization in Table 1 between water and the solvent mixtures. For ammonia in 50% methanol pK_a = 8.7, 0.5 units less than pK_a = 9.2 in water [26]. The combined effect of 1.4 log units for the two kinds of deprotonation is to multiply the k_{12}/k_{13} ratio determined in 50% methanol by 25 to provide an estimate for k_{12}/k_{13} in water. The ratio k_{13}/k_{12} = 3 suggested above for 50%

methanol yields an estimated $k_{12}/k_{13} \simeq 8$ in water. This value is
identical to that determined above in evaluating the microconstant
values that appear in Table 2. Though the arguments leading to this
last ratio are necessarily approximate, the discussion does serve to
emphasize the significant role played by dielectric media in even
reversing the favored deprotonation. We conclude from the NMR chemical shift studies that for tetracycline in water the HO12 ionization
is strongly favored over the dimethylammonium group deprotonation.
As the dielectric constant of the medium is lowered, the two deprotonations become more competitive attaining near equality in 50%
DMSO. The dimethylammonium group deprotonation is favored in 50%
methanol. As the water content of a nonaqueous solvent mixture
decreases further, we anticipate the outright predominance of the
dimethylammonium group deprotonation. These conclusions resolve a
great deal of previously conflicting results and are important for
metal ion binding, which has also been studied in a variety of solvent media. The conclusions are also in agreement with the structural information reviewed in Sec. 2.1.

2.2.3. Summary

The preceding section furnishes a comprehensive and consistent analysis of the relative acidities of the three acidic groups in tetracycline. The conclusions reached should also be applicable to its
related molecules including oxytetracycline and chlorotetracycline.
All the microconstants in the scheme of Fig. 2 have been evaluated
for tetracycline and the values are listed in Table 2. The first
deprotonation occurs from the A ring to give the zwitterion shown
in Fig. 1. Model compound comparisons indicate that in water the
predominant deprotonation from the zwitterion occurs at HO12 and
this process may be associated with pK_2 in Table 1. The resulting
molecule with a single net negative charge contains a positively
charged dimethylammonium group and one negative charge delocalized
on the A ring and the second negative charge at O12. The third
deprotonation in water to give the dianion occurs from the dimethyl-

ammonium group and is identified with pK_3 in Table 1. Ionization of HO10 does not take place at pH <12.

Thus in water the successive deprotonations occur across the top of the scheme in Fig. 2. Starting in acid water solution and raising the pH, the most basic site moves from the A ring to O12 and finally to N4 at high pH.

As the dielectric constant of the medium decreases, the HO12 ionization becomes less favored relative to the dimethylammonium group deprotonation. In 50% methanol the latter process becomes dominant and in more anhydrous media the tetracycline anion consists predominantly of electrically neutral amine and HO12 groups with the negative charge delocalized about the O1 to O3 axis of the A ring.

2.3. Metal Ion Binding

Before considering any experimental results on metal ion binding to tetracyclines, it is worthwhile to examine our expectations for metal ion-binding sites based on our newly acquired knowledge of proton-binding sites in the preceding section. The experimental results are then economically interpreted in the framework of these expectations.

We assume that significant metal ion binding occurs only at basic sites, i.e., sites that have been deprotonated. Two factors in addition to basicity influence the extent of metal ion binding at a basic site. Binding will be strengthened if five- or six-membered ring chelation can occur with a basic site as an anchor for chelation. Many of the metal ions to be considered prefer oxygen to nitrogen donors, and this preference will sometimes need to be considered. On the basis of these criteria tetracycline cations should not sig-nificantly bind metal ions.

By far the major tetracycline species of net zero charge in water is the (A^-, BH, NH^+) zwitterion (Fig. 2). It is favored over the species with all neutral groups (AH, BH, N) by $k_1/k_3 = 10^3$. With $pk_1 = 3.3$ the A ring is a weaker base than formate ($pK_a = 3.7$) and only weak metal binding is anticipated in unidentate modes at O3, O1,

or O (amide). There is, however, a metal ion-binding chelation
possibility between O3 and O (amide). Due to the pronounced delocalization in the deprotonated A-ring system O1-O (amide) chelation may
also occur. O1-HO12-(HO12a) bidentate and tridentate chelation also
appear favorable. Unless deprotonated, an amide nitrogen is not a
metal ion-binding site [37].

Mixed solvent systems of low water content or anhydrous solvents
favor the species (AH, BH, N) with all groups uncharged. The most
basic center is now the deprotonated amino nitrogen N4, which may
bind metal ions in a unidentate mode. Chelation between N4 and HO12a
appears attractive sterically. In 5-hydroxytetracycline N4-HO5 chelation is also possible. It is possible, however, that the neutral
molecule with all groups uncharged adopts in its metal ion complexes
the alternative folded conformation produced by rotation about the
C4a-C12a bond (Sec. 2.1). In this folded conformation the N4-HO12a
chelate is lost.

The predominant anionic species in water is (A^-, B^-, NH^+) which
is favored over (A^-, BH, N) by about $k_{12}/k_{13} = 8$. In the former spe-
cies both the A ring and O12 are ionized with O12 the most basic site.
Three bidentate chelation modes are possible: O12-O1, O12-HO12a, and
O12-O11. Two tridentate chelation modes involving O12 are attractive:
O12-HO12a-O1 and O12-O11-O1.

Mixed solvent systems with low water content may favor the less
polar anion species (A^-, BH, N), where the most basic site is now the
amino nitrogen N4. In addition to the N4-HO12a chelation mentioned
above for the neutral species in mixed solvents, N4-O3 chelation
becomes a new possibility. In 5-hydroxytetracycline N4-HO5 chelation
may again occur. In the folded conformation (Sec. 2.1) the N4-HO12a
and N4-O3 opportunities are replaced by O1-HO12a.

In dianionic tetracycline all three groups are in their basic
form with the amino N4 group about eight times more basic than O12
in water. Since O12 appears to offer more favorable chelating characteristics, metal ions which do not bear strong preferences for
nitrogen donors may well be found bound to O12 and chelated to a

neighboring oxygen in ways already described. A proton remains at
HO10 and no metal ion binding is anticipated at this site.

Two crystal structure determinations of metal ion complexes
give binding modes in accord with the above expectations. In a
zwitterionic oxytetracycline complex of $HgCl_2$ the shortest Hg(II)
bonds are to the two undisplaced Cl^-, and only weak Hg-O bonds
chelate to 01-0 (amide) of the A ring [28]. There are no interactions with still protonated HN4 or any substituents on the B, C,
or D rings.

A di K^+ salt of dianionic oxytetracycline was crystallized
from a strongly basic solution [28]. There are three K^+ sites and
each of the K^+ interacts with at least two different dianionic oxytetracycline ligands. Though deprotonated the amino nitrogen remains
uncoordinated to any K^+, which favors oxygen donors. For each K^+ the
shortest K^+-oxytetracycline bond is always made to 012, the most
basic oxygen. One K^+ chelates to 012-011 of one ligand, to 01-0
(amide) of a second, and coordinates to 03 of a third. The last two
K^+ each form a tridentate chelate with the same three oxygen atoms
from a pair of oxytetracycline dianions. One K^+ chelates 012-HO12a-01
and the second 012-011-01 from each of two ligands.

Table 3 compiles literature values of stability constants for
tetracycline in water. Several of the investigations include values
for other tetracyclines. The values for oxotetracycline and chlorotetracycline are closely similar to those of tetracycline. Though
the experimental protocol is not always described in the references,
all the stability constants in Table 3 appear to refer to coordination of the metal ion with the LH or (A^-, B^-, NH^+) species. Since
the most basic unprotonated site in this species is 012, all the
stability constants refer to chelation at 012-011 with N4 still
protonated. This conclusion is supported strongly by results for
tetracycline betaine (pK_2 = 7.50) where N4 bears three methyl groups.
The stability constant logarithms in parentheses for Mg^{2+} (3.8),
Ca^{2+} (4.0), Mn^{2+} (4.0), Ni^{2+} (5.9), Cu^{2+} (8.0), and Zn^{2+} (4.7) at
25°C and 0.1 M ionic strength [38] are similar to or greater than

TABLE 3

Tetracycline Stability Constants

Metal	$\log K_1$	Temp., I[a]	Ref.
Mg^{2+}	3.8	25°, 0.10	17
	3.5	24°, 0.15	18
	3.6	37°, 0.15	39
Ca^{2+}	3.0	24°, 0.15	18
	$(3.9)^b$	37°, 0.15	8
Mn^{2+}	4.4	20°, 0.01	13
	4.1	25°, 0.10	17
Fe^{2+}	5.3	20°, 0.01	13
Co^{2+}	5.4	20°, 0.01	13
	5.5	25°, 0.10	17
Ni^{2+}	6.0	20°, 0.01	13
	5.8	25°, 0.10	17
	6.1	30°, 0.01	19
Cu^{2+}	7.8	20°, 0.01	13
	7.8	25°, 0.01	16
	7.6	25°, 0.10	17
	7.5	30°, 0.01	19
	7.6	30°, 0.10	20
Zn^{2+}	4.9	20°, 0.01	13
	4.6	25°, 0.10	17
	5.1	30°, 0.01	19
Cd^{2+}	3.3	25°, 0.10	17
	3.2	30°, 0.10	20
Pb^{2+}	3.8	30°, 0.10	20
Al^{3+}	7.4	20°, 0.01	13
Fe^{3+}	9.9	20°, 0.01	13

[a]See same note in Table 1.
[b]Inferred from product of first and second stability constants.

the values in Table 3. Thus even the most nitrogen-seeking metal ion in Table 3, Cu^{2+}, prefers chelation at O12-O11 to coordination at N4 in the (A⁻, BH, N) tautomer. Selective broadening in ^{1}H NMR spectroscopy by Gd^{3+} indicates that it also chelates at O12-O11 in tetracycline anion [29]. Literature values for stability constants ascribed to tetracycline dianion L or (A$^-$, B$^-$, N) probably represent instead metal chelation at O12-O11 and hydroxo and hydroxo-bridged complexes in many cases [38].

Ca^{2+} [8] and Mg^{2+} [39] complexes of tetracycline, oxytetracycline, and doxycycline (6-deoxy-5β-hydroxytetracycline) have been characterized by a sophisticated computer analysis of potentiometric titration data. This kind of study is potentially very valuable, but there are some puzzling conclusions that are difficult to understand. Of the six combinations for two metal ions and three ligands a MLH complex was found for half: Mg^{2+} and tetracycline and both metal ions with doxycycline. A ML complex was also found for three of the six combinations: Ca^{2+} and doxycycline and both metal ions with oxytetracycline. There does not seem to be any pattern for these variations among the three closely related antibiotics. The treatment has been extended to mixed Ca^{2+} and Mg^{2+} complexes of the antibiotics [5]. These three papers should be used with caution.

From the stability constants in Table 3, under conditions existing in blood plasma, pH 7.4, 1.3 mM Ca^{2+}, and 0.9 mM Mg^{2+}, most of the tetracycline antibiotics occur as alkaline earth metal ion complexes. Inside cells at pH 6.6 where the Ca^{2+} concentration drops to less than 0.1 μM and the Mg^{2+} rises to about 2 mM, most of the tetracyclines occur as Mg^{2+} complexes. In both cases there is about 10% unbound ligand. Metal ions influence absorption of tetracyclines from the gut.

Ca^{2+} and Mg^{2+} binding to tetracyclines has been studied in a variety of solvent systems by circular dichroism (CD) among other methods. Tetracycline and desdimethyltetracycline exhibit similar CD spectra with both metal ions indicating that N4 is not involved in binding [40]. This result confirms our expectations that the metal ions chelate O12-O11. As indicated in Sec. 2.1, tetracycline appears to occur in most aqueous solutions in the extended conformation and in solvents of very low water content in the folded conformation [12]. Addition of Mg^{2+} [41] or Ca^{2+} [42] to a DMSO solution of tetracycline in the folded conformation transforms the CD spectra in the direction of that displayed by the extended conformation in aqueous solutions. Thus both metal ions apparently shift the conformational equilibrium toward the extended conformation of the zwitterion form by binding preferentially to this species. This conclusion is also in accord with our expectations for O12 has undergone ionization in the zwitterion but not in the neutral tautomer of the folded conformation. Whether some differences in CD between Ca^{2+} and Mg^{2+} tetracycline solutions is one of degree [42] or of kind [43] still seems unresolved. NMR results indicate that with Mg^{2+} there is slow exchange of tetracycline between different environments [41,42]. The differences between properties of Mg^{2+} and Ca^{2+} have been contrasted [44].

A series of papers report effects of both paramagnetic and diamagnetic metal ions on chemical shifts and broadening of neutral tetracycline NMR peaks in 100% dimethylsulfoxide (DMSO). In this solvent the (AH, BH, N) ligand species predominates over the zwitterion (A^-, BH, NH^+), the favored tautomer in water. For most of the metal ions the most affected ^1H NMR peaks are H4, H4a, H(amide), and H(dimethyl) [45]. As expected these sites are all on the A ring where either O3 or N4 is deprotonated, depending on the tautomer. In ^{13}C NMR addition of the same metal ions perturbs the A-ring carbon peaks including C12a [46]. The marked broadening reported for C12a and C4 by Nd^{3+} may be due to the N4-HO12a chelate described above for the favored (AH, BH, N) tautomer in 100% DMSO or to interactions at O3 and O1 in the zwitterion (A^-, BH, NH^+) tautomer. Lanthanides prefer oxygen donors [44] and it is likely that the Nd^{3+}, never present in excess, reacts preferentially with the zwitterion. This view is supported by the similarity of Nd^{3+} perturbations of tetracycline and its methiodide, which possesses a trimethylammonium group at C4 that enforces the zwitterion [47]. Removal of the tetracycline dimethylamino group to yield desdimethylaminotetracycline gave an apparently inactive ligand [47]. However, the O3 basicity in this compound is about 300 times greater than that of tetracycline and its methiodide [15]. In the 100% DMSO medium the nitrogen base

added is unable to remove the proton from the more basic O3 and no
basic sites are available to a metal ion. This is another example
of the reversal of amino group and O3 basicities in low dielectric
media described in Sec. 2.2.2.

Addition of Mn^{2+} was found to perturb the ^{13}C NMR of tetracycline A-ring carbons from pH 7-8 in 80% DMSO [48]. Dimethylamino
carbon and C11 resonances were relatively unaffected. Once again the
results indicate metal ion interaction with amino protonated ligand.

When present in excess, Na^+ may flush other ions present in
lesser amounts from tetracycline-binding sites [49,50].

3. DAUNORUBICIN

3.1. Structure

Daunorubicin (daunomycin) and doxorubicin (adriamycin) are closely
similar anthracycline antibiotics with an amino sugar, daunosamine.
As shown in Fig. 3, the structures differ in only one ring substituent; doxorubicin is 14-hydroxydaunorubicin. This difference appears
insignificant for metal ion-binding capabilities, and the two antibiotics are considered together in this chapter. Whether the name

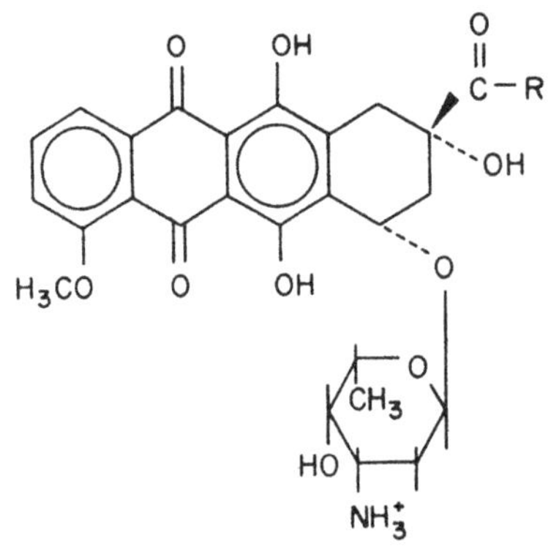

FIG. 3. Structures of daunorubicin (R = CH$_3$) and doxorubicin (R =
CH$_2$OH). Quinizarin (1,4-dihydroxyanthraquinone) consists of the
three leftmost rings without the methoxy group. (From Ref. 54.)

daunorubicin or doxorubicin is used in the text depends on which
compound the investigators happened to use in the paper under review.

Both antibiotics display high cytotoxicity against both normal
and neoplastic cells and are used clinically. Daunorubicin was the
first antibiotic found to exhibit activity against acute leukemia in
humans. Doxorubicin appears especially effective against solid
tumors. The development of the antibiotics was the result of a
systematic search for wide-spectrum antitumor compounds in a group
of microbial metabolites, the anthracycline glycosides [51]. The
antibiotics are thought to function by binding to DNA and inhibiting
replication and transcription. The anthracycline portion of daunorubicin has been found by x-ray diffraction to intercalate between
stacked bases in a double-stranded DNA fragment [52].

3.2. Proton Binding

Daunorubicin (or doxorubicin) contains three potentially acidic
groups: the ammonium group on the sugar ring and the two phenolic
hydroxy groups on the anthracycline ring system. We may utilize the
microconstant scheme in Fig. 2 for daunorubicin by designating the
ammonium group deprotonation by the subscript 3 and the two phenolic
deprotonations by subscripts 1 and 2. Because the ammonium group in
daunorubicin is remote from both phenolic groups (Fig. 3), we cannot
distinguish experimentally between them, and all subscripts 1 and 2
referring to phenolic groups in Fig. 2 are interchangeable. A similar but real symmetry also occurs for the two terminal carboxylic
acid groups in the microconstant scheme for citric acid [53].

Daunorubicin macroconstants and microconstants, evaluated by
a combination of potentiometric and spectrophotometric titrations
of the hydrochloride salt, are listed in Table 2 [54]. Between
$9 < pH < 11.5$ the absorption spectrum of daunorubicin is sensitive
to deprotonation of a single phenolic group and 14 experimental
intensities were employed in the microconstant analysis. In the
more acid region of the range, the spectral change is perturbed by

the ammonium group deprotonation. The second phenolic group does
not undergo deprotonation until pH > 13. The microconstant evaluation [54] indicates that deprotonation of the three acidic groups in
daunorubicin proceeds almost sequentially across the bottom of the
scheme in Fig. 2. To a good approximation in daunorubicin and doxorubicin the ammonium group deprotonates with pK_1 = 8.6, the first
phenolic group with pK_2 = 10.0, and the second phenolic group with
pK_3 = 13.7.

The preceding assignments of deprotonation sites and acidity
constants receive support from other studies. At unspecified temperature and ionic strength only a few points were used to estimate
for doxorubicin pK_1 = 8.15, pK_2 = 10.16, and pK_3 = 13.2 [55]. These
results are in agreement with those of Table 2. After the ammonium
group is protonated, with increasing acid, the next protonation is
that of a carbonyl group with pK_a = -5.9, in very acid sulfuric acid
solutions [55]. A later claim of a protonation with pK_a = 2.1 [56]
is without any support [54].

The results quoted in Table 1 receive additional support from
an investigation of quinizarin or 1,4-dihydroxyanthraquinone [54].
This molecule is closely related to the chromophore in daunorubicin
(Fig. 3). Quinizarin is much more stable than daunorubicin. The
results of potentiometric and spectrophotometric titrations in 50%
ethanol at 20°C and 0.15 M ionic strength give, with high precision,
pK_1 = 9.92 ± 0.01 and pK_2 = 13.7 ± 0.1 [54]. These values are nearly
identical to the pK_2 and pK_3 values for daunorubicin in Table 2. The
near identity of the pairs of values provides strong support for both
the assignment to the phenolic groups and acidity constant values of
pK_2 and pK_3 for daunorubicin in Table 2.

A recent paper offers alternative pK_a value assignments to
doxorubicin deprotonations. Values of pK_a of 8.94 and 9.95 are both
assigned to phenolic deprotonations and one of 6.8 to the ammonium
group deprotonation [57]. Similar to the results presented in Table
2 there is good agreement that the predominant spectral change is
associated with a phenolic deprotonation that occurs with pK_a = 10.0.

Though the data analysis differs, there is also agreement that a group associated with pK_a = 8.61-8.94 appears as a perturbation on the greater spectral change that occurs with the pK_a = 10.0 phenolic deprotonation. The authors identify their pK_a = 8.94 as the first phenolic deprotonation [57] rather than as the ammonium group deprotonation for pK_1 in Table 2. Potentiometric titration of quinizarin cited above reveals only one deprotonation between 5 < pH < 11 and makes it extremely unlikely that both the pK_a = 8.6-8.9 and pK_a = 10.0 values should be assigned to phenolic deprotonations in daunorubicin or doxorubicin. The authors adopt the two phenolic deprotonation assignment because they find a deprotonation with pK_a = 6.8, which they assign to the ammonium group [57]. Other investigators found no evidence for a deprotonation in this range in either daunorubicin or doxorubicin [54,55]. In these compounds the ammonium group on the sugar appears as a substituted ethanolamine. Ethanolamine itself exhibits pK_a = 9.5 [26] and an ammonium group deprotonation with pK_a = 6.8 is an inexplicable 2.7 log units more acidic. Dimerization or stacking of daunorubicin cannot account for such a discrepancy and in fact works in the wrong direction for the positively charged antibiotics as the studies with the pK_a = 6.8 deprotonation were performed at the lowest concentration. Explanation of the anomalous pK_a = 6.8 deprotonation is uncertain. It corresponds well to deprotonation of dihydrogen phosphate $H_2PO_4^-$, which might appear as the counterion for the positively charged antibiotic. Phosphate buffer is used in the isolation of doxorubicin, but it is removed in a subsequent extraction into chloroform at pH 8.6 [51].

3.3. Metal Ion Binding

Stability constants for daunorubicin determined spectrophotometrically from intensity measurements in the visible region are listed in Table 4. The usual stability constant K_M refers to the reaction $M + LH \rightleftarrows MLH$, where LH designates a ligand with one protonated and

TABLE 4

Daunorubicin Stability Constant Logarithms[a]

Metal ion	log K_M	log K_7 [b]
H^+	10.0	—
Mg^{2+}	3.7	0.7
Ca^{2+}	3.3	0.3
Fe^{3+}	11.0	8.0
Cu^{2+}	7.3[c]	4.3
Zn^{2+}	4.5	1.5
Tb^{3+}	7.2	4.2
Th^{4+}	10.3	7.3

[a] In aqueous solutions at 20°C and 0.15 M ionic strength. From Ref. 54.
[b] Apparent stability constant at pH 7.0.
[c] Predicted, see text.

one deprotonated phenolic group. The ammonium group is not involved in bonding to any metal ion and its remoteness from the anthracycline ring center renders its protonation state of little importance. The listed K_M values were calculated from results at a specific pH by using pK_a = 10.0 for the phenolic group where the metal ion is bound. We now reverse the procedure and from K_M calculate the apparent stability constant at any pH from

$$K_X = \frac{K_M}{1 + (H)/K_a}$$

At pH 7.0 with pK_a = 10.0 we obtain

$$\log K_7 = \log K_M - 3.0$$

Values of the apparent stability constant at pH 7.0 appear in the last column of Table 4.

From the stability constants it is possible to show that in body fluids only small fractions of daunorubicin and doxorubicin bear a Mg^{2+} or Ca^{2+} ion. Fe^{2+} is oxidized to Fe^{3+} by daunorubicin

[58]. There is no observable complex formation at pH 1 between $Hg(ClO_4)_2$ and the anthracycline chromophore of daunorubicin [54].

From the pH dependence of the apparent stability constants it is evident that one proton per daunorubicin ligand is removed for each metal ion bound. The strong shift in the absorption spectrum establishes metal ion binding at the anthracycline chromophore. That closely similar stability constants are obtained with quinizarin (1,4-dihydroxyanthraquinone) suggests that the amine group of daunorubicin is not chelated to any metal ion [54]. This conclusion is supported by selective broadening by Gd^{3+} of only the anthracycline and not the daunosamine portion of daunorubicin [59]. Relative broadenings suggest that the less hindered 11, 12 site (top of Fig. 1) binds more strongly than the 5, 6 site. All these results concur and lead to the conclusion that the primary daunorubicin metal ion-binding site is a deprotonated phenolic oxygen and a carbonyl oxygen to form a six-membered chelate ring.

The stability constant logarithms for metal ion and proton binding to daunorubicin are almost equal to the values for quinizarin [54]. These values in turn are almost half the values for binding of the same metal ions to tiron (1,2-dihydroxybenzene-3,5-disulfonate). In daunorubicin and quinizarin metal ion chelation occurs at a phenolic site with $pK_a \simeq 10.0$, while in tiron chelation occurs at a catecholate function with a sum $pK_1 + pK_2 = 7.7 + 12.6 = 20.3$. The generalization is exploited to predict a stability constant for Cu^{2+} binding to daunorubicin as listed in Table 4. The value of $\log K_M = 7.3$ should be reliable to ± 0.5 log units. Another investigation reports that the Cu^{2+} complex of daunorubicin is not stable in blood plasma [60].

Strongly binding metal ions such as the trivalent lanthanide ions and Cu^{2+} permit observation of 2:1 daunorubicin-metal ion complexes even in equimolar solutions. Careful analysis of data for addition of 0.4 to 10 eq of Tb^{3+} to daunorubicin at pH 7 reveals a second stability constant about 50 times greater than the first (recalculated from Ref. 54). Similar results have been obtained with Yb^{3+}, where the second to first stability constant ratio appears to be 100 or greater [61]. The nature of the system with a greater second stability constant and free ligand association with an uncertain self-association constant (below) make difficult precise determination of a stability constant ratio. In contrast to the daunorubicin results, for quinizarin in 50% ethanol the binding of a second ligand molecule to Tb^{3+} occurs with a sixfold typical reduction in the second compared to the first stability constant [54]. This difference may be ascribed to appreciable self-association of daunorubicin in the aqueous medium and to negligible self-association of quinizarin in 50% ethanol. Considering both ligands together daunorubicin self-association results in a 300 times greater second stability constant than is expected without self-association.

Attempts to express quantitatively the tendency of daunorubicin to self-associate in water have led to a range of results. Analysis of circular dichroism and absorbance intensity variations with daunorubicin concentration led to an equilibrium constant for dimerization of 6.4 mM^{-1} at pH 7.3 and 25°C in 10 mM phosphate buffer [62]. This value should be increased to about 10.0 mM^{-1} at 0.2 M ionic strength. An identical dimerization constant value was obtained indirectly from an analysis of Fe^{3+} binding to daunorubicin at pH 0.70 and 20°C [54].

It is also possible to formulate daunorubicin self-association in terms of an isodesmic model of infinite association with identical equilibrium constants for association of each additional monomer [63]. It is a necessary consequence of the simple isodesmic model and the usual assumption that perturbation of a molecule within a stack is twice that of a molecule at the end of a stack for intensity and chemical shift measurements, that the value of the association equilibrium constant is exactly double the value of the dimerization equilibrium constant [64]. Results of equilibrium sedimentation measurements suggest that daunorubicin association proceeds beyond the dimer stage and that the isodesmic model is more appropriate [63]. If the highest dimerization constant value of 10 mM^{-1} is used, then at 0.1 mM concentration 50% of daunorubicin molecules are monomeric and 50% appear as dimers. For the corresponding conditions in the isodesmic model, 25% of the molecules appear as monomers, 50% occur at the ends of stacks, and 25% within stacks.

The nature of the complexes formed with doxorubicin and Cu^{2+} depends on the ligand to metal ion ratio and pH. In solutions with a 2:1 ratio a 2:1 complex predominates from 5 < pH < 8 [57,65]. In equimolar solutions an equimolar complex occurs at pH 7. From the conservative visible CD spectrum occurring in an equimolar solution at pH 5.8, it is inferred that a pair of ligands are stacked in a 2:1 complex [57]. N-trifluoroacetyl daunorubicin gives results similar to daunorubicin [65]. This comparison supports inferences from other studies that the daunosamine amino group is not involved significantly in metal ion binding. Formation of the neutral pH complexes is slow, taking about 30 min for the CD bands to reach maximal amplitudes [57]. This result is disquieting and suggests formation of polymeric and hydroxo complexes. For most metal ions hydroxo and doxorubicin complex formation are competitive. Hydrolysis of Cu^{2+} complexes needs to be considered at pH >5 [66].

With excess doxorubicin complexes with up to three ligands for each Fe^{3+} occur at pH 7 when the Fe^{3+} is stabilized against hydrolysis with acetohydroxamate [58]. Probably due to lack of consideration of the effects of Fe^{3+} hydrolysis, unreasonably large doxorubicin stability constants have been reported [56]. The log K_M value quoted in Table 4 is consistent with results from quinizarin and tiron [54]. The log K_7 value in Table 4 refers to pH 7 but due to hydrolysis the constant cannot be measured directly at that pH in the absence of a stabilizing ligand. The quoted log K_7 value is derived from the log K_M value determined near pH 1. Even in the presence of a stabilizing ligand at pH 7 the effects of hydroxo complex formation still must be considered with Fe^{3+}.

An Fe^{3+} complex of doxorubicin catalyzes the reduction of oxygen by thiols such as cysteine and glutathione yielding superoxide and hydrogen peroxide [58]. The last two agents are thought to be responsible for destruction of human erythrocyte ghost membranes upon binding of an Fe^{3+}-doxorubicin complex in the presence of glutathione [58]. This study was directed toward understanding the cardiac toxicity of

an Fe^{3+}-doxorubicin complex. A ternary complex doxorubicin-Fe^{3+}-ADP
is reduced by P450 reductase and causes decomposition of unsaturated
fatty acids [67].

With large excesses of metal ion M_2L complexes of daunorubicin
were detected with Mg^{2+}, Ca^{2+}, and Tb^{3+}. The stability constants for
binding of the second metal ion are 1.1-1.4 log units less than the
values tabulated in Table 4 [54]. In these M_2L complexes the second
phenolic group deprotonates and the antibiotic binds two metal ions
in two separated six-membered chelate rings.

In the absence of a stabilizing ligand such as citrate, the pH
range of the Fe^{3+} experiments is limited by metal ion hydrolysis at
pH >1 and by complex ion formation with anions at pH <0.5. Daunorubicin binds Fe^{3+} so weakly at pH <1 that metal ion hydrolysis begins
before MLM complex formation. Daunorubicin or doxorubicin cannot bind
three Fe^{3+} ions at pH 7 to form the proposed simple complex triferric
doxorubicin (quelamycin) used in some cancer therapy [68].

Several different structures may be envisaged for each of the
daunorubicin-metal ion compositions present in solution. The basic
monomeric structure is that of a ligand coordinated through two oxygens, one of which has undergone deprotonation, in a six-membered
chelate ring. In equimolar neutral and basic solutions the second
phenolic oxygen may become substituted by a metal ion and a polymeric
complex can form with metal ions at both six-membered chelate ring
sites and each metal ion bound to two ligands. Since there is a
strong tendency of the ligand to self-associate, one can visualize
mononuclear 2:1 complexes in which only one or both ligands have
undergone deprotonation. As long as there is some preferred orientation in the former complex, both kinds of 2:1 complexes should
yield a conservative circular dichroism spectrum. Finally, the 2:1
complexes may themselves be arranged in longer stacks.

Many studies have reported on the binding of daunorubicin (or
doxorubicin) to DNA [69]. Since the nucleic acids contain several
metal ion-binding sites [70], it is likely that ternary complexes
form in a system containing the antibiotic, DNA, and a metal ion.

Metal ions do affect interactions of the antibiotic with DNA [71-73] including DNA strand scission [74]. The Cu^{2+} complex of the related antibiotic iremycin elongates DNA more than iremycin itself or daunorubicin [75].

ACKNOWLEDGMENTS

I thank the authors of Refs. 5, 57, and 59 for sending manuscripts in advance of publication.

REFERENCES

1. W. Dürckheimer, *Angew. Chem. Internat. Ed. Engl., 14,* 721 (1975).

2. J. P. White and C. R. Cantor, *J. Mol. Biol., 58,* 397 (1971).

3. F. Fey, M. Reiss, and H. Kersten, *Biochemistry, 12,* 1160 (1973).

4. T. R. Tritton, *Biochemistry, 16,* 4133 (1977).

5. L. Lambs, M. Brion, and G. Berthon, *Agents and Actions, 14,* 743 (1984).

6. L. Saxen, *Science, 149,* 870 (1965).

7. M. M. Mull, *Amer. J. Dis. Child., 112,* 483 (1966).

8. M. Brion, G. Berthon, and J. Fourtillan, *Inorg. Chem. Acta, 55,* 47 (1981).

9. J. J. Stezowski, *J. Am. Chem. Soc., 98,* 6012 (1976).

10. G. J. Palenik and J. A. Bentley, *J. Am. Chem. Soc., 100,* 2863 (1978).

11. R. Prewo and J. J. Stezowski, *J. Am. Chem. Soc., 99,* 1117 (1977).

12. L. J. Hughes, J. J. Stezowski, and R. E. Hughes, *J. Am. Chem. Soc., 101,* 7655 (1979).

13. A. Albert and C. W. Rees, *Nature, 177,* 433 (1956).

14. C. R. Stephens, K. Murai, K. J. Brunings, and R. B. Woodward, *J. Am. Chem. Soc., 78,* 4155, 6425 (1956).

15. L. J. Leeson, J. E. Krueger, and R. A. Nash, *Tetrahedron Lett., 18,* 1155 (1963).

16. L. Z. Benet and J. E. Goyan, *J. Pharm. Sci., 54,* 983 (1965).

17. J. J. R. Frausto da Silva and M. H. M. Dias, *Rev. Port. Quim., 14,* 159 (1972).

18. S. R. Martin, *Biophys. Chem.*, *10*, 319 (1979).

19. J. T. Doluisio and A. N. Martin, *J. Med. Chem.*, *6*, 16 (1963).

20. N. P. Sachan and C. M. Gupta, *Talanta*, *27*, 457 (1980).

21. N. E. Rigler, S. P. Bag, D. E. Leyden, J. L. Sudmeier, and C. N. Reilley, *Anal. Chem.*, *37*, 872 (1965).

22. G. L. Asleson and C. W. Frank, *J. Am. Chem. Soc.*, *98*, 4745 (1976).

23. J. J. Stezowski, *J. Am. Chem. Soc.*, *99*, 1122 (1977).

24. M. C. Lim, *J. Chem. Soc. Dalton*, 726 (1978).

25. M. L. Eidinoff, *J. Am. Chem. Soc.*, *67*, 2072 (1945).

26. R. A. Robinson and R. H. Stokes, *Electrolyte Solutions*, 2nd ed., Butterworths, London, 1959, Appendix 12.1.

27. C. V. McDonnell, Jr., M. S. Michailidis, and R. B. Martin, *J. Phys. Chem.*, *74*, 26 (1970).

28. K. H. Jogun and J. J. Stezowski, *J. Am. Chem. Soc.*, *98*, 6018 (1976).

29. M. Celotti and G. V. Fazakerley, *J. Chem. Soc. Perkin II*, 1319 (1977).

30. B. Noszal and R. B. Martin, unpublished experiments performed in 1980.

31. D. L. J. Clive, *Quart. Rev.*, *22*, 435 (1968).

32. R. B. Martin, J. T. Edsall, D. B. Wetlaufer, and B. R. Hollingworth, *J. Biol. Chem.*, *233*, 1429 (1958).

33. T. L. Sayer and D. L. Rabenstein, *Can. J. Chem.*, *54*, 3392 (1976).

34. R. B. Martin, *Metal Ions Biol. Syst.*, *9*, 1 (1979).

35. R. B. Martin, *Introduction to Biophysical Chemistry*, McGraw-Hill, New York, 1964, Chap. 4.

36. R. B. Martin, *Science*, *139*, 1198 (1963).

37. H. Sigel and R. B. Martin, *Chem. Rev.*, *82*, 385 (1982).

38. M. H. M. Dias, J. J. R. Frausto da Silva, and A. V. Xavier, *Rev. Port. Quim.*, *21*, 5 (1979).

39. G. Berthon, M. Brion, and L. Lambs, *J. Inorg. Biochem.*, *19*, 1 (1983).

40. E. C. Newman and C. W. Frank, *J. Pharm. Sci.*, *65*, 1728 (1976).

41. G. W. Everett, Jr., J. Gulbis, and J. Shaw, *J. Am. Chem. Soc.*, *104*, 445 (1982).

42. J. Shaw and G. W. Everett, Jr., *J. Inorg. Biochem.*, *17*, 305 (1982).

43. A. H. Caswell and J. D. Hutchison, *Biochem. Biophys. Res. Commun.*, *43*, 625 (1971).

44. R. B. Martin, *Metal Ions Biol. Syst.*, *17*, 1 (1984).

45. D. E. Williamson and G. W. Everett, Jr., *J. Am. Chem. Soc.*, *97*, 2397 (1975).

46. J. Gulbis and G. W. Everett, Jr., *J. Am. Chem. Soc.*, *97*, 6248 (1975).

47. J. Gulbis and G. W. Everett, Jr., *Tetrahedron*, *32*, 913 (1976).

48. J. Y. Lee and G. W. Everett, Jr., *J. Am. Chem. Soc.*, *103*, 5221 (1981).

49. J. Gulbis, G. W. Everett, Jr., and C. W. Frank, *J. Am. Chem. Soc.*, *98*, 1280 (1976).

50. C. Coibion and P. Laszlo, *Biochem. Pharmacol.*, *28*, 1367 (1979).

51. F. Arcamone, *Doxorubicin*, Academic Press, New York, 1981.

52. G. J. Quigley, A. H. Wang, G. Ughetto, G. van der Marel, J. H. van Boom, and A. Rich, *Proc. Natl. Acad. Sci.*, *77*, 7204 (1980).

53. R. B. Martin, *J. Phys. Chem.*, *65*, 2053 (1961).

54. R. Kiraly and R. B. Martin, *Inorg. Chim. Acta*, *67*, 13 (1982).

55. R. J. Sturgeon and S. G. Schulman, *J. Pharm. Sci.*, *66*, 958 (1977).

56. P. M. May, G. K. Williams, and D. R. Williams, *Inorg. Chim. Acta*, *46*, 221 (1980).

57. H. Beraldo, A. Garnier-Suillerot and L. Tosi, *Inorg. Chem.*, *22*, 4117 (1983).

58. C. E. Myers, L. Gianni, C. B. Simone, R. Klecker, and R. Greene, *Biochemistry*, *21*, 1707 (1982).

59. Y. H. Mariam, personal communication.

60. K. Mailer and D. H. Petering, *Biochem. Pharmacol.*, *25*, 2085 (1976).

61. Y. H. Mariam, W. Wells, and B. Wright, personal communication.

62. S. R. Martin, *Biopolymers*, *19*, 713 (1980).

63. J. B. Chaires, N. Dattagupta, and D. M. Crothers, *Biochemistry*, *21*, 3927 (1982).

64. P. R. Mitchell and H. Sigel, *Eur. J. Biochem.*, *88*, 149 (1978).

65. F. T. Greenaway and J. C. Dabrowiak, *J. Inorg. Biochem.*, *16*, 91 (1982).

66. H. Sigel and R. B. Martin, *Chem. Rev.*, *82*, 385 (1982), Section XVI B.

67. K. Sugioka and M. Nakano, *Biochem. Biophys. Acta, 713,* 333 (1982).

68. M. Gosalvez, M. F. Blanco, C. Vivero, and F. Valles, *Eur. J. Cancer, 14,* 1185 (1978).

69. J. B. Chaires, N. Dattagupta, and D. M. Crothers, *Biochem., 21,* 3933 (1982).

70. R. B. Martin and Y. H. Mariam, *Met. Ions. Biol. Syst., 8,* 57 (1979).

71. P. Mikelens and W. Levinson, *Bioinorganic Chem., 9,* 441 (1978).

72. D. R. Phillips and G. A. Carlyle, *Biochem. Pharmacol., 30,* 2021 (1981).

73. M. Spinelli and J. C. Dabrowiak, *Biochemistry, 21,* 5862 (1982).

74. A. Someya and N. Tanaka, *J. Antibiotics, 32,* 839 (1979).

75. H. Fritzsche, H. Triebel, J. B. Chaires, N. Dattagupta, and D. M. Crothers, *Biochemistry, 21,* 3940 (1982).

Chapter 3

INTERACTION OF METAL IONS WITH STREPTONIGRIN AND
BIOLOGICAL PROPERTIES OF THE COMPLEXES

Joseph Hajdu[*]
Department of Chemistry
Boston College
Chestnut Hill, Massachusetts

[*]Present affiliation: Department of Chemistry, California State
University, Northridge, California

1. INTRODUCTION

The aminoquinone antibiotic streptonigrin, a metabolite of *Strepto-*
myces flocculus [1], is one of the most effective agents for the
treatment of human cancers [2]. It has been shown to be active
against lymphoma, melanoma, cancers of breast, cervix, head and
neck, as well as viruses. At the same time, however, streptonigrin
is known to have a number of undesirable side effects, including
severe bone marrow depression [3]. Its high toxicity precludes the
clinical use of the antibiotic; the tolerated dose is limited to
2 mg per treatment [4].

Evidence to date indicates that streptonigrin exerts its antitumor action via (1) interference with cell respiration and (2) disruption of cell replication [4]. Both mechanisms involve participation of metal ions [5] as addition of chelating agents inhibits the
biological activity of the drug. Although the chemistry of streptonigrin-metal complexes has not been delineated, Cu^{2+} and Fe^{2+} are
known to accelerate streptonigrin-induced DNA scission while Co^{2+}
appears to inhibit the same process [6].

Structural and mechanistic elucidation of the metal complexes
of streptonigrin is a key step toward understanding the mode of
action of the antibiotic and a prerequisite for developing a rational
approach to improve its chemotherapeutic properties.

This chapter intends to illustrate how, by using a combination
of experimental methods involving spectroscopic, physical-organic,
and bioinorganic studies, one can begin to understand the structural
properties and chemical reactivities of the metal complexes of streptonigrin, while at the same time provide insight into the possible
mechanism of biological functioning of this potent antitumor antibiotic.

2. STRUCTURAL AND CHEMICAL PROPERTIES OF STREPTONIGRIN AND ITS METAL COMPLEXES

2.1. Streptonigrin as a Substituted Picolinic Acid

Streptonigrin (1) belongs to the class of antitumor antibiotics possessing the aminoquinone moiety as a common structural element.

(1)

Other members in the group include mitomycin C, porfiromycin, actino-mycin, rifamycin, and geldanamycin [4]. The compound was first iso-lated in the United States from a strain of *S. flocculus* [1].

   The chemical structure of the antibiotic was established by Woodward and co-workers [7] and its crystal structure was determined in Lipscomb's laboratory [8]. The compound is best described as a highly substituted picolinic acid. Analysis of its single-crystal x-ray diffraction pattern shows the A, B, and C rings in a nearly coplanar arrangement, while the fourth D ring is nearly perpendicular to the plane defined by the ABC triad. The coplanarity of the latter appears to be held by a hydrogen bond between the quinoline nitrogen and the amino group of the adjacent pyridine ring. It is not clear, however, if a similar conformational arrangement exists in solution.

   Streptonigrin forms dark brown crystals which can be obtained from ethyl acetate. The compound is only sparingly soluble in water and in lower alcohols; it readily dissolves in dimethyl formamide,

dimethyl sulfoxide, pyridine, dioxane, and tetrahydrofuran [9].
Consequently, most of its physicochemical properties were determined
in mixed solvent systems. A monobasic acid, streptonigrin has a pK_a
of 6.5 in 50% aqueous dioxane [10]. Its ^{13}C NMR spectrum was originally determined by Lown [11]; subsequently a number of resonances
have been reassigned on the basis of results obtained in biosynthetic
studies conducted by Gould and Weinreb [9].

The proton NMR of streptonigrin has also been reported [9],
however due to the few unsubstituted ring positions the spectrum
has merely confirmed the structures of the substituents.

Because of the complexity and rather large number of substituents in the molecule, streptonigrin is best discussed in terms of
constituent structural units responsible for activity and binding.
Specifically, the 5,8-quinolinedione moiety containing the redox-
active aminoquinone function has been shown to be the structural
element essential for antitumor activity. A series of reduction
products derived from this fragment have been observed in vitro and
implicated in vivo [4] while streptonigrin derivatives, in which the
p-quinone moiety was blocked, showed no antitumor activity [12,13].

The importance of structural integrity of the functional groups
in the quinone moiety has been well documented [4,6]. As one notable
exception, however, it has been recently reported [9] that acylation
of the quinone amino group produces an antitumor-active streptonigrin
derivative. At the same time, however, the relatively high effective
dose requirement seems to raise some doubts as to whether the modified analog or regenerated streptonigrin (formed by in vivo hydrolysis) is in fact the active entity [9].

The substituents comprising rings B and C provide an assembly
of functional groups and heteroatoms with excellent metal-binding
properties. Due to the conformational flexibility allowing free
rotation around the carbon-carbon bond at the pyridine-quinoline
ring junction, a number of alternative chelating sites become available. For example, the orientation and proximity between the heterocyclic nitrogen of ring B and the 2'-carboxypyridine portion of ring

C provide a bipyridyl-carboxylate fragment, well known to give tight-binding complexes with a series of metal ions [14].

Alternatively, the quinoline nitrogen in conjunction with the amino group of the pyridine ring bridged by a metal ion provides a stable coordinating arrangement, analogous to the conformation found in the crystal structure of streptonigrin [8]. Significantly, both conformational arrangements place the metal ion in close proximity to the aminoquinone-oxygen, potentiating direct coordination, thus catalytic interaction, between the redox-active quinone moiety and the cationic center of the complex.

The specific role of ring D in the streptonigrin molecule is less obvious at the present. One might surmise that the hydrophobic dimethoxyphenol moiety is likely to be involved in noncovalent binding interactions between the antibiotic and target biomolecules (DNA and/ or proteins) [4] in conjunction with the biological functioning of the antibiotic.

2.2. Complexation Between Streptonigrin and a Series of Metal Ions

Although the importance of metal ion participation in the antitumor action of streptonigrin has been well recognized, for nearly 20 years following its discovery no systematic studies of their interaction with the antibiotic have been reported. The first complexation studies between streptonigrin and metal ions [10,15] were inspired by discoveries reported by White [16] and Lown [4,6], demonstrating that addition of *specific* metal ions enhanced the bactericidal and DNA-cleaving potency of the antibiotic. It became apparent that elucidation of the interactions between metal ions and streptonigrin would lead to creative chemistry, in terms of differential activation of the coordinating ligands, and provide important clues as to how specific metal substitution might improve the chemotherapeutic properties of the drug [10].

A series of streptonigrin complexes have been prepared utilizing a number of biologically occurring transition metal ions which
are likely to interact with the antibiotic in vivo. Formation of
1:1 complexes of streptonigrin with Zn^{2+} and Cu^{2+} was first demonstrated by potentiometric titration of compound (1) in presence and
in absence of equimolar amounts of the corresponding metal. Adapting
the solvent system originally used for the physicochemical characterization of streptonigrin (dioxane-water 1:1) [9], a pK_a of 6.5 has
been obtained for the free antibiotic. This dissociation constant
is lowered by 2.3 and 3.3 pK_a units in presence of equimolar Zn^{2+}
and Cu^{2+} respectively [10]. The complexation reaction between these
metal ions and streptonigrin results in the release of 1 mol H^+ per
mol metal bound [Eq. (1)]:

$$HSN + M^{2+} \rightleftharpoons {}^-SN \cdot M^{2+} + H^+ \tag{1}$$

Although the stoichiometry of the substitution reaction could
readily be established, the site at which the proton is being released
from the complexing ligand molecule has not yet been elucidated.

Independent evidence for the formation of streptonigrin-metal
complexes came from spectroscopic studies of the reactions. Addition
of anhydrous metal halide to an acetonitrile solution of streptonigrin produces a deep brown solution of the corresponding metal
complex. The spectra of the resulting metal derivatives differ from
that of the antibiotic in a number of characteristic ways (Fig. 1).

In case of the zinc complex, for example, the long-wavelength
UV absorption is red-shifted (375 → 400 nm) and its intensity increased in comparison with the spectrum of streptonigrin. The 245-nm
band of the parent antibiotic is split into a doublet, yielding a
maximum at 235 nm and a shoulder at 255 nm, both decreased in intensity with respect to the absorption of the free ligand. Complexation
with Cu^{2+} results in substantial broadening of the long-wavelength
absorption (attributed to the quinolinequinone moiety) which is now
centered at 415 nm. The pattern of the short-wavelength UV maxima
remains quite similar to that of the parent antibiotic, except for
a substantial decrease in the intensity of the 294-nm absorption
band.

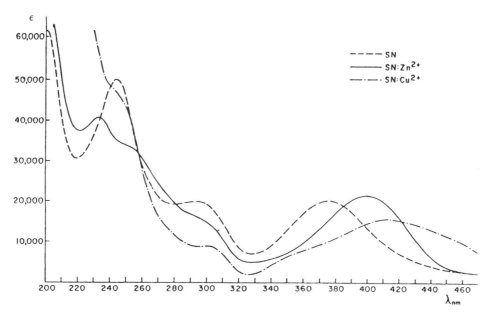

FIG. 1. The spectra of streptonigrin and its copper(II) and zinc(II) chloride complexes in acetonitrile at 25°C. (Reprinted from Ref. 10 by permission of the American Chemical Society.)

Complexation between streptonigrin and metal ions has been found to be sensitive to the solvent. In aqueous media displacement of the streptonigrin-proton by the complexing metal ion is dependent on the pH. The dissociation constants presented in Table 1 have been determined spectrophotometrically. The double-reciprocal plots ($1/\Delta$ OD vs. $1/M^{2+}$) were obtained by following the spectral changes on increasing the concentration of the metal ion added to a solution of streptonigrin (Fig. 2). Due to the low aqueous solubility of streptonigrin the K_D values were determined in solvent-systems containing up to 10% acetonitrile. Under the experimental conditions employed, no complexation between the buffer and the metal ions was observed.

The stoichiometry of the reaction between streptonigrin and the corresponding metal ions (i.e., the metal-to-ligand ratio) has been determined by spectrophotometric and potentiometric titrations.

TABLE 1

Dissociation Constants of Streptonigrin-Metal
Complexes in Aqueous Solutions[a]

Metal	Buffer	pH	K_D (M)
Zn^{2+}	0.1 M Acetate	5.1	1.3×10^{-4}
	0.1 M Tris	6.75	1.3×10^{-4}
	0.1 M Tris	8.26	1.3×10^{-4}
Cu^{2+}	0.1 M Acetate	5.1	5.6×10^{-5}
	0.005 M Hepes[b]	7.01	5.3×10^{-5}
	0.005 M Pipes[b]	7.07	5.1×10^{-5}
Co^{2+}	0.1 M Acetate	5.1	1.1×10^{-4}
	0.005 M Hepes[b]	7.01	4.1×10^{-5}
	0.005 M Pipes[b]	7.07	4.7×10^{-5}
Cd^{2+}	0.1 M Acetate	5.1	2.3×10^{-5}
	0.005 M Hepes[b]	7.01	2.3×10^{-5}
	0.005 M Pipes[b]	7.07	2.3×10^{-5}
Mn^{2+}	0.005 M Hepes[b]	7.01	2.2×10^{-4}

[a]Unpublished results from the author's laboratory.
[b]Ionic strength adjusted to 0.1 with NaCl.

Utilizing streptonigrin concentrations above the dissociation constants, the difference spectra obtained on stepwise addition of metal ion plotted against the number of equivalents of the titrant produced sharp break points for the entire series of complexes. For Zn^{2+}, Cu^{2+}, Cd^{2+}, and Mn^{2+} 1:1 metal-to-ligand ratios were obtained, the only exception being Co^{2+}, for which this ratio turned out to be 1:2, indicating that cobalt(II) is capable of binding 2 mol of antibiotic per mol of metal. Similar results emerged from potentiometric titrations measuring the number of moles of metal required for the release of the first mol of protons from the ligand.

It is important to point out that in the course of spectrophotometric titrations it was discovered that streptonigrin is capable of additional metal binding, beyond the point at which the first proton displacement is completed [15]. Specifically, when the addition of metal ion is continued past the break point, the difference spectrum does not level off but rather increases. Although the magnitude of

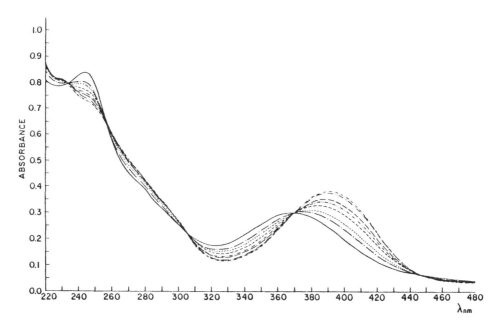

FIG. 2. Complexation between streptonigrin and zinc chloride in
0.1 M Tris of pH 6.76 containing 8% acetonitrile. The curves corre-
spond to (—•—) 2.1×10^{-5} M, (---) 5.6×10^{-5} M, (———) 1.4×10^{-4} M,
(••-••) 2.2×10^{-4}M, (——) 3.6×10^{-4}M, (•••-•••) 7.2×10^{-4} M, and
(•-••-•-••) 1.1×10^{-3} M, concentrations of $ZnCl_2$ in the solution.
(Reprinted from Ref. 10 by permission of the American Chemical Society.)

the absorption increment is greatly diminished after the first break
point, the increase is definite, and indicates that the antibiotic
molecule has multiple binding sites. Considering the streptonigrin
structure, this is indeed not unexpected. The existence of multiple
binding sites might have a specific role in the biological functioning of the antibiotic [16-18].

2.3. Proposed Structures of the Complexes

Definite structural assignments of the streptonigrin-metal complexes
have not yet been established. Proposals regarding possible structures have relied on spectroscopic observations and changes in the

chemical properties of the antibiotic as a result of metal-ligand
interactions.

Complexation with Zn^{2+} in acetonitrile has been shown to alter
the absorption spectrum of the antibiotic in a manner substantially
different from the corresponding changes observed on complex formation between streptonigrin and copper(II). Figure 1 clearly demon-
strates that both the long-wavelength absorption as well as the spectrum in the 245-nm peak region are perturbed differently by the two
metals. These observations suggest that the coordinating ligand
moieties in the corresponding metal complexes are likely to be different. Considering the conformational flexibility of the molecule,
two metal derivatives might arise from complexation involving two
different streptonigrin conformers (2a) and (2b).

(2a)   (2b)

Hajdu [10] proposed that for metal ions which are known to
function as strong Lewis acids, including Cu^{2+}, the amine nitrogen
of the pyridine C ring may serve as a donor atom, while the alternative binding site, including the picolinic acid moiety, may be functional in anchoring metal ions such as Zn^{2+}, where charge-charge
interactions are necessary to provide tight complexation. Additional
support for the proposition that the complexes of these two metal
ions are structurally dissimilar has come from the observed differences in their chemical and electrochemical properties (see Sec.
2.5).

2.4. Evidence for Formation of Ternary Complexes Involving DNA and Its Components

In addition to complexation between streptonigrin and metal ions, recent studies have definitively established the formation of ternary complexes involving the antibiotic, metal ion, and DNA. The first evidence for metal ion-promoted binding of streptonigrin by DNA was provided by White [16] in conjunction with studies demonstrating the effect of specific transition metal ions enhancing the bactericidal action of the antibiotic. Subsequently, Rao prepared a ternary complex consisting of streptonigrin-Zn^{2+}-DNA in the ratio of 1:7:25,* stable to dialysis and to gel filtration chromatography [15].

These findings provided important evidence toward clarification of earlier conflicting reports [19-21] regarding the nature of reversible vs. irreversible binding of streptonigrin to DNA. The structural role of metal ions in these complexes has not yet been elucidated. Furthermore, preliminary evidence has been obtained [22] to the effect that while metal ions potentiate tight complexation between streptonigrin and polynucleotides, there is only weak interaction using Zn^{2+} as a complexing metal at the mononucleotide level. It appears, therefore, that elucidation of the preferential nucleotide binding site on DNA will have to be determined using homopolynucleotides.

Involvement of ternary complexation has been suggested by Bachur et al. [17,18] as part of the overall mechanism of antitumor action of streptonigrin (see Sec. 3) utilizing the metal ion for targeting the antibiotic, or its semiquinone, to react directly with DNA.

2.5. Perturbation of the Chemical Properties of Streptonigrin as a Result of Complex Formation

The chemistry of streptonigrin-metal complexes was developed in an attempt to delineate the role of metal ions in the biological functioning of the antibiotic. As evidence became available regarding

---

*The ratio of 25 relates to the number of nucleotides [15].

the cofactors required for the antitumor action of streptonigrin,
followed by the identification of reactive intermediates [17,18],
it became possible to formulate specific questions toward the
elucidation of the chemical properties of the complexes.

Degradative damage of DNA by streptonigrin has been shown to
depend on (1) reductive activation of the antibiotic, (2) molecular
oxygen, and (3) metal ions. Furthermore, direct evidence has been
obtained for the formation of free radical intermediates, such as
streptonigrin semiquinone (3), which has been identified by EPR
spectroscopy [4,17,18]. Furthermore, administration of streptonigrin

(3)

has been shown to effect elevated in vivo levels of the inducible
enzyme superoxide dismutase, implicating the involvement of superoxide [23] in the overall sequence of reactions. Finally, it has
been observed that the bactericidal action of streptonigrin is enhanced by a series of transition metal ions, such as Zn^{2+}, Cu^{2+},
Fe^{2+}, Cd^{2+}, and Mn^{2+}, that are also known to promote binding of the
antibiotic to DNA [16].

Thus mechanistic elucidation of the reactions involved in the
biological activity of streptonigrin seemed to require (1) development of the redox chemistry of streptonigrin-metal complexes as it
relates to formation of reactive intermediates, (2) delineation of
the individual steps involved in oxygen activation, including identification of the resulting oxygen derivatives generated, and (3)
description of the role of metal ions in the formation of ternary
complexes including the antibiotic, the metal ion, and DNA.

Elucidation of the catalytic role of metal ions in activation of streptonigrin was first addressed by Lown [4,6]. Utilizing an in vitro assay system, developed for monitoring streptonigrin-induced cleavage of PM2 covalently closed circular DNA, addition of Cu^{2+} and Fe^{2+} was shown to result in rate acceleration, while addition of Co^{2+} appeared to inhibit degradation of the nucleic acid. As NADH (or borohydride) was found to be essential for streptonigrin-induced cleavage of DNA in vitro, Lown and Sim suggested [24] that the catalytic step is preceded by a two-electron reduction of the antibiotic to produce the corresponding dihydroquinone (4). This compound (4) then undergoes metal ion-catalyzed autoxidation to yield streptonigrin semiquinone [Eq. (2)]. This step in turn is followed by

electron transfer to O_2 leading to the formation of superoxide, which indeed has been detected in the reaction mixture [4,6,24].

Support for this hypothesis came from model studies using a series of synthetic 5,8-quinolinequinones, structurally related to streptonigrin, which exhibited both in vivo antitumor action and in vitro degradation of covalently closed circular DNA [4]. It has been pointed out, however, that since the process requires both reduction of the quinone and subsequent reoxidation of the hydroquinone, the relationship between the redox potentials (of the synthetic models vs. that of streptonigrin) and metal ion catalysis might be more complicated [4].

Studies initiated in the author's laboratory for elucidation of the chemistry of the complexes began focusing on the first reduction

TABLE 2

Reduction of Streptonigrin and Its Metal Complexes
by N-benzyldihydronicotinamide[a]

				$t_{1/2}$ (sec)	
				MeCN	0.005 M Hepes[b] pH 7.5
$[SN]_M$	$[CuCl_2]_M$	$[ZnCl_2]_M$	$[Py:H]_M$		
1.10×10^{-5}	0	0	2.20×10^{-4}	8.41×10^3	2.47×10^3
1.13×10^{-5}	4.5×10^{-4}	0	2.15×10^{-4}	<2	2.75×10^2
1.15×10^{-5}	0	4.5×10^{-4}	1.92×10^{-4}	N.R.[c]	2.63×10^2
0	4.3×10^{-3}	0	1.98×10^{-4}	N.R.[c]	N.R.[c]
0	0	4.5×10^{-4}	2.20×10^{-4}	N.R.[c]	N.R.[c]

[a]Unpublished results from the author's laboratory.
[b]Ionic strength adjusted to 0.1 with NaCl.
[c]N.R., no reduction observed.

step of the overall process. Specifically, Hajdu obtained evidence
[10] that in acetonitrile Cu^{2+} activates the antibiotic toward reduction by dihydronicotinamides (NADH and its N-alkyl analogs) whereas
complexation with Zn^{2+} inhibits the reaction. As the results in
Table 2 demonstrate, the copper(II) complex is reduced in acetonitrile by N-benzyldihydronicotinamide more than 1,000 times faster
than the metal-free antibiotic. Furthermore, the activation by Cu^{2+}
and inhibition by Zn^{2+} are strongly dependent on the solvent.

Comparison of the acceleratory effect of copper(II) in acetonitrile to that in water, corresponding to a decrease of more than two
orders of magnitude, clearly indicates that solvation has a major
role in influencing the catalytic reaction. The change in the reactivity of the zinc(II) complex from inhibition to a modest acceleration by the metal ion provides further evidence for solvent participation in the catalytic process. These observations, together with

the spectroscopic data indicating significant changes in the UV absorption of the complexes transferred from dipolar aprotic to hydroxylic media [10] (Figs. 1 and 2) strongly suggest that the nature of the microenvironment is an important feature of the mechanism of metal-ligand interactions in the activation of streptonigrin. It appears that efficient catalytic interactions might require formation of inner sphere complexes via direct metal-quinone coordination [25,26]. Specifically, in a dipolar aprotic solvent, such as acetonitrile, direct coordination to the quinone-oxygen is more readily achievable than in aqueous media, where solvent competition for the first coordination sphere of the metal decreases the efficacy of catalysis. Chemical precedent for a similar type of inhibition of catalysis by protic solvents comes from a closely related reaction involving metal ion-promoted dihydronicotinamide-dependent reduction of a chelating aldehyde [25]. For that system it has been demonstrated that addition of hydroxylic solvent (i.e., water) completely abolishes the catalytic reaction by preferentially coordinating to the complexing metal ion [25,26].

Additional evidence in support of the kinetic results has emerged from electrochemical studies [10]. Specifically, the cyclic voltammogram of streptonigrin is shifted on complexation with Cu^{2+} by 0.28 V toward less negative potentials while retaining its double-wave character. Complexation with Zn^{2+}, on the other hand, results in a completely irreversible reduction pattern shifted by -0.7 V toward more negative reduction potentials [10].

Hajdu proposed that coordination of the quinone oxygen by an activating metal such as copper(II) could readily account for the enhanced rate of chemical reduction via stabilization of the developing negative charge. Similar interaction could explain the greater ease with which the electrochemical reduction of the copper complex takes place.

The inhibitory effect of Zn^{2+} toward streptonigrin reduction, on the other hand, might be explicable in terms of a metal ion-assisted tautomeric shift [Eq. (3)] transforming the p-quinone into an aza-substituted o-quinoid structure (3b'), isoelectronic with the

biologically inactive isopropylidene azastreptonigrin (5) [12,13] and its related o-quinoid analogs. While this as yet tentative assignment is certainly consistent with the spectral, chemical, and electrochemical reduction data obtained for the streptonigrin zinc

(5)

complex, precedents for such tautomerism have been implied in previous investigations involving related aminoquinone derivatives
[Eq. (4)] [27].

(6)   (7)

$$(4)$$

While the zinc-catalyzed tautomeric shift in acetonitrile could
readily account for the inhibition of the chemical and electrochemical
reduction of the complex, experiments designed to test this hypothesis
should also provide important information regarding the mechanism of
action of the antibiotic, as the biological functioning of streptonigrin depends on its reduction. The question can now be addressed
directly by the study of the redox chemistry of the zinc complexes—
including streptonigrin derivatives and streptonigrin models—with
and without the potentially tautomeric amino protons in the molecule.

    The in vivo mechanism of streptonigrin activation by metal ions
remains to be elucidated. Formation of streptonigrin semiquinone,
for example, may involve either two-electron reduction followed by
metal ion-catalyzed autoxidation, or a single-step one-electron

reduction. Activation by the complexing metal ion is expected to assist stabilization of the developing negative charge in either case, yet the role of metal ions in the subsequent reactions involving streptonigrin semiquinone and leading to the eventual degradation of DNA is less obvious at present. Significantly, the question of their direct involvement in oxygen activation has not yet been addressed. Metal ions are likely to participate in (1) oxygen binding, (2) promoting electron transfer toward formation of superoxide, or (3) in catalysis of disproportionation of dioxygen species [4,9]. Alternatively, they might promote DNA binding by streptonigrin or streptonigrin semiquinone [15-18]. Through further characterization of the metal complexes of streptonigrin, these questions might be elucidated.

3. BIOLOGICAL ACTIVITY AND SUGGESTED MECHANISTIC SCHEMES FOR THE ANTITUMOR ACTION OF STREPTONIGRIN

3.1. Antitumor Activity

Streptonigrin is currently regarded as one of the most effective agents for the treatment of human cancers [27]. Exhibiting a broad spectrum of antitumor activities, it inhibits various tumors [28] including carcinoma 755, sarcoma 180, Lewis lung carcinoma, Walker 256 carcinosarcoma, and Ridgeway osteogenic sarcoma. The human tumor H.S. No. 1 grown in rats is particularly sensitive [29] and viral tumors such as Rauscher and Friend leukemia are inhibited [30,31].

Clinically, streptonigrin has been shown to be effective for the treatment of lymphoma, melanoma, and cancers of the breast, cervix, head, and neck [32,33]. During the past several years the compound has also been used in combination with other chemotherapeutic agents. In the United States these combinations included vincristine, prednisone, and bleomycin for the treatment of lymphosarcoma and reticulum cell sarcoma [34,35]. Combination chemotherapy

in Europe has been shown to be beneficial in treating primary malig-
nant melanoma [36] and non-Hodgkin lymphoma in children [37].

3.2. Toxicity

The clinical use of streptonigrin is severely limited at present
because of its high toxicity. Specifically, the compound (1) is the
most potent bone marrow depressant known [38]. Additional side
effects associated with the antibiotic include depression, nausea,
vomiting, diarrhea, and alopecia [3]. The maximum dose tolerated
is approximately 2 mg per treatment [4].

3.3. Target Sites for Antitumor Action

The evidence to date indicates that streptonigrin has as its principal
target site the nucleic acids and exerts its antitumor action by extensive degradation of DNA, causing disruption of the replicative
mechanism of the cell [5]. More recently it has been reported that
the antibiotic interferes with the cell respiratory mechanism as well
[4].

Streptonigrin causes chromosomal damage when added to cultures
of human leukocytes [39,40], inhibits DNA synthesis in tissue culture
cells [41-43] and in bacteria [19,42,44,45]. It has been found to
uncouple oxidative phosphorylation in human leukemic leukocytes [46].
The antibiotic decreases the ATP level as well as protein synthesis
in intact cells and causes strand breakage of extractable DNA [46].

3.4. Functional Groups Essential for
     Anticancer Activity

Considerable evidence has been gathered indicating that the 5,8-
quinolinequinone moiety is the structural element principally responsible for the expression of the antitumor character of streptonigrin

[4,47]. Other functional groups, such as the carboxyl of ring C,
appear to influence the *degree* of activity. Thus the methyl ester
of streptonigrin has only 1% activity of the parent compound in
tissue culture cells [48], 0.2% activity against Meloney leukemia
virus replication, and 33% activity in inhibiting DNA polymerase of
C-type RNA virus [49].

Intense effort has been expended in the area of chemical modification of streptonigrin [12,13], primarily to improve its chemotherapeutic properties. This goal, however, has not been achieved and the
modified derivatives were found to be considerably less effective as
antineoplastic agents than the parent antibiotic [9].

In recent years a new series of streptonigrin analogs became
available as a result of efforts aimed at the total synthesis of (1).
Among these derivatives compound (8) was found to be twice as active
as streptonigrin against *B. subtilis* [9] while compound (9) proved

to be inactive against KB cells [9]. These results seem to suggest
that in addition to the 5,8-quinolinequinone fragment the *metal-
binding site* involving the functional groups of ring C *is necessary
for significant biological potency.*

3.5. Cellular Activation and Inhibition

In a series of studies with *E. coli* White [41,45] demonstrated that
the requirements for the bactericidal activity of streptonigrin
include (1) electron source (i.e., a bioreductant such as glucose)

and (2) oxygen. Cyanide and carbonylcyanide phenylhydrazone have
enhanced lethality, while phenazine methosulfate (functioning as an
autoxidizable electron scavenger) has been shown to inhibit the bactericidal action of the antibiotic [45]. Furthermore, it has been
observed that *E. coli* and *B. megaterium* are capable of reducing
streptonigrin to the corresponding semiquinone (3), whose EPR spectrum was found to be identical with that of the chemically prepared
free radical [50,51] [Eq. (5)].

Subsequently, the enzyme xanthine oxidase was shown to reduce
streptonigrin [20], and more recently Bachur et al. obtained evidence
for the formation of streptonigrin semiquinone (3) in NADPH-dependent
reduction of the antibiotic by mammalian microsomes under aerobic
conditions [17,18].

Furthermore, it has been demonstrated that streptonigrin is
more toxic (bactericidal) in presence of oxygen and appears to function as an intracellular source of superoxide ($O_2^{-\cdot}$) [23]. In accord
with this view, increased levels of the inducible enzyme superoxide
dismutase were observed in streptonigrin-treated *E. coli* which exhibited resistance toward the antibiotic [23].

While these observations have implicated the participation of
both superoxide and streptonigrin semiquinone in the mechanism of
action of the antibiotic, it remains to be ascertained if the antitumor action occurs via superoxide-mediated path or if the semiquinone itself functions as a DNA-targeted site-specific free radical.

3.6 Suggested Mechanisms of Antitumor Action

The single-strand scission of DNA induced by streptonigrin is reminiscent of that produced by ionizing radiation [52,53]. The precise
mechanism by which degradation of the nucleic acid occurs is presently
not known. There are, however, two suggested schemes. The first one,
proposed by Lown [4], based largely on in vitro studies utilizing
streptonigrin and a series of synthetic 5,8-quinolinequinone model
compounds, implies superoxide- or hydroxyl radical-mediated disruption of DNA; the second one, advanced by Bachur et al. [17,18], based
on enzymatic studies involving mammalian microsomes, suggests that
streptonigrin semiquinone, generated in vivo by one of the microsomal
enzymes (possibly by NADPH cytochrome P-450 reductase), is the active
agent targeted at the DNA as a "site-specific free radical" without
the intervention of superoxide.

Lown's mechanism [4] describes the following sequence of events:

1. Streptonigrin is reduced in vivo to the dihydro compound
    which undergoes metal ion-catalyzed autoxidation.

$$(i) \quad SN + NADH \xrightarrow{enzyme} SNH_2 + NAD^+$$

$$(ii) \quad SNH_2 + O_2 \longrightarrow SNH\cdot + HO_2^{\cdot}$$

$$(iii) \quad HO_2^{\cdot} \rightleftharpoons O_2^{\bar{\cdot}} + H^+$$

2. At this point, superoxide might be subject to enzymatic
    conversion to hydrogen peroxide first, catalyzed by an
    inducible protective enzyme, superoxide dismutase, which
    step is then followed by catalytic disproportionation to
    oxygen and water, catalyzed by the enzyme catalase.

$$(iv) \quad 2O_2^{\bar{\cdot}} + 2H^+ \xrightarrow[dismutase]{superoxide} H_2O_2 + O_2$$

$$(v) \quad 2H_2O_2 \xrightarrow{catalase} 2H_2O + O_2$$

In absence of these enzymes, however, the superoxide
radical generated in step (iii) may further react with
hydrogen peroxide, which forms on rapid spontaneous dis-

mutation of O_2^-, as in reaction (iii), to produce hydroxyl radicals and these OH radicals in turn attack DNA.

$$\text{(vi)} \quad O_2^- + H_2O_2 \longrightarrow \cdot OH + {}^-OH + O_2$$

This scheme is consistent with the protective effect of superoxide dismutase [23] as well as with the synergistic protection by superoxide dismutase and catalase, observed in vitro [53] against degradation of covalently closed circular DNA (cccDNA) by streptonigrin and model 5,8-quinolinequinones, in presence of dihydropyridine or borohydride reductants. Furthermore, xanthine oxidase-generated superoxide [53] resulted in similar degradation of cccDNA, suggesting that in each case a similar type of intermediate might be involved.

The mechanism of antitumor action of streptonigrin according to the scheme developed by Bachur et al. [17,18] involves the streptonigrin radical (generated on enzymatic one-electron reduction of the parent antibiotic) as the reactive species which interacts directly with DNA. The formation of streptonigrin semiquinone in microsomal reduction of the antibiotic was evidenced by the observed EPR signal [17,18]. (A series of other quinone-containing antibiotics gave free radicals as well.) It was suggested that the resulting semiquinone has sufficient stability and selective affinity to enter the nucleus and bind to DNA. Unlike superoxide, which is subject to enzymatic conversion to hydrogen peroxide by normally available cellular detoxification utilizing superoxide dismutase [23], the antibiotic semiquinone could readily reach the DNA and cause its degradation.

Clearly, both mechanisms are incomplete concerning a number of details involved in the sequence of events, particularly because the chemistry of streptonigrin semiquinone and especially of the semiquinone-metal complexes are not yet understood at present.

In addition to the need to elucidate the catalytic mechanisms involved in the metabolic interconversion of the antibiotic and its metal complexes, there is the yet unanswered question of how streptonigrin or its semiquinone interacts with DNA. As metal ions are obligatory components in the antitumor apparatus, understanding of their

role in the system—both concerning oxidation-reduction and delivery/
affinity to DNA—is likely to advance the level of our knowledge of
the mechanism by which the system operates.

4. CONCLUSIONS

Since its discovery nearly 25 years ago, streptonigrin has been the
subject of intensive investigations regarding its potential use as an
anticancer drug, its mechanism of action, its chemical modification
and total synthesis. Despite its highly effective antitumor activity
against a wide spectrum of human cancers, the clinical use of the
antibiotic is still severely limited because of its toxicity.

Early efforts relying on structural modification of streptonigrin as well as syntheses of antibiotic analogs have failed to
produce an effective anticancer agent with tolerably low cytotoxicity.
More recently, three total syntheses [54-56] have been reported, and
attention has been directed toward elucidation of its biosynthetic
pathway [9] as well as development of microbial syntheses of antitumor-active analogs with tolerably low toxicity [9]. Despite this
resurgence of activity, none of these efforts has been able to change
the situation. It appears, therefore, that an approach based on
understanding the chemistry involved in the mechanism of action of
the drug should be a more promising route to improve its chemotherapeutic properties.

As it has been recognized that the metal complexes constitute
the antitumor-active form of the antibiotic, detailed characterization of the complexes and elucidation of the reactions involved in
their biological functioning should be of the highest priorities in
this regard. Specifically, delineation of the catalytic mechanisms
involved in the oxidation-reduction reactions as well as in the interaction of the complexes with DNA should provide important clues as to
how one might be able to affect their reactivities (i.e., with regard
to redox properties as well as to specificity/selectivity toward the
antitumor target) in order to produce the desired chemistry. It is

obvious from this chapter that the work which has been accomplished toward these goals is in its initial stages and, as such, it has contributed more to defining the problems rather than providing the answers for them. Nevertheless, the results that have already emerged clearly indicate that this direction of research provides a new and exciting way to modify the chemical and biological properties of this interesting and potent antitumor antibiotic.

ACKNOWLEDGMENTS

The persevering efforts of the author's associates, Ellen Armstrong and Donald Chace at Boston College, are gratefully acknowledged. The author is also indebted to Professor Steven J. Gould of Oregon State University and Professor Steven M. Weinreb of Pennsylvania State University for stimulating discussions, and Dr. John D. Douros of Drug Research and Development, Chemotherapy, NCI, for a generous gift of streptonigrin. Work in the author's laboratory was supported by the Research Corporation and by Grant No. 1424-C-1 of the American Cancer Society, Massachusetts Chapter.

REFERENCES

1. K. V. Rao and W. P. Cullen, *Antibiot. Ann.*, 950 (1959).

2. H. L. Davis, D. D. VonHoff, J. T. Henney, and M. Rozencweig, *Cancer Chemother. Pharmacol.*, *1*, 83 (1978).

3. W. L. Wilson, C. Labra, and E. Barrist, *Antibiot. Chemother.* (Basel), *11*, 147 (1961).

4. J. W. Lown, in *Bioorganic Chemistry* (E. E. Van Tamelen, ed.), Acad. Press, New York, 1977, pp. 95-121.

5. B. K. Bhuyan, in *Antibiotics* (I. D. Gottlieb and P. D. Shaw, eds.), Springer-Verlag, New York, 1967, p. 175.

6. J. W. Lown and S. K. Sim, *Can. J. Chem.*, *54*, 2563 (1976).

7. K. V. Rao, K. Biemann, and R. B. Woodward, *J. Am. Chem. Soc.*, *85*, 2532 (1963).

8. Y. H. Chiu and W. N. Lipscomb, *J. Am. Chem. Soc.*, *97*, 2525 (1975).

9. S. J. Gould and S. M. Weinreb, *Fortschr. Chem. Org. Natur.*, *41*, 77 (1982).

10. J. Hajdu and E. C. Armstrong, *J. Am. Chem. Soc.*, *103*, 232 (1981).

11. J. W. Lown and A. Begleiter, *Can. J. Chem.*, *52*, 2331 (1974).

12. W. B. Kremer and J. Laszlo, *Cancer Chemother. Rep.*, *19*, 51 (1967).

13. J. S. Driscoll, G. F. Hazard, H. H. Wood, and A. Goldin, *Cancer Chemother. Rep.*, *4*, Pt. 2, No. 2, 1-362 (1974).

14. R. W. Hay and C. R. Clark, *J. Chem. Soc.*, *1977*, 1866.

15. K. V. Rao, *J. Pharm. Sci.*, *68*, 853 (1979).

16. J. R. White, *Biochem. Biophys. Res. Commun.*, *77*, 387 (1977).

17. N. R. Bachur, S. L. Gordon, and M. V. Gee, *Cancer Res.*, *38*, 1745 (1978).

18. N. R. Bachur, S. L. Gordon, M. V. Gee, and H. Kon, *Proc. Natl. Acad. Sci. USA*, *76*, 914 (1979).

19. N. S. Mizuno and D. P. Gilboe, *Biochem. Biophys. Acta*, *224*, 319 (1970).

20. H. L. White and J. R. White, *Mol. Pharmacol.*, *4*, 564 (1968).

21. Yu. V. Dudnik, G. G. Gauze, V. L. Karpov, L. I. Kozmyan, and C. Padron, *Antibiotiki*, *18*, 968 (1973), *Chem. Abstr.*, *80*, 105000e.

22. J. Hajdu and D. Chace, unpublished results.

23. R. M. Hassan and I. Fridovich, *J. Bacteriol.*, *129*, 1574 (1977).

24. J. W. Lown and S. K. Sim, *Can. J. Biochem.*, *54*, 446 (1976).

25. For detailed discussion of a closely related system, see D. S. Sigman, J. Hajdu, and D. J. Creighton, in *Bioorganic Chemistry* (E. E. Van Tamelen, ed.), Academic Press, New York, 1978, Vol. IV, pp. 385-407.

26. D. J. Creighton, J. Hajdu, and D. S. Sigman, *J. Am. Chem. Soc.*, *98*, 4619 (1976).

27. T. K. Liao, W. H. Nyberg, and C. C. Cheng, *J. Heterocycl., Chem.*, *13*, 1063 (1976).

28. H. C. Reilly and K. Sugiura, *Antibiot. Chemother.*, *11*, 174 (1961).

29. M. N. Teller, S. F. Wagshul, and G. W. Wooley, *Antibiot. Chemother.*, *11*, 165 (1961).

30. T. J. McBride, J. J. Oleson, and D. Woolf, *Cancer Res.*, *26A*, 727 (1966).

31. P. S. Ebert, M. A. Chirigos, and P. A. Ellsworth, *Cancer Res.*, *28*, 363 (1968).

32. D. D. VonHoff, M. Rozencweig, W. T. Soper, L. J. Hellman, J. S. Penta, H. L. Davis, and F. M. Muggia, *Cancer Treatm. Rep.*, *61*, 759-768 (1977).

33. R. B. Livingston and S. K. Carter, *Single Agents in Cancer Chemotherapy,* Plenum Press, New York, 1970, pp. 389-392.

34. N. I. Nissen, T. F. Pajak, I. Glidewell, H. Blom, M. Flaherty, D. Hayes, O. R. McIntyre, and J. F. Holland, *Cancer Treatm. Rep., 61,* 1097 (1977).

35. R. J. Forcier, O. R. McIntyre, N. I. Nissen, T. F. Pajak, D. Glidewell, and J. F. Holland, *Med. Pediatr. Oncol., 4,* 351 (1978).

36. P. Banzet, C. Jacquillat, J. Civatte, J. Maral, C. Chastang, L. Israel, S. Belaich, J. C. Jourdarn, M. Weil, and G. Anclerc, *Cancer, 41,* 1240 (1978).

37. M. Gout-Lemerle, C. Rodary, and D. Sarrazin, *Arch. Fr. Pediatr., 33,* 527 (1976).

38. C. A. Hackethal, R. B. Golbey, C. T. C. Tan, D. A. Karnefsky, and J. H. Burchenal, *Antibiot. Chemother., 11,* 176 (1961).

39. M. M. Cohen, *Cyrogenetics, 2,* 271 (1963).

40. M. M. Cohen, M. W. Shaw, and A. P. Craig, *Proc. Natl. Acad. Sci. USA, 50,* 16 (1963).

41. H. L. White and J. R. White, *Mol. Pharmacol., 4,* 549 (1968).

42. N. S. Mizuno, *Biochem. Biophys. Acta, 108,* 394 (1965).

43. C. W. Young and S. Hodas, *Biochem. Pharmacol., 14,* 205 (1965).

44. M. Levine and M. Bartwick, *Virology, 21,* 568 (1963).

45. H. L. White and J. R. White, *Biochem. Biophys. Acta, 123,* 648 (1966).

46. D. S. Miller, J. Laszlo, K. S. McCarthy, W. R. Guild, and P. Hochstein, *Cancer Res., 27,* 632 (1967).

47. K. V. Rao, *Cancer Chemother. Rep., 4,* 11 (1974).

48. N. S. Mizuno, *Biochem. Pharmacol., 16,* 933 (1967).

49. M. A. Chirigos, J. W. Pearson, T. S. Papas, W. A. Woods, H. B. Wood, Jr., and G. Spahn, *Cancer Chemother. Rep., 57,* 305 (1973).

50. K. Ishizu, H. H. Dearman, M. T. Huang, and J. R. White, *Biochem. Biophys. Acta, 165,* 283 (1968).

51. J. R. White and H. H. Dearman, *Proc. Natl. Acad. Sci. USA, 54,* 887 (1965).

52. G. N. Gale, E. Cundliffe, P. E. Reynolds, M. M. Richmond, and M. J. Waning, *The Molecular Basis of Autibiotic Action,* Wiley-Interscience, New York, 1972, pp. 188-312.

53. R. Cone, S. K. Hasan, J. W. Lown, and A. R. Morgan, *Can. J. Biochem., 54,* 219 (1976).

54. A. S. Kende, P. Lorah, and J. Boatman, *J. Am. Chem. Soc., 103,* 1271 (1981).

55. S. M. Weinreb, F. Z. Basha, S. Hibino, N. A. Khatori, D. Kim,
     W. E. Pye, and T. T. Wu, *J. Am. Chem. Soc., 104,* 536 (1982).

56. D. L. Boger and J. S. Panek, *J. Org. Chem., 48,* 621 (1983).

Chapter 4

BLEOMYCIN ANTIBIOTICS: METAL COMPLEXES AND THEIR BIOLOGICAL ACTION

Yukio Sugiura
Faculty of Pharmaceutical Sciences
Kyoto University
Kyoto, Japan

Tomohisa Takita and Hamao Umezawa
Institute of Microbial Chemistry
Tokyo, Japan

1. INTRODUCTION

Bleomycins are a family of metalloglycopeptide antitumor antibiotics
clinically used in the treatment of Hodgkin's lymphoma, carcinomas of
the skin, head, and neck, and tumors of the testis mostly in combination with radiation or other chemotherapeutic agents [1]. The bleomycin complexes radiolabeled with ^{55}Co, ^{99m}Tc, and ^{111}In are also
useful in the tumor detection [2]. Cleavage of cellular DNA by bleomycin accounts for the antineoplastic action of bleomycin [3]. The
bithiazole and terminal amine residues contribute toward the interaction of DNA, and then the β-aminoalanine-pyrimidine-β-hydroxyhistidine portion is capable of dioxygen activation by the chelation with
ferrous ion. Indeed, the bleomycin-Fe(II)-O_2 complex system cleaves
isolated DNA specifically at G-C(5'→3') and G-T(5'→3') sequences [4],
and also can preferentially degrade the DNA sequences in open chromatin
within isolated nuclei [5].

The base-specific DNA binding of the drug is responsible for the
sequence specificity in the DNA cleavage. The specific nucleotide
recognition and the sequence-specific cleavage of DNA by bleomycin
antibiotics are a typical example of macromolecular receptor(DNA)-
drug(bleomycin) interaction in the field of chemotherapy. Certain
biosynthetic intermediates and synthetic analogs of bleomycin are
able to mimic the metal binding, dioxygen activation, and DNA cleavage
[6]. On the basis of various experimental results, we describe the
molecular mechanism of bleomycin action mechanism.

2. STRUCTURAL AND SYNTHETIC ASPECTS OF BLEOMYCIN

Various bleomycins produced by a bleomycin-producing microorganism,
such as *Streptomyces verticillus,* are different from one another in
the terminal amine moiety. Bleomycin consists of a linear hexapeptide and disaccharide (see Fig. 1). In the course of structural
studies of bleomycin and its Cu(II) complex, the natural form

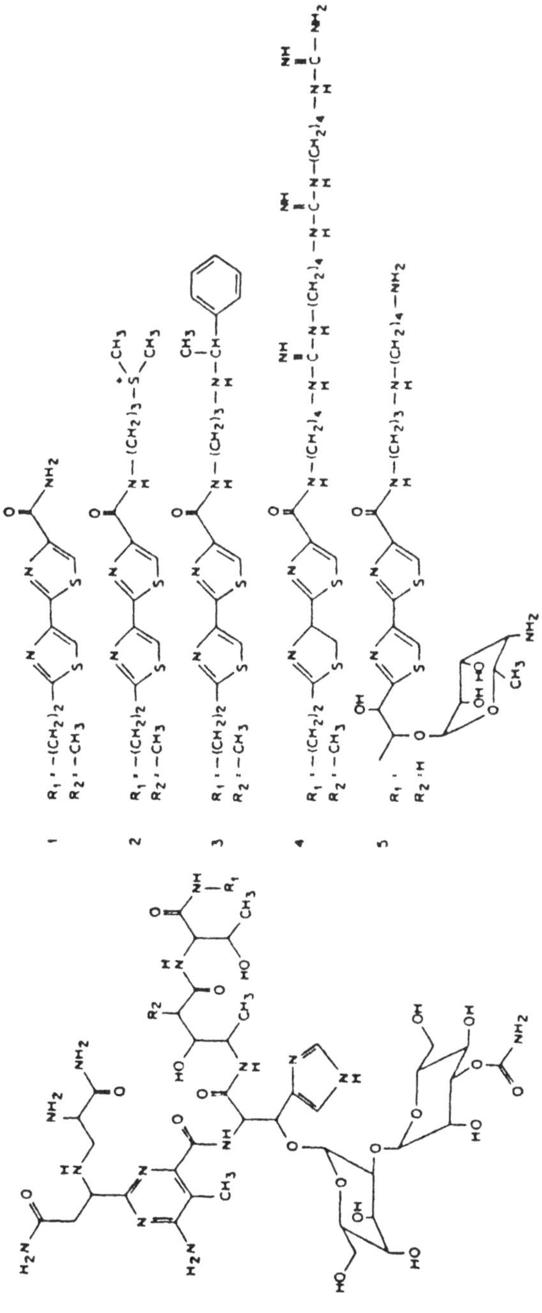

FIG. 1. Bleomycin and its analogues: (1), bleomycin B$_1$; (2), bleomycin A$_2$; (3), peplomycin; (4), phleomycin G; and (5), tallysomycin B.

produced by fermentation [7,8], Umezawa and his colleagues found
several specific reactions which are caused by neighboring group
participation, metal coordination, and enzymes: isomerization, epi-
merization, deamidation, depyruvamidation, decarbamoylation, and
deglycosidation. Therefore, several derivatives of bleomycin are
known: isobleomycin, having the carbamoyl group attached to 02' of
mannose; depyruvamide bleomycin, missing the alanine amide portion
of the antibiotic, and deamidobleomycin, having a hydrolyzed amide
of the aminoalanine-amide moiety of the drug.

Selective removal of the terminal amine moiety has also been
successfully achieved by chemical and enzymatic methods [9,10]. This
led to new artificial bleomycins, which were also prepared by fermentation by addition of artificial amines [11]. Peplomycin was selected
from hundreds of such artificial bleomycins and has been used clinically. In 1981, Takita et al. succeeded in the total synthesis of
bleomycin for the first time [12] and in 1982, the improved total
synthesis was established [13].

On the basis of this synthesis, we have planned the reconstruction of the bleomycin molecule from the fragments derived from bleomycin. As shown in Chart 1, the C-terminal tetrapeptide (tetrapeptide S), the sugar moiety, and pyrimidobleomic acid (1) were yielded

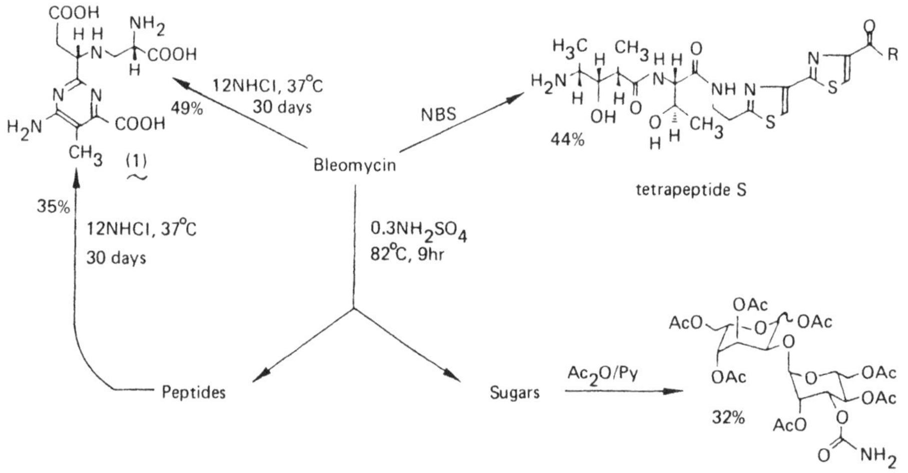

Chart 1

under specified conditions. Boc-pyrimidoblamic acid, which is a
key compound to synthesize bleomycin, has been derived from (1) by
esterification, Cu complexation, followed by butoxycarbonylation and
ammonolysis. The key reaction is a selective hydrolysis by Cu(II)
complexation. The stereoselective synthesis of erythro-β-hydroxy-
histidine, the central part of bleomycin molecule, was also achieved
by Cu(II)-mediated coupling of 4-formylimidazole and N-pyruvylidene-
glycinato-Cu(II) (see Chart 2). The reconstruction of the bleomycin
molecule from the above-mentioned fragments has been achieved by
utilization of the coupling reactions used in the total synthesis.
Establishment of the reconstruction of bleomycin enabled us to pre-
pare the desirable derivatives, in which a part of the bleomycin
molecule is replaced by other groups.

3. CHARACTERISTICS OF BLEOMYCIN-METAL COMPLEXES

Potentiometric studies revealed that the divalent metal complexes of
bleomycin have substantially a similar coordination core, though their
stabilities are in the order Fe(II) < Co(II) < Ni(II) < Cu(II) > Zn(II)
[14]. The bleomycin-Cu(II) complex has a square-pyramidal configuration in which the secondary amine nitrogen, pyrimidine ring nitrogen,

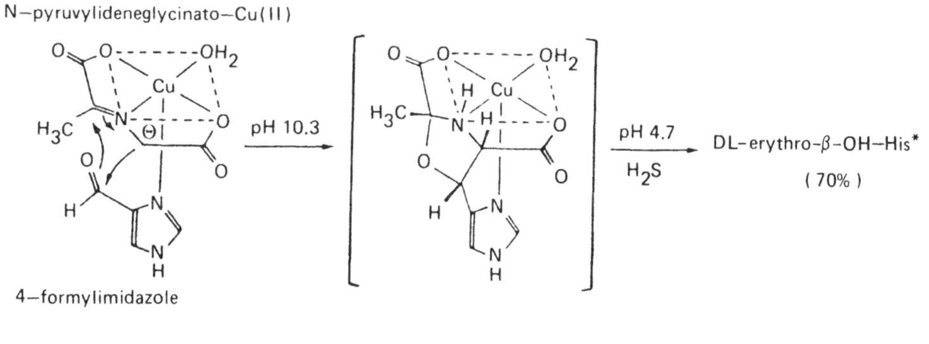

N—pyruvylideneglycinato—Cu(II)

4—formylimidazole

*The threo-isomer was not found in the reaction product.

Chart 2

deprotonated peptide nitrogen of histidine residue, and histidine
imidazole nitrogen bind to Cu(II) as planar ligand donors, and the
α-amino nitrogen as axial donor. Of special interest is the fact
that the planar Cu(II) coordination sites between bleomycin and human
serum albumin are similar, and the bleomycin-Cu(II) complex is more
stable than the albumin-Cu(II) complex. It is well known that human
serum albumin has a specific Cu(II)-binding site with histidine as
the third amino acid residue, Asp-Ala-His, and its Cu(II) complex is
a stable complex with donor set of $N_A(N_p)_2N_{Im}$ [15].

Indeed, the x-ray crystallographic result for the Cu(II) complex
of P-3A, a biosynthetic intermediate of bleomycin, demonstrated that
the Cu(II) site is a distorted square-pyramidal structure with four
chelate rings of 5-5-5-6 ring members coordinated by the α-amino,
secondary amine, pyrimidine ring, deprotonated peptide of histidine
residue, and histidine imidazole nitrogens, and that the Cu(II) ion
is displaced about 0.20 Å from the basal plane in the direction of
the axial α-amino nitrogen ligand [16]. The cyclic voltammogram of
the bleomycin-Cu(II) complex for Cu(II)/Cu(I) exhibited redox wave
with an $E_{1/2}$ value of -319 mV vs. the neutral hydrogen electrode
(NHE) [17]. Based on ^{1}H NMR and potentiometric pH titration data,
the bleomycin-Cu(I) complex is shown to have a geometry very different from that of the corresponding Cu(II) complex [18]. In addition,
if anaerobic reduction of the Cu(II)-bleomycin is carried out in the
presence of Fe(II), iron is chelated by the drug [19].

The bleomycin-Co(II) complex has the ESR features ($g_\perp$ = 2.272,
$g_\parallel$ = 2.025, $A_\parallel^{Co}$ = 92.5 G, and $A_\parallel^N$ = 13 G) typical of a low-spin
square-pyramidal structure with $(d_{xy})^2(d_{yz})^2(d_{zx})^2(d_{z^2})^1$ electronic
configuration. With oxygenation, the bleomycin-Co(II) complex forms
a monooxygenated low-spin Co(II)-dioxygen adduct complex and its ESR
spectrum undergoes drastic change [20]. The effective g values
($g_\perp$ = 2.007 and $g_\parallel$ = 2.098), the relationship of $g_\parallel > g_\perp \simeq 2.00$, and
the relatively smaller A_{iso}^{Co} value (15 G) of the bleomycin-Co(II)-O_2
complex strongly suggest that the unpaired spin density no longer
resides on the cobalt metal center, but instead resides on the

dioxygen moiety. The ESR characteristics closely resemble those
of typical monooxygenated low-spin Co(II) complexes such as Schiff
base-Co(II)-O_2, oxy-Co(II)-myoglobin, and oxy-Co(II)-peroxidase.
Electron spin echo envelope spectroscopy identified imidazole as a
metal ligand for the Co(II)-bleomycin as well as for the Cu(II)-drug
[21]. Two molecules of the mononuclear superoxo Co(III) complex
react together, with the loss of oxygen, to yield the dinuclear
μ-peroxo-Co(III) complex; the dimerization follows a second order
rate law with k = 200 ± 50 M^{-1} sec^{-1} at 25°C [22]. Meares and his
collaborators have suggested that the green Co(III)-bleomycin complex contains a hydroperoxide group bound to cobalt with unusual
stability [23].

The bleomycin-iron complexes with CO, NO, C_2H_5NC, OH^-, N_3^-,
CN^-, and CH_3NH_2 were characterized by electronic, ESR, ^1H NMR, and
Mössbauer spectroscopies and the findings were compared with the
corresponding hemoprotein complexes [24-26]. Table 1 summarizes the
spectroscopic characteristics of these iron complexes. The bleomycin-
Fe(II) complex has the Mössbauer parameters typical of a high-spin
ferrous ion. The quadrupole splitting (ΔE_Q) and the isomer shift (δ)
are considerably larger than those (ΔE_Q = 2.40 and δ = 0.90 mm/sec)
of reduced hemoglobin. The visible absorption spectra of the bleomycin-
Fe(II) complex with CO, C_2H_5NC, and NO differ markedly from that of
the 1:1 bleomycin-Fe(II) complex, and the absorption maxima are shifted
to a longer wavelength in the order C_2H_5NC > NO > CO. Burger et al.
[27] reported that a short-lived bleomycin-Fe(II)-O_2 complex has λ_{max} =
385 nm (ϵ 3,000).

Carbon monoxide, nitric oxide, dioxygen, and ethyl isocyanide
are potential π-acceptor ligands and hence these adducts are expected
to show a net shift of electron density from the iron to the ligands.
The Mössbauer features of the CO and C_2H_5NC adducts of the bleomycin-
Fe(II) complex are consistent with an S = 0 ferrous assignment, although the ΔE_Q value of the bleomycin-Fe(II)-CO complex is larger
than that of the corresponding hemoglobin complex (ΔE_Q = 0.36 and
δ = 0.18 mm/sec). Upon CO or C_2H_5NC binding to the bleomycin-Fe(II)
complex, the paramagnetic shifted protons completely disappeared,

TABLE 1

Electronic, ESR, ^{1}H-NMR, and Mössbauer Parameters of Bleomycin-Fe Complexes with Oxygen Analogous Ligands

Complex	Electronic spectra λ_{max}, nm (ε)	ESR spectra (77 K)			^{1}H NMR spectra (25°C) (ppm from TSP)	Mössbauer spectra (110 K)		Iron spin state
		g_1	g_2	g_3		δ (mms^{-1})	ΔE_Q (mms^{-1})	
Bleomycin-Fe(II)	476 (380)	no ESR signals			47.4, 45.1, 42.9, 34.7, 23.8, 20.6, 13.9, 9.9, -4.1, -5.2, -6.7, -11.8, -15.1	1.10	2.99	S = 2
Bleomycin-Fe(II)-NO	470 (2300)	2.041	1.976	2.008	broad (48.0, 33.1, 29.8, -17.1, -28.8)	0.36	0.78	S = 1/2 (·NO)
Bleomycin-Fe(II)-CO	380 (3000)	no ESR signals			no paramagnetic signals	0.19	0.66	S = 0
Bleomycin-Fe(II)-C$_2$H$_5$NC	495 (2700)	no ESR signals			no paramagnetic signals	0.18	0.62	S = 0
Bleomycin-Fe(III)-OH	365 (2000) 384 (1900)	1.893	2.185	2.431	11.6	0.37	2.53	S = 1/2
Bleomycin-Fe(III)-CN	390 (2000) 465 (800)	g = 4 (broad, 4 K)			64.5, 13.8, 12.6, -6.2, -10.8, -14.9, -19.4, -27.3, -41.9, -46.5	0.48	0.89	S = 5/2
Bleomycin-Fe(III)-N$_3$	355 (2400) 375 (2200)	1.857	2.223	2.492	17.2	0.32	2.95	S = 1/2
Bleomycin-Fe(III)-CH$_3$NH$_2$	350 (sh. 2700)	1.847	2.179	2.540	25.1	0.30	2.36	S = 1/2

Source: Reprinted from Ref. 25 by permission of Elsevier Biomedical Press B.V., Amsterdam.

indicating the presence of a diamagnetic Fe(II) ion (S = 0) in these adducts.

The OH^-, CH_3NH_2, and N_3^- adducts of the bleomycin-Fe(III) complex show the ESR, ^{1}H NMR, and Mössbauer spectra typical of a low-spin Fe(III) (S = 1/2). In particular, the ESR parameters for the N_3^- and OH^- adducts of the bleomycin-Fe(III) complex are close to those for the corresponding adducts of hemoproteins. The magnitude of the proton chemical shifts over ±50 ppm indicates a high-spin ferric type for the bleomycin-Fe(III)-CN^- complex [28]. The Mössbauer and ESR spectra of the CN^- adduct differ substantially from those of typical low-spin hemoprotein-cyanide complexes. Except for the CN^- adduct, the spectroscopic and crystal field parameters of these bleomycin-iron complexes are similar to those of the corresponding hemoprotein complexes.

In the bleomycin-Fe(II)-CO adduct complex, Oppenheimer et al. [29] suggested from the ^{1}H NMR chemical shifts that the six coordination sites of the Fe(II) ion are occupied by the imidazole, propionamide, pyrimidine, β-aminoalanine amide, and mannose carbamoyl groups in addition to the carbon monoxide. On the basis of various physicochemical results, we demonstrated five nitrogen coordinations of the α-amino, secondary amine, pyrimidine, deprotonated peptide, and imidazole groups for the bleomycin-iron complexes. The ^{13}C NMR study also indicates the Fe(II)-chelation by the amine-pyrimidine-imidazole portion [30].

4. OXYGEN ACTIVATION AND REDOX CYCLE OF THE BLEOMYCIN-IRON COMPLEX

Although the pink colored 1:1 bleomycin-Fe(II) complex is ESR negative at 77 K, the exposure of this iron complex to molecular oxygen yielded ESR positive species [24]. At the initial reaction stage, the unique low-spin Fe(III) complex signal with small rhombic splitting of the g values (g_1 = 1.937, g_2 = 2.171, and g_3 = 2.254) was evidently detected. At the final reaction stage, another stable bleomycin-Fe(III)-OH^- complex species was formed and characterized by the large rhombic splitting (g_1 = 1.893, g_2 = 2.185, and g_3 = 2.431). The intermediate bleomycin-Fe(III) complex was also obtained by the addition of KO_2 or H_2O_2 to the stable bleomycin-Fe(III)-OH^- complex and its ESR parameters are close to those of the complexes [hemoglobin-Fe(III)-O_2H^-] (g_1 = 1.94, g_2 = 2.16, and g_3 = 2.30) and [peroxidase-Fe(III)-O_2H^-] (g_1 = 1.95, g_2 = 2.16, and g_3 = 2.31), prepared by ^{60}Co γ irradiation of the corresponding oxyhemoproteins [31,32].

A similar ferric-peroxide structure has also been proposed in the single-electron reduced product of oxycytochrome P-450 [33]. Therefore, the intermediate bleomycin-Fe(III) complex is considered to be [bleomycin-Fe(III)-O_2H^-]. The Mössbauer spectrum (ΔE_Q = 3 ± 0.2 and δ = 0.10 ± 0.07 mm/sec) also gives convincing evidence that the iron of the intermediate bleomycin is low-spin ferric [34]. It is important to note that potential reactive dioxygen radicals such as •OH are generated from the bleomycin-Fe(II)-O_2 complex system [35]. In addition, the reduction potential of the bleomycin-Fe(III)-OH^- complex was estimated to be +0.13 V vs. NHE by using a micro-coulometric and optical absorption method [36]. The result reveals that the ferric bleomycin complex easily undergoes the reduction to the bleomycin-Fe(II) complex by the biological reducing agents such as NAD(P)H (E_p = -0.32 V vs. NHE at pH 7), glutathione (-0.23 V), and superoxide (-0.33 V). This high reduction potential appears to be one characteristic feature in bleomycin-iron redox reaction. On the basis of these observations, the following hypothetical mechanism for the cyclic function of the bleomycin-iron complex has been proposed [32] (see Fig. 2). Of interest is the similar cyclic function between the bleomycin-iron complex and hemoxygenases.

5. INTERACTION OF BLEOMYCIN AND ITS IRON COMPLEX WITH DNA

The quenching of the bithiazole fluorescence on addition of DNA has been used as an assay of DNA-drug complex formation. The binding constant for DNA of bleomycin and the saturation ratio of bleomycin

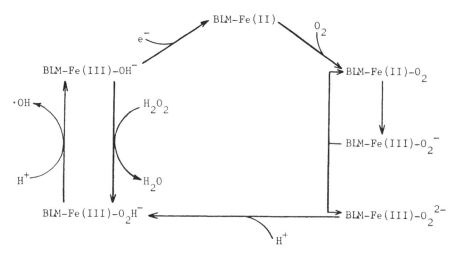

FIG. 2. Hypothetical mechanism for cyclic function of bleomycin-iron complex. Reproduced from Ref. 32 by permission of Academic Press, Inc.

equivalents per DNA nucleotide pair were thus determined to be 10^{-5} M^{-1} and 0.2, respectively [37]. The terminal "tripeptide S" moiety which comprises the terminal amine, aminoethylbithiazole, and threonine residues, is primarily required for the DNA binding because the apparent equilibrium constant for the binding of tripeptide S to DNA is similar to that of the parent antibiotic [38]. The planar, aromatic bithiazole ring binds intercalatively DNA, and then the terminal amine with a positive charge binds through electrostatic interaction with DNA. The specific broadening of the proton signals originated from the two bithiazole protons also suggests the participation of the groups in the DNA binding [37].

The kinetics and mechanism of binding of Cu(II)-bleomycin and Fe(III)-bleomycin to DNA were studied by using fluorometry, equilibrium dialysis, electric dichroism, temperature jump, and stopped-flow spectrophotometry [39]. The DNA affinity of Cu(II)-bleomycin is greater than that of metal-free bleomycin but less than that of Fe(III)-bleomycin. DNA is lengthened by 4.6 Å/molecule of bound Cu(II)-bleomycin and by 3.2 Å per bound Fe(III)-bleomycin, suggesting that both bleomycin-metal complexes indeed intercalate. When calf

thymus DNA was added to the bleomycin-Fe(II)-NO complex, the ESR
spectrum of the ferrous-NO complex clearly changed to become more
rhombic [40,41]. DNA induces a large shift of the g_x and g_y values
which are attributed to the in-plane anisotropy in the iron site.
In contrast, RNA does not entirely affect the original ESR parameters.
Similarly, the bleomycin-Fe(II)-CO complex binds to poly(dA-dT)·poly
(dA-dT) via its bithiazole group to change the envirónment of the
Fe(II) ion [42]. The binding affinity of bleomycin or its metal
complexes toward ribohomopolymers is in decreasing order poly G >
poly A > poly U > poly C, indicating higher affinity for guanine
bases [38]. The base-specific DNA binding of the antibiotic is
presumably responsible for the nucleotide sequence specificity in
DNA cleavage by the bleomycin-iron complex system.

6. SEQUENCE-SPECIFIC DNA CLEAVAGE BY THE BLEOMYCIN-IRON COMPLEX

The development of a new method for sequencing DNA [42] allowed us
to determine the base sequences in DNA which are preferentially
cleaved by bleomycin. Figure 3 shows the nucleotide sequences at

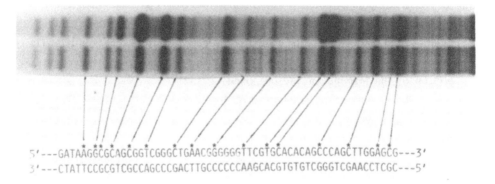

5'---GATAAGGCGCAGCGGTCGGGCTGAACGGGGGGTTCGTGCACACAGCCCAGCTTGGAGCG---3'
3'---CTATTCCGCGTCGCCAGCCCGACTTGCCCCCCAAGCACGTGTGTCGGGTCGAACCTCGC---5'

FIG. 3. Base release from double-stranded restriction fragment of
plasmid pBR 322 DNA by iron complex systems of bleomycin (upper)
and deglyco-bleomycin (lower).

the specific sites on a pBR 322 DNA fragment (327 base pairs) cleaved
by the bleomycin-Fe(II) complex system, together with the deglyco-
bleomycin-Fe(II) complex system [6]. Cleavage by the bleomycin-iron
system evidently occurred preferentially at G-C(5'→3') and G-T (5'→3')
sequences, and in particular the pyrimidine bases at the positions 30,
33, 41, and 59 were the most preferred cleavage sites. The guanine-
pyrimidine (5'→3') sequence specificity is consistent with the observations by D'Andrea and Haseltine [43] and Takeshita et al. [4], who
used lactose operon pL J3 and bacteriophage φX 174 DNA fragments,
respectively. The frequency of bases released were $C \geq T > A > G$,
and hence the preferred "recognition site" for bleomycin in DNA
involves $G_p C$ or $G_p T$ sequences.

Interestingly, the Fe(II) complex system of bleomycin and
deglyco-bleomycin showed almost identical nucleotide sequence-
specific mode for the DNA cleavages [6]. The result indicates that
the gulose-mannose sugar moiety does not produce a noticeable effect
on the specificity for the binding and cleavage of DNA by bleomycin
antibiotics. The binding specificity of the four bleomycin congeners
was also compared, and the result revealed no significant contribution
of the terminal amine to the nucleotide sequence specificity [44].
On the other hand, several bithiazole derivatives were tested for
their ability to inhibit the bleomycin-DNA interaction [45]. It was
found that although the DNA degradation was diminished, the specificity of DNA cleavage by iron-bleomycin was not altered by any of
the bithiazoles tested.

In contrast, we observed that certain DNA-binding agents such
as distamycin A (DMC) and actinomycin D(AMD) clearly alter the sequence specificity of DNA breakage by the bleomycin-iron complex
system [46] (see Fig. 4). Distamycin A masked the cleavage at certain G-T and G-A sequences, and produced higher specificity for G-C
sequences than that of iron-bleomycin only. This is reasonably
explained by the specific A and T binding of distamycin A, having
planar pyrrole rings connected to each other through peptide bonds.
Actinomycin D as well as ethidium bromide (ETB) also influenced the
nucleotide sequence specificity of DNA cleavage. In this case, the

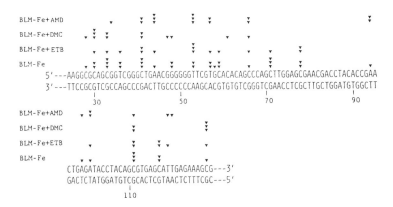

FIG. 4. Base release from the DNA fragment by iron-bleomycin in the absence or presence of certain intercalators. The release of bases and the relative intensity of bands are indicated by the symbol and its numbers, respectively.

preferred cleavage sites were shifted from G-C sequences to G-A and G-T sequences. Partial wobble of the DNA duplex by the phenoxazone ring and/or the peptide loop of actinomycin D may induce a delicate change for DNA binding by the bithiazole group of bleomycin.

The DNA-cleaving activity of bleomycin was substantially modified by cis-dichlorodiammineplatinum(II), and a number of specific new cutting sites in guanine-rich parts of the sequence were activated by the platinum complex [47]. The result emphasizes the possibility that the synergism found when cis-dichlorodiammineplatinum(II) and bleomycin are used in combination chemotherapy may be due to interactions at the level of DNA-drug binding. In contrast with cis-dichlorodiammineplatinum(II) which induces intrastrand crosslinking of guanine N7 atoms, the monocovalent guanine modifiers, such as aflatoxin B_1, dimethyl sulfate, and mitomycin C, did not give remarkable alteration of the DNA nucleotide cleavage mode by bleomycin [48].

These results clearly show the difference of DNA cleavage modes by bleomycin among the monocovalent, crosslinked, and intercalated modifications of guanine residues. Certainly, the preferred "recognition site" of bleomycin in DNA involves G_pC and G_pT sequences and the nucleotide sequences with reversed polarity (C_pG and T_pG) are not recognized. The guanine-pyrimidine (5'→3') specificity of the iron-bleomycin system appears to be mainly due to the interaction between the bleomycin bithiazole group and DNA guanine bases.

The ^{32}P-labeled DNA cleavage experiments showed that the biological activity of the inactive bleomycin-Fe(III)-OH⁻ complex is evidently induced by addition of H_2O_2 or by irradiation of UV light [32]. Similar light-induced nicking of DNA is observed in the bleomycin-Co(III) complex [49]. Using a 5'-end-labeled DNA fragment of defined sequence, the sites of DNA damaged by the light-activated Co(III)-bleomycin are found to be very similar to those caused by the iron-bleomycin complex system.

7. METAL COMPLEXES AND DNA CLEAVAGE OF BLEOMYCIN ANALOGS: PEPLOMYCIN, TALLYSOMYCIN, AND PHLEOMYCIN

Peplomycin is a bleomycin antibiotic which contains a secondary amine and a terminal phenyl group as the side chain in the terminal amine portion. Tallysomycin also resembles bleomycin but contains an additional amino-sugar moiety and a methyl group linked to the backbone of the molecule. Phleomycin is structurally related to bleomycin, differing in that one of the two thiazole ring moieties is partially saturated. The metal-binding properties and antitumor mechanism of these antibiotics appear to be similar to those of bleomycin. In fact, the tallysomycin-metal complexes give the ESR features similar to the corresponding bleomycin complexes: tallysomycin-Cu(II) complex ($g_\perp$ = 2.057, $g_\parallel$ = 2.210, and $A_\parallel$ = 178 G); tallysomycin-Co(II) complex ($g_\perp$ = 2.270, $g_\parallel$ = 2.025, $A_\parallel$ = 92.2 G, and A^N = 13 G), and tallysomycin-Co(II)-O_2 complex ($g_\perp$ = 2.008, $g_\parallel$ = 2.101, and $A_\parallel$ = 20.0 G) [50].

Using ^{13}C NMR, Dabrowiak et al. suggested the region of iron binding for both bleomycin and tallysomycin which do apparently not involve the 2,4'-bithiazole moiety [30]. Identical ESR behavior

to the bleomycin-Fe(III) complex systems has also been observed for
the tallysomycin-Fe(III) complex systems [30]. Similarly, Cu(I)-
bleomycin and Cu(I)-tallysomycin have been shown to be air-sensitive
and readily oxidize in the atmosphere to produce radicals and a
Cu(II)-antibiotic [50]. The ^{13}C NMR study of the tallysomycin-Zn(II)
complex also indicated that the metal-binding region for tallysomycin
was the same as that found for bleomycin, although it was not possible
to identify specific tallysomycin donor atoms from the observed carbon
resonance shifts [51]. Therefore, the metal-binding properties of
these antibiotics, bleomycin, peplomycin, tallysomycin, and phleomycin, are considered to be substantially the same.

On the other hand, bleomycin, peplomycin, and phleomycin preferentially attack G-C (5'→3') and G-T (5'→3') sequences, and display
generally similar sequence specificities in DNA cleavages [44,52,53].
However, phleomycin shows considerably less preference for purines
than does any of the bleomycin, reflecting the increased difficulty
encountered by the partly nonplanar phleomycin ring in forming an
intercalation complex. Povirk et al. [39] indicated that Cu(II) and
Fe(III)-bleomycin complexes intercalate while the Cu(II)-phleomycin
complex does not. It is unlikely that intercalative and noninter-
calative binding could cause precisely the same specific cleavage
of G-C and G-T sequences. In order to clarify the question, the
detailed interaction between phleomycin and DNA must be investigated.

The base specificity of tallysomycin differs quantitatively
from that of the bleomycin analogs in that it shows the most marked
preference for the G-T sequence and a general preference for adenine
over cytosine. As discussed above, the terminal amine does not contribute significantly to base specificity. Therefore, the altered
specificity of tallysomycin must be attributed to the presence of
the additional amino sugar and/or methyl moieties. Synthesis of
other analogs modified in this region of the bleomycin molecule may
be of considerable interest.

Peplomycin is a modified drug of bleomycin which was developed
to enhance the therapeutic activity and to minimize the pulmonary
and renal toxicities caused by the bleomycin drug. Such modification

of the side chains does not give remarkable alteration on base-
specific DNA cleavage but is expected to alter transport, organ
distribution, and perhaps toxicity of the drug.

8. METAL COMPLEXES AND OXYGEN ACTIVATION BY SYNTHETIC ANALOGS AND BIOSYNTHETIC INTERMEDIATES OF BLEOMYCIN

Although the structure of the bleomycin-metal complexes as seen in
Fig. 5 seems to be most reasonable [24], some possible transition
metal-binding sites have been proposed on the basis of the spectroscopic investigations. The coordination of the sugar carbamoyl group
for Cu(II) [54] and Fe(II) [29] complexes of bleomycin, and the bind-
ings of the diaminopropionamide and the β-aminopropionamide groups
for Co(III) complexes [55] have been proposed. A recent ^{1}H nuclear
relaxation study of the bleomycin-Mn(II) complex also suggested the
bithiazole group as a metal ligand [56].

In order to confirm the metal-binding sites and to clarify the
role of the respective functional groups, we have studied metal binding, dioxygen interaction, and DNA cleavage of the synthetic analogs
and biosynthetic intermediates of bleomycin as shown in Fig. 6 [6].
N-[6-[[[(S)-2-amino-2-(carbamoyl)-ethyl]amino]methyl]pyridine-2-
carbamoyl]-L-histidine(PYML-1) and its methyl ester(PYML-2) are
synthetic analogs corresponding to the metal-binding site shown in
Fig. 5. N-(2-aminoethyl)-6-[[(2-aminoethyl)amino]methyl]pyridine-2-
carboxamide(PEML) has a simple amino group instead of an imidazole
group.

Table 2 summarizes the ESR parameters for the divalent metal
complexes of these ligands [6]. Except for the PEML-Fe(II) complex,
the Fe(II) complexes of PYML, P-3A, deglyco-bleomycin, and bleomycin
react with carbon monoxide, ethyl isocyanide, and nitric oxide to
form these adduct complexes. ^{57}Fe Mössbauer spectra of the PYML-1-
Fe(II) complex (ΔE_Q = 3.00 and δ = +1.05 mm/sec) and its CO adduct
(ΔE_Q = 0.51 and δ = +0.18 mm/sec), measured at 110 K in zero magnetic field, are remarkably close to those of the corresponding

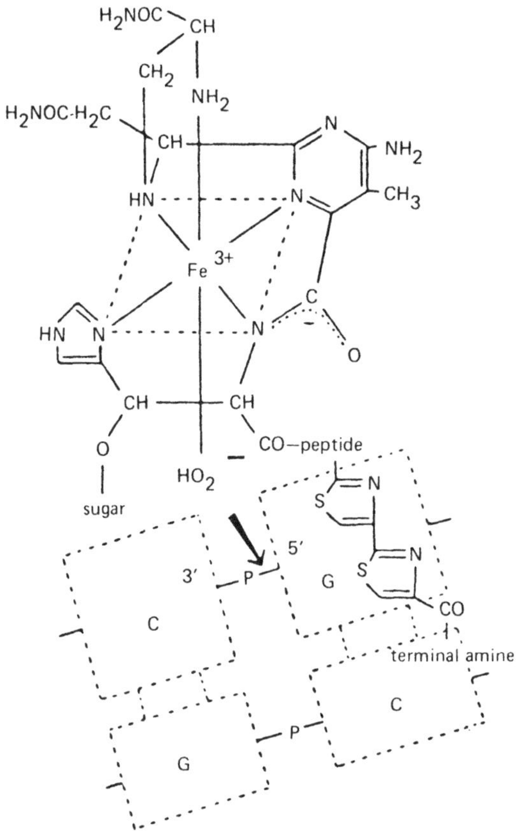

FIG. 5. Active iron-oxygen species of bleomycin for DNA cleavage. Reproduced from Ref. 24 by permission of the American Chemical Society.

bleomycin complexes. The ^{1}H NMR behaviors of the PYML-1-Fe(II) complex and its CO adduct are also in accordance with the results of the ^{1}H NMR spectra of the bleomycin-Fe(II) complex and its CO adduct which indicate the presence of high-spin Fe(II) ion (S = 2) and diamagnetic Fe(II) ion (S = 0), respectively [57]. In addition, the ESR spin-trapping experiments showed that hydroxyl radicals are generated from the PYML-1-Fe(II)-O_2 complex system. Carbon monoxide competitively interferred the dioxygen activation by the PYML-1-Fe(II) complex. The relative radical spin concentration of the

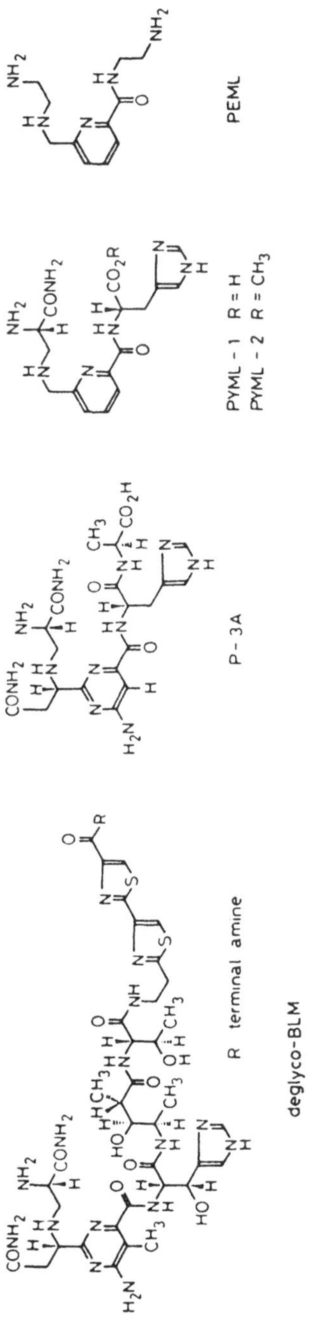

FIG. 6. Biosynthetic intermediates and synthetic analogs of bleomycin.

TABLE 2

ESR Parameters for Divalent Metal Complexes of PYML, PEML, P-3A, Deglyco-BLM, and BLM

Complex	$g_\parallel$ (g_z)	$g_\perp$ ($g_x g_y$)	$A_\parallel$, G	A_v, G	Nitrogen-hyperfine splitting (line)
PYML-1-Cu(II)	2.206	2.048	179.4		
PYML-1-Co(II)	2.022	2.255	92.5	13	3
PYML-1-Co(II)-O_2	2.093	2.005	22.5		
PYML-1-Fe(II)		No ESR signals			
PYML-1-Fe(II)-^{14}NO	2.009	2.036 1.972		25.6	3
PYML-1-Fe(II)-^{15}NO	2.009	2.036 1.972		35.6	2
PYML-1-Fe(II)-^{14}NO + DNA	2.009	2.036 1.972		25.6	3
PYML-2-Cu(II)	2.204	2.052	179.8		
PYML-2-Co(II)	2.020	2.252	92.0	13	3
PYML-2-Co(II)-O_2	2.093	2.005	22.5		
PYML-2-Fe(II)		No ESR signals			
PYML-2-Fe(II)-^{14}NO	2.009	2.036 1.970		25.5	3
PEML-Cu(II)	2.237	2.050	170.8		
PEML-Co(II)[a]	2.018	2.260	84.3	13	3
PEML-Co(II)-O_2		Not detected			
PEML-Fe(II)[a]		No ESR signals			
PEML-Fe(II)-^{14}NO		Not detected			

Complex	g_1	g_2	g_3	A_1	A_2	n
P-3A-Cu(II)	2.214	2.133	2.078	167.3		3
P-3A-Co(II)	2.027	2.275		93.8	13	3
P-3A-Co(II)-O_2	2.102	2.007		22.4		
P-3A-Fe(II)	No ESR signals					
P-3A-Fe(II)-^{14}NO	2.007	2.038	1.968		24.8	3
P-3A-Fe(II)-^{15}NO	2.007	2.038	1.969		35.0	2
P-3A-Fe(II)-^{14}NO + DNA	2.007	2.038	1.968		24.8	3
Deglyco BLM-Cu(II)	2.214	2.131	2.077	167.0		
Deglyco BLM-Co(II)	2.027	2.277		95.0	13	3
Deglyco BLM-Co(II)-O_2	2.100	2.009		22.5		
Deglyco BLM-Fe(II)	No ESR signals					
Deglyco BLM-Fe(II)-^{14}NO	2.007	2.038	1.969	24.8		3
Deglyco BLM-Fe(II)-^{15}NO	2.007	2.038	1.970	33.0		2
Deglyco BLM-Fe(II)-^{14}NO + DNA	2.007	2.046	1.963	25.1		3
BLM-Cu(II)	2.211	2.055		183.0		
BLM-Co(II)	2.025	2.272		92.5	13	3
BLM-Co(II)-O_2	2.098	2.007		20.2		
BLM-Fe(II)	No ESR signals					
BLM-Fe(II)-^{14}NO	2.008	2.041	1.976		23.6	3
BLM-Fe(II)-^{15}NO	2.008	2.040	1.976		31.6	2
BLM-Fe(II)-^{14}NO + DNA	2.006	2.060	1.962		24.0	3

[a] The complex was detected only under the fully anaerobic condition which was achieved by using a vacuum line.

PYML-1 (or P-3A)-Fe(II) and deglyco-bleomycin-Fe(II) complex systems
was estimated to be approximately 20 and 40% of that of the corresponding bleomycin-Fe(II) complex system, respectively.

These results provide the most reliable evidence for the proposed metal-binding sites in which (1) the β-aminoalanine-pyrimidine-
β-hydroxyhistidine portion of the bleomycin ligand is substantially
important for the Fe(II), Co(II), and Cu(II) interactions and (2)
the gulose-mannose, methylvalerate, and bithiazole groups in bleomycin are not participating as direct ligands toward the metal coordination. This result is consistent with the structural assignment
by x-ray analyses for the P-3A-Cu(II) complex [16] and the Co(III)
complex of pseudotetrapeptide A of bleomycin [58]. In the total
structure of bleomycin, consequently, the portion corresponding to
PYML is considered to be responsible for metal binding and dioxygen
activation, and the sugar moiety seems to play an important role as
an environmental factor in the effective dioxygen activation, just
as the pivalimidophenyl groups in the picket fence porphyrins [59].
The presence of the imidazole group, in particular the trans position
of pyrimidine (or pyridine) and imidazole groups in the Fe(II) coordination, appears to be essential for the effective binding and activation of molecular oxygen by the Fe(II) complexes of bleomycin-
related ligands.

The Fe(II) complexes of PYML and P-3A showed the effective
dioxygen activation but were remarkably less active than the corresponding bleomycin and deglyco-bleomycin complexes in the DNA cleavage
reaction. Therefore, the DNA-binding molecule to deliver a metal ion
to the site of the DNA helix, where activated molecular oxygen attacks
the DNA, is required for the effective DNA strand scission. Indeed,
DNA was clearly cleaved by a combination of PYML and a DNA-interacting
site consisting of 1-methylpyrrole systems. Dervan et al. also reported interesting DNA cleaving molecules, methidium-propyl-EDTA-
Fe(II) [60], distamycin-EDTA-Fe(II) [61], and penta-N-methylpyrrole-
carboxamide-EDTA-Fe(II) [62].

9. MOLECULAR MECHANISM OF BLEOMYCIN ACTION

The cytotoxic activity of bleomycin antibiotics results from DNA cleavage, which is also accomplished in vitro by a reaction system containing Fe(II), drug, and molecular oxygen. The base sequence-specific DNA cleavage by bleomycin is considered to be due to (1) selective DNA binding by bithiazole-terminal amine portion and (2) metal chelation and dioxygen activation by the β-aminoalanine-pyrimidine-β-hydroxyhistidine moiety. The iron ligand donors in bleomycin are arranged in a rigid square-pyramidal configuration with a 5-5-5-6 ring membered system, as seen in the case of heme. Probably, the aromatic nitrogen-containing and electron-rich structure formed by the trans coordination of pyrimidine (or pyridine) and imidazole nitrogen ligands creates the same properties in the iron electronic state as found in hemoproteins.

The species attacking DNA formed effectively from the redox cycle of the bleomycin-Fe(II)-O_2 complex system is considered to be a ferric-peroxide species of bleomycin (see Figs. 2 and 5). Iron-bleomycin induces the release of free bases from DNA glycosidic linkage and also the cleavage of the polymer backbone at the deoxyribose C3-C4 bond [63,64]. The guanine-pyrimidine (5'→3') sequence specificity such as GCGC in DNA cleavage of bleomycin indicates that a guanine base portion of the DNA molecule plays a typical role as a receptor having a specific interaction with the bithiazole moiety of bleomycin. Indeed, a computer graphic model-building study of 3-(2'-phenyl-2,4'-bithiazole-4-carboxamido)propyldimethylsulfonium iodide, an analog of the DNA-binding portion of bleomycin A_2, has shown that the phenyl group and the second thiazole ring can be intercalated between the base pairs of the double-stranded deoxydinucleoside phosphate $d(C_pG)$, and also that the sulfonium cation can interact with a backbone phosphate group [65]. It is also known that double-stranded DNA is a much better substrate for bleomycin cleavage than is single-stranded DNA [52]. As illustrated in Fig. 5, "the site-specific activated dioxygen species" produced from iron-bleomycin is able to account for the action mechanism of DNA cleavage by bleomycin.

10. CONCLUSIONS

The antitumor and antimicrobial antibiotic bleomycin cleaves DNA in a reaction that is dependent on the presence of Fe(II) ion and molecular oxygen. Bleomycin is a bifunctional compound consisting of the binding site to DNA and the reaction site with DNA. The β-aminoalanine-pyrimidine-β-hydroxyhistidine portion of bleomycin is essential for the metal binding and dioxygen activation, and then the bithiazole moiety contributes to the specific binding to the guanine base of DNA.

The bleomycin-iron complexes with CO, NO, C_2H_5NC, OH^-, N_3^-, and CN^- were characterized by electronic, ESR, ^1H NMR, and Mössbauer spectroscopies. Of special interest is the fact that the physico-chemical properties of the bleomycin-iron complexes and the corresponding hemoprotein complexes are remarkably similar.

In the DNA strand scission, the bleomycin-Fe(II) complex system preferentially attacked the pyrimidine bases which were located in G-C (5'→3') and G-T (5'→3') sequences.

Certain oligopeptides are able to mimic the metal binding and dioxygen activation by bleomycin, but they cannot induce the effective DNA cleavage. For the effective DNA strand scission, a DNA-binding molecule must deliver a metal ion to the site of the DNA helix attacked by the activated dioxygen species. The nucleotide sequence-specific binding and cleavage of DNA by bleomycin antibiotics are also of interest as an example of a specific interaction with the biological receptor of the drug.

REFERENCES

1. H. Umezawa, *Lloydia, 40,* 67 (1977).

2. O. E. Nieweg, H. Beekhuis, A. M. J. Paans, D. A. Piers, W. Vaalburg, J. Welleweers, T. Wiegman, and M. G. Woldring, *Eur. J. Nucl. Med., 7,* 104 (1982).

3. H. Umezawa, in *Bleomycin: Current Status and New Developments* (S. K. Carter, S. T. Crooke, and H. Umezawa, eds.), Academic Press, New York, 1978, p. 15 ff.

4. M. Takeshita, A. P. Grollman, E. Ohtsubo, and H. Ohtsubo, *Proc. Natl. Acad. Sci. USA, 75,* 5983 (1978).

5. M. Kuo, *Cancer Res., 41,* 2439 (1981).

6. Y. Sugiura, T. Suzuki, M. Otsuka, S. Kobayashi, M. Ohno, T. Takita, and H. Umezawa, *J. Biol. Chem., 258,* 1328 (1983).

7. T. Takita, Y. Muraoka, T. Nakatani, A. Fujii, Y. Umezawa, H. Naganawa, and H. Umezawa, *J. Antibiot., 31,* 801 (1978).

8. T. Takita, Y. Muraoka, T. Nakatani, A. Fujii, Y. Iitaka, and H. Umezawa, *J. Antibiot., 31,* 1073 (1978).

9. T. Takita, A. Fujii, T. Fukuoka, and H. Umezawa, *J. Antibiot., 26,* 252 (1973).

10. H. Umezawa, Y. Takahashi, A. Fujii, T. Saino, T. Shirai, and T. Takita, *J. Antibiot., 26,* 117 (1973).

11. A. Fujii, T. Takita, N. Shimada, and H. Umezawa, *J. Antibiot., 27,* 73 (1974).

12. T. Takita, Y. Umezawa, S. Saito, H. Morishima, N. Naganawa, H. Umezawa, T. Tsuchiya, T. Miyake, S. Kageyama, S. Umezawa, Y. Muraoka, M. Suzaki, M. Otsuka, M. Narita, S. Kobayashi, and M. Ohno, *Tetrahedron Lett., 23,* 521 (1982).

13. S. Saito, Y. Umezawa, T. Yoshioka, T. Takita, H. Umezawa, and Y. Muraoka, *J. Antibiot., 36,* 92 (1983).

14. Y. Sugiura, K. Ishizu, and K. Miyoshi, *J. Antibiot., 32,* 453 (1979).

15. K. S. Iyer, S. J. Lau, S. H. Laurie, and B. Sarkar, *Biochem. J., 169,* 61 (1978).

16. Y. Iitaka, H. Nakamura, T. Nakatani, Y. Murata, A. Fujii, T. Takita, and H. Umezawa, *J. Antibiot., 31,* 1070 (1978).

17. K. Ishizu, S. Murata, K. Miyoshi, Y. Sugiura, T. Takita, and
     H. Umezawa, *J. Antibiot.*, *34*, 994 (1981).

18. N. J. Oppenheimer, C. Chang, L. O. Rodriguez, and S. M. Hecht,
     J. Biol. Chem., *256*, 1514 (1981).

19. J. H. Freedman, S. B. Horwitz, and J. Peisach, *Biochemistry, 21*,
     2203 (1982).

20. Y. Sugiura, *J. Am. Chem. Soc.*, *102*, 5216 (1980).

21. R. M. Burger, A. D. Adler, S. B. Horwitz, W. B. Mims, and
     J. Peisach, *Biochemistry, 20*, 1701 (1981).

22. J. P. Albertini and A. Garnier-Suillerot, *Biochemistry, 21*,
     6777 (1982).

23. C.-H. Chang, J. L. Dallas, and C. F. Meares, *Biochem. Biophys.
     Res. Commun.*, *110*, 959 (1983).

24. Y. Sugiura, *J. Am. Chem. Soc.*, *102*, 5208 (1980).

25. Y. Sugiura, T. Suzuki, H. Kawabe, H. Tanaka, and K. Watanabe,
     Biochim. Biophys. Acta, 716, 38 (1982).

26. Y. Sugiura, J. Kuwahara, and T. Suzuki, *J. Chem. Soc. Chem.
     Commun.*, 908 (1982).

27. R. M. Burger, S. B. Horwitz, J. Peisach, and J. B. Wittenberg,
     J. Biol. Chem., *254*, 12299 (1979).

28. Y. Sugiura, S. Ogawa, and I. Morishima, *J. Am. Chem. Soc., 102*,
     7944 (1980).

29. N. J. Oppenheimer, L. O. Rodriguez, and S. M. Hecht, *Proc. Natl.
     Acad. Sci. USA, 76*, 5616 (1979).

30. J. C. Dabrowiak, F. T. Greenaway, F. S. Santillo, and S. T.
     Crooke, *Biochem. Biophys. Res. Commun.*, *91*, 721 (1979).

31. H. Kuramochi, K. Takahashi, T. Takita, and H. Umezawa, *J. Anti-
     biot.*, *34*, 576 (1981).

32. Y. Sugiura, T. Suzuki, J. Kuwahara, and H. Tanaka, *Biochem.
     Biophys. Res. Commun.*, *105*, 1511 (1982).

33. R. C. Blake and M. J. Coon, *J. Biol. Chem.*, *255*, 4100 (1980).

34. R. M. Burger, T. A. Kent, S. B. Horwitz, E. Munck, and J.
     Peisach, *J. Biol. Chem.*, *258*, 1559 (1983).

35. Y. Sugiura and T. Kikuchi, *J. Antibiot.*, *31*, 1310 (1978).

36. D. L. Melnyk, S. B. Horwitz, and J. Peisach, *Biochemistry, 20*,
     5327 (1981).

37. M. Chien, A. P. Grollman, and S. B. Horwitz, *Biochemistry, 16*,
     3641 (1977).

38. H. Kasai, H. Naganawa, T. Takita, and H. Umezawa, *J. Antibiot.*,
     31, 1316 (1978).

39. L. F. Povirk, M. Hogan, N. Dattagupta, and M. Buechner, *Biochemistry*, *20*, 665 (1981).

40. Y. Sugiura and K. Ishizu, *J. Inorg. Biochem.*, *11*, 171 (1979).

41. Y. Sugiura, T. Takita, and H. Umezawa, *J. Antibiot.*, *34*, 249 (1981).

42. A. M. Maxam and W. Gilbert, *Methods Enzymol.*, *65*, 499 (1980).

43. A. D. D'Andrea and W. A. Haseltine, *Proc. Natl. Acad. Sci. USA*, *75*, 3608 (1978).

44. M. Takeshita, L. S. Kappen, A. P. Grollman, M. Eisenberg, and I. H. Goldberg, *Biochemistry*, *20*, 7599 (1981).

45. J. Kross, W. D. Henner, W. A. Haseltine, L. O. Rodriguez, M. D. Levin, and S. M. Hecht, *Biochemistry*, *21*, 3711 (1982).

46. Y. Sugiura and T. Suzuki, *J. Biol. Chem.*, *257*, 10544 (1982).

47. P. K. Mascharak, Y. Sugiura, J. Kuwahara, T. Suzuki, and S. J. Lippard, *Proc. Natl. Acad. Sci. USA*, *80*, 6795 (1983).

48. T. Suzuki, J. Kuwahara, and Y. Sugiura, *Biochem. Biophys. Res. Commun.*, *117*, 916 (1983).

49. C.-H. Chang and C. F. Meares, *Biochemistry*, *21*, 6332 (1982).

50. Y. Sugiura, *Biochem. Biophys. Res. Commun.*, *90*, 375 (1979).

51. F. T. Greenaway, J. C. Dabrowiak, R. Grulich, and S. T. Crooke, *Org. Mag. Res.*, *13*, 270 (1980).

52. J. Kross, W. D. Henner, S. M. Hecht, and W. A. Haseltine, *Biochemistry*, *21*, 4310 (1982).

53. C. K. Mirabelli, C.-H. Huang, S. M. Hecht, and W. A. Haseltine, *Biochemistry*, *22*, 300 (1983).

54. R. D. Bereman and M. E. Winkler, *J. Inorg. Biochem.*, *13*, 95 (1980).

55. C. M. Vos, G. Westera and D. Shipper, *J. Inorg. Biochem.*, *13*, 165 (1980).

56. R. P. Sheridan and R. K. Gupta, *J. Biol. Chem.*, *256*, 1242 (1981).

57. M. Otsuka, M. Yoshida, S. Kobayashi, M. Ohno, Y. Sugiura, T. Takita, and H. Umezawa, *J. Am. Chem. Soc.*, *103*, 6986 (1981).

58. J. C. Dabrowiak and M. Tsukayama, *J. Am. Chem. Soc.*, *103*, 7543 (1981).

59. J. P. Collman, J. I. Brauman, K. M. Doxsee, T. R. Halbert, and K. S. Suslick, *Proc. Natl. Acad. Sci. USA*, *75*, 564 (1978).

60. R. P. Hertzberg and P. B. Dervan, *J. Am. Chem. Soc.*, *104*, 313 (1982).

61. P. G. Schultz, J. S. Taylor, and P. B. Dervan, *J. Am. Chem. Soc.*, *104*, 6861 (1982).

62. P. G. Schultz and P. B. Dervan, *Proc. Natl. Acad. Sci. USA, 80,*
     6834 (1983).

63. R. M. Burger, A. R. Berkowitz, J. Peisach, and S. B. Horwitz,
     J. Biol. Chem., 255, 11832 (1980).

64. R. M. Burger, J. Peisach, and S. B. Horwitz, *J. Biol. Chem.,*
     257, 3372 (1982).

65. R. Kuroda, S. Neidle, J. M. Riordan, and T. T. Sakai, *Nucl.*
     Acids Res., 10, 4753 (1982).

Chapter 5

INTERACTION BETWEEN VALINOMYCIN AND METAL IONS

K. R. K. Easwaran
Molecular Biophysics Unit
Indian Institute of Science
Bangalore, India

1. INTRODUCTION

1.1. Transmembrane Ion Transport

Understanding the transport processes across cellular membranes has
been found difficult not only because of the complexity of the process, which involves several stages [1-6], but also because of the
complexity of the system, namely, the membrane across which the
transport occurs [5,6]. Although some of the biochemical nutrients
can permeate the membrane unaided, transport of cations is a process
mediated by membrane proteins [7]. Progress achieved so far in understanding the mechanism of transmembrane cation transport is mainly due
to the discovery of macrocyclic and linear antibiotics (ionophores)
[8,9], also known as complexones [10], which selectively enhance the
cation permeability across natural [11-13] and model membranes [14-20].

1.2. Ionophores

Ionophores are compounds extracted from microorganisms that are capable of selectively inducing cation passage across membranes by enhancing the cation permeability of the latter [8]. These compounds have
moderate molecular weights (200-2000), were initially identified by
their effect of stimulating energy-dependent transport across mitochondria [21], and their antibiotic action [22] is linked to the
transport process carried out by these molecules [23,24].

Ionophores are classified as carriers and channels based on
their mechanism of ion transport (see Refs. 8 and 9 for a list of
available ionophores and their sources.) Among the carrier ionophores there are two kinds: (1) neutral ionophores, which do not
contain any ionizable groups and therefore carry cations as charged
complexes (e.g., valinomycin, enniatins, etc.), and (2) carboxylic
or charged ionophores, which contain ionizable carboxylic groups and,
depending on the state of ionization of the carboxylic group, charge
of the cation, and stoichiometry of the complex, the transported
complex can be either neutral or charged (e.g., lasalocid A, A23187,

etc.) [25]. Among the channel-forming antibiotics are the linear polypeptides gramicidin A, alamethicin, etc.

1.3. Carriers and Channels

The transmembrane cation transport by the ionophores is thought to be brought about by an interplay of two mechanisms, namely, the carrier and channel mechanisms. In the carrier mechanism, the cation is complexed by the membrane-soluble substance, the carrier (e.g., valinomycin), at one of the membrane-water interfaces of the lipid bilayer and is carried to the other interface by diffusion. Such a process generally involves conformational changes for the carrier molecule upon complexation and decomplexation and possibly during the diffusion process. In the channel mechanism, the membrane-soluble substance, very often an aggregate of linear polypeptides such as gramicidin A, spans across the membrane from one interface to the other creating a channel for the cation to pass through. The channel is static, in the sense that it does not undergo diffusion and does not involve any significant conformational changes upon complexation.

1.4. Valinomycin as a Carrier

Among the neutral ionophores, valinomycin, a cyclic dodecadepsipeptide with the sequence cyclo(L-val-D-hyi-D-val-L-lac)$_3$, has emerged as an extraordinary molecule (see Fig. 1) due to its high specificity for K^+ over Na^+ and a high selectivity for K^+ over Na^+ ion (by a factor of over 10^4) in transport across natural [11-13] and model [14-20] membranes. Valinomycin was first isolated from cultures of *Streptomyces fulvissimus* [21] and its chemical structure was correctly identified and confirmed by synthesis by Shemyakin et al. [26]. It is readily soluble in organic solvents and sparingly soluble in water.

On lower level microorganisms it is mildly toxic but either highly toxic or has strong physiological effects on higher organisms

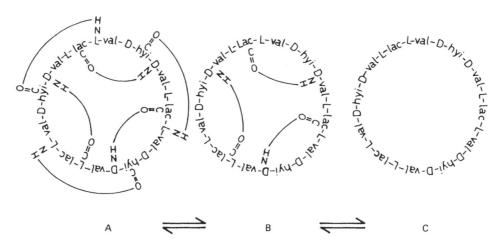

FIG. 1. A, B, and C conformations of valinomycin in solution.
(Reproduced by permission from Ref. 10.)

[27]. It is highly active against acid-fast bacteria, yeast, fungi,
and the like [28,29]. A large number of analogs of valinomycin differing in ring size, in the number, nature, and position of amino
and hydroxy residues have been synthesized, and their antibiotic
activity and structure analysed in comparison with the parent valinomycin molecule [10]. Extensive studies using a variety of physicochemical and spectroscopic methods have been carried out on valinomycin and its cation complexes [13,30,31]. These studies have indicated that the structural and conformational properties of this molecule under varying conditions play a significant role in understanding
the mechanism of transport at the molecular level mediated by this
ionophore. In this chapter we discuss valinomycin and its interaction with metal ions.

2. STRUCTURE AND CONFORMATION OF FREE VALINOMYCIN

2.1. Solution Conformations

The conformations of free valinomycin in different solvents have been
extensively studied using several physicochemical techniques such as

CD [10,30,32], NMR [33-42], IR [43,44], ultrasonic absorption [45], Raman spectroscopy [46], and energy calculations [47,48]. Based on these studies, three distinct conformational models, A, B, and C, which differ in the number of intramolecular hydrogen bonds have been proposed (Fig. 1). The A conformation is observed in nonpolar solvents such as chloroform, hexane, and so on. The molecule forms a "bracelet" structure with six intramolecular hydrogen bonds between the NH of valine residues and amide carbonyls. The B conformation is observed in solvents of medium polarity such as a 3:1 mixture of CCl_4 and DMSO in which only three D-val amide protons are hydrogen bonded. This conformation is also known as the "propellar" conformation. The C conformation, observed in highly polar and good hydrogen bond-accepting solvents like DMSO, methanol, etc., is a relatively more flexible and stretched conformation with no intramolecular hydrogen bonds. In the A conformation, the depsipeptide chain forms fused systems of six 10-membered rings which are alternating type II and type II' β turns with intramolecular hydrogen bonds between an amide carbonyl with the neighboring amide proton in the direction of acylation. The resulting shape represents a compact molecule resembling a bracelet about 8 Å diameter and 4 Å high. Conformation B, stabilized by three intramolecular 4→1 hydrogen bonds between D-val amide protons and L-lac carbonyls, is a more flexible conformation than the conformation A. This is described by a hydrophobic core of the D-val and L-lac side chains. The 10-membered hydrogen-bonded rings are in the periphery making form B resemble a propellar. Both forms A and B are C_3-symmetric. In all the solvents, forms A, B, and C are in equilibrium with one another, the relative populations of the forms depending on the polarity of the solvent.

Grell and Funck [45], from ultrasonic absorption studies, have suggested conformations for free valinomycin ranging from a conformation in which all six possible hydrogen bonds involving amide carbonyls through conformations with five, four, three, two, and one of the possible six hydrogen bonds to a final completely flexible conformation with no hydrogen bonds. All these conformations are simultaneously present and are in dynamic equilibrium, the proportion of each conformer being determined by the solvent.

Raman spectroscopic studies of valinomycin in various solvents [46] have suggested that ester carbonyls could also be involved in hydrogen bonding in nonpolar solvents. Increasing the polarity of the solvent shifts the conformational equilibrium of valinomycin to forms containing fewer hydrogen bonds.

2.2. Solid State Conformations

The crystal structures of free valinomycin [49-51] show no C_3 symmetry but do have a pseudosymmetric center (Fig. 2). In the crystal structure, valinomycin has six intramolecular hydrogen bonds of which four are 4→1 type formed by the amide carbonyls and NH groups. The remaining two are of the 5→1 type between ester carbonyls and NH groups.

Raman spectroscopic studies on valinomycin recrystallized from n-octane, carbon tetrachloride, chloroform, 1-chloroethane, acetone,

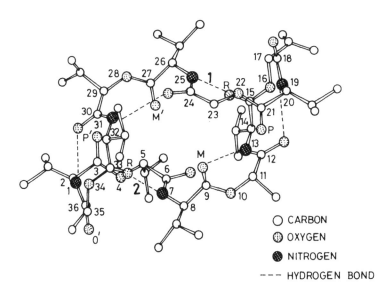

FIG. 2. Crystal structure of free valinomycin. (Reproduced from Ref. 50 by permission of the American Chemical Society.)

and acetonitrile indicate a structure resembling that obtained from
x-ray crystallography [46]. However, valinomycin recrystallized from
o-dichlorobenzene and p-dioxane exhibits a considerably different
structure. This conformation lacks strongly hydrogen-bonded ester
carbonyl groups and resembles the conformation which has been observed
in nonpolar solvents.

3. STRUCTURE AND CONFORMATION OF VALINOMYCIN-METAL ION COMPLEXES

3.1. Alkali Metal Ion Complexes

3.1.1. Solution Conformations

Most of the studies carried out on the conformations of valinomycin
with cations have been with alkali metal ions. UV [52], IR [10,42,
53], CD [54,55], Raman [56], and NMR [10,41,57-59] methods have been
widely employed for a detailed characterization of these complex
conformations.

From UV and CD studies on the complexes of valinomycin with
Na^+, K^+, and Cs^+ in methanol, Stark et al. [52] found that complexation generally led to a decrease in UV absorption above 190 nm and
that CD spectra showed a positive cotton effect around 210 nm with
the exception of Na^+ which showed a negative cotton effect. From
UV temperature jump and ultrasonic absorption measurements, it has
been shown [53] that complexation is a multistep process characterized by two relaxation times with ligand conformational changes which
occur during the stepwise substitution of solvent molecules around
the cations. The CD of valinomycin-Li^+ complex in acetonitrile is
characterized by a negative cotton effect around 217 nm [60] (Fig. 3).
The analysis of CD data gave an equimolar stoichiometry for the valino-
mycin-Li^+ complex with a stability constant of 48 M^{-1} as against
values of 10^3 and 3.16×10^5 M^{-1} for Na^+ and K^+ complexes, respectively, in acetonitrile [55]. Li^+ and Na^+ showed CD spectra with
negative cotton effects while K^+, Rb^+, and Cs^+ show positive cotton

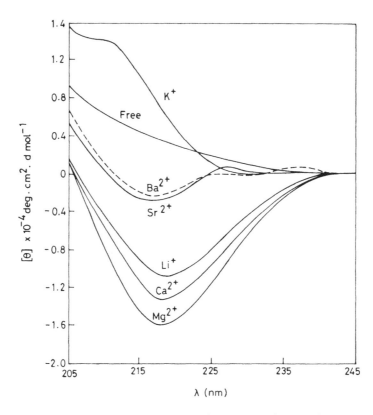

FIG. 3. Representative CD spectra of valinomycin and its metal
ion complexes.

effects. Although all the five cations form equimolar complexes,
the CD data show that the conformations of valinomycin with Li[+] and
Na[+] are different from the other three.

The IR spectra of free valinomycin in nonpolar solvents clearly
indicated the existence of intramolecular hydrogen bonds involving
amide NH's and carbonyls, ester carbonyls being practically free [10,
43,44]. The IR data of valinomycin with Na[+], K[+], Rb[+], and Cs[+] showed
that complexation was accompanied by conformational changes [10,43].
The IR data are summarized in Table 1. From the shifts in the IR
frequencies of the amide and ester carbonyls, it is clear that all
the six ester carbonyls are involved in coordination with the metal
ion with all the six amide carbonyls hydrogen-bonded to amide protons.

TABLE 1

IR Data of Valinomycin-Alkali Metal Ion Complexes

Cation	Anion	Solvent	Ester stretch (cm^{-1})		Amide I (cm^{-1})	Ref.
			Free	Complexed		
-	-	CH_3CN	1750	-	1650	61
-	-	$n-C_7H_{16}$	1750	-	1651	43
-	-	CH_3OH	1752	-	1676	43
-	-	CH_3CN/CCl_4	1755	-	1665	10
Li^+	ClO_4^-	CH_3CN	1750	1738	1640^a	61
Na^+	ClO_4^-	CH_3OH	1748^a		1686	43
Na^+	Cl^-	$CH_3CN-CCl_4$ (1:2)	1755	1747	1657	10
K^+	Cl^-	CH_3OH	-	1745	1658	43
Rb^+	Cl^-	CH_3OH	-	1744	1655	43
Cs^+	Cl^-	CH_3OH	-	1745	1659	43

^{a}Broad.

IR studies on the Na^+ complex showed a broad ester stretch in methanol
[43] and an asymmetric ester stretch in CCl_4/CH_3CN (2:1) mixture [10].
This has been interpreted as being due to the asymmetrically placed
Na^+ in the cavity formed by the six ester carbonyls folded inside the
bracelet. IR data on valinomycin-Li^+ complex (Table 1) have shown
that Li^+ forms a complex with only one set of ester carbonyls coordinating to the cation on one side of the bracelet with the other
side containing the other three ester carbonyls being free [61]. IR
spectral changes observed for the Na^+ complex can also be interpreted
in this way.

^1H and ^{13}C NMR studies on K^+, Rb^+, and Cs^+ complexes in various
solvents [10,41,57-59] have provided a wealth of information corrobo-
rating the results obtained through UV, CD, and IR studies. NMR
studies clearly indicated a C_3 symmetry for the complexed molecule.
The solvent perturbation and deuterium exchange studies on the amide
proton and temperature coefficient of chemical shift studies on the

TABLE 2

^{1}H NMR Chemical Shifts (in ppm) of Valinomycin-Alkali Metal Ion Complexes

Cation	Anion	Solvent	Amide protons		C^α protons				Ref.
			D-val	L-val	D-val	L-val	D-hyi	L-lac	
-	-	$CDCl_3$	7.80	7.67	4.13	4.00	5.02	5.29	42
-	-	CD_3CN	7.60	7.58	4.13	4.16	4.98	5.24	60
-	-	CCl_4-DMSO-d_6 (3:1)	7.48	8.67	4.79	4.08	4.95	5.43	42
-	-	CD_3OD + 2% D_2O	8.17	8.24	4.31	4.42	4.93	5.19	42
Li^+	ClO_4^-	CD_3CN	7.46		4.27	4.19	4.79	5.07	60
Na^+	SCN^-	CH_3OH	7.99	7.99	-	-	-	-	31
Na^+	SCN^-	CCl_4-CH_3CN (1:1)	7.88	7.94	4.07	4.13	4.78	5.07	31
K^+	SCN^-	CH_3CN	8.34	8.42	3.80	3.85	4.64	4.95	31
K^+	SCN^-	CCl_4-CH_3CN (1:1)	8.30	8.37	3.79	3.84	4.57	4.89	31
Cs^+	Cl^-	$CDCl_3$	7.89	7.98	3.81	3.81	4.72	5.08	31

K^+ complex indicated that both D-val and L-val NH's are intramolecularly hydrogen-bonded. The changes in the chemical shift of C^α protons (Table 2) and the vicinal $^3J_{HNC^\alpha H}$ proton-coupling constants of L-val and D-val residues (Table 3) indicated substantial changes in the conformation accompanying complexation. The significantly large ^{13}C chemical shift changes of the ester carbonyls as compared to amide carbonyls [42] showed that the ester carbonyls are involved in metal ion binding. By using ^{13}C- and ^{15}N-enriched analogs of valinomycin, the multiple resonance techniques and heteronuclear coupling constant-dihedral angle correlations, Bystrov et al. [42] gave a detailed description for the K^+ complex. The proposed solution conformation for the K^+ complex has a bracelet structure with all six NH's intramolecularly hydrogen bonded and with the K^+ ion in the center of the cavity and liganding to all six ester carbonyls. The

TABLE 3

Coupling Constants and dδ/dT data of Valinomycin-Alkali
Metal Ion Complexes

| Cation | Anion | Solvent | $^3J_{HNC^\alpha H}$ | | dδ/dT (ppb/°C) | | Ref. |
			D-val	L-val	D-val NH	L-val NH	
-	-	CDCl$_3$	8.80	6.60			42
-	-	CD$_3$CN	8.40	7.50	3.0	3.5	60
-	-	CCl$_4$-DMSO-d$_6$ (3:1)	10.10	7.60	-		42
-	-	CD$_3$OD	5.60	5.60	-	-	42
Li$^+$	ClO$_4^-$	CD$_3$CN	6.20	6.60	2.1		60
Na$^+$	SCN$^-$	CH$_3$OH	6.61	6.61	6.6	6.6	31
Na$^+$	SCN$^-$	CCl$_4$-CH$_3$CN (1:1)	5.50	5.50	0.9	0.9	31
K$^+$	SCN$^-$	CH$_3$CN	4.77	4.77	-	-	31
K$^+$	SCN$^-$	CCl$_4$-CH$_3$CN (1:1)	4.95	5.05	1.5	1.8	31
Cs$^+$	Cl$^-$	CDCl$_3$	5.05	5.05	-	-	31

conformations of the K[+], Rb[+], and Cs[+] complexes of valinomycin are
similar in almost all solvents [37,53] and also with different anions
[57] reflecting the high stability of the complex. Also the conformations of the K[+] complex in solution and solid state (see Fig. 4) are
similar.

The Na[+] complex of valinomycin differed from the K[+] complex in
terms of the observed anion and solvent dependence [53,57]. It has
been shown that Na[+] forms a weaker complex with a reduced number of
coordinating ester carbonyl oxygens and intramolecular hydrogen bonds.
Detailed NMR investigations of the Li[+] complex [60] showed that Li[+]
forms an equimolar complex with only the D-val ester carbonyls coordinating to the cation with the L-val ester carbonyls on the other

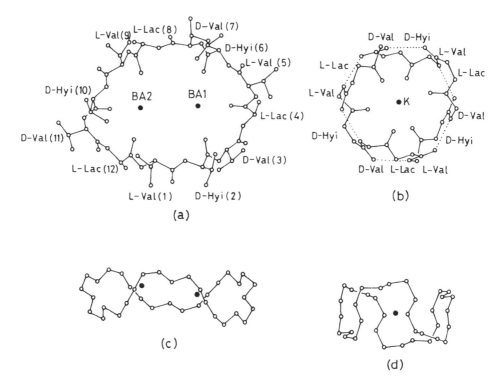

FIG. 4. Top view of the valinomycin molecule in its (a) barium
perchlorate complex and (b) potassium iodide complex. Parts (c)
and (d) represent the corresponding side views. (Reproduced by
permission from Ref. 73.)

side of the bracelet being free. The cation is not inside the bracelet in this conformation but is complexed on the periphery by D-val
ester carbonyls that are folded out. The model for the 1:1 complex
of valinomycin-Li[+] complex is shown in Fig. 5. The NMR data of the
Na[+] complex are also compatible with this model.

The possibility of valinomycin forming nonequimolar complexes
with K[+] was suggested at high absolute concentrations of valinomycin
[62]. At high concentrations of valinomycin using UV and CD, it was
shown that valinomycin forms a 2:1 (valinomycin:K[+]) ion sandwich
complex in ethanol. It was also shown by Ovchinnikov and Ivanov
[63] that analogs of valinomycin, namely, valinomycin with L-val
replaced by L-lys and L-glu, could form charged or neutral cation
complexes similar to the ion sandwich of valinomycin-K[+]

3.1.2. *Solid State Conformations*

The crystalline complex of valinomycin-K[+] [64-67] has a conformation
similar to the one described for the complex in solution. It has six

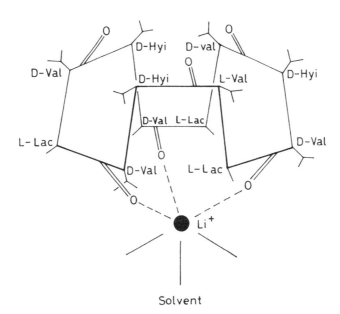

FIG. 5. Model for the 1:1 valinomycin-Li[+] complex.

intramolecular hydrogen bonds all of 4→1 type between the six amide
NH's and carbonyls (Fig. 4). The cation is held at the center of
the cavity of the bracelet structure by ion-dipole interactions
between the metal ion and six carbonyl ligands. The cation is
effectively shielded by the hydrogen bond system and the pendant
isopropyl chains. The hydrophobic exterior of this complex explains
its high solubility in organic solvents and membranes. When the
anion is picrate, there is a weak but definite interaction between
picrate and K^+ [66].

The single-crystal x-ray structure of the valinomycin-sodium
picrate complex has been solved recently [68]. Sodium forms a 1:1
complex with the three D-val ester carbonyls coordinating to the
cation on the periphery of the bracelet molecules. The structure
revealed a water molecule in the center of the cavity of the bracelet usually occupied by K^+ in the K^+ complex. Na^+ is displaced by
2.3 Å from the position of K^+. This conformation is similar to the
one proposed for the Li^+ complex in acetonitrile [60,61].

3.2. Alkaline Earth Metal Ion Complexes

Even though valinomycin is highly specific for K^+, the conformation
of the K^+ complex does not represent all the stages during K^+ transport. For a better understanding of the complexation-decomplexation
reactions at the interface and the mode of transport after complexation, a number of spectroscopic and crystallographic studies on
valinomycin complexes with cations of various sizes and charges have
been carried out mainly by our group at Bangalore. The results on
valinomycin-divalent cation complexes are summarized below.

3.2.1. Solution Conformations

The stoichiometries and solution conformations of the complexes of
valinomycin with Ca^{2+} [70,71], Mg^{2+} [61,68], Ba^{2+} [72-75], and Sr^{2+}
[61,69] were studied by UV, CD, IR, and NMR methods. The conforma-

tions of the Sr^{2+} complexes are similar to the Ba^{2+} complexes and the valinomycin-Mg^{2+} system is similar to the Ca^{2+} system.

The UV studies on the valinomycin-Ba^{2+} system with both perchlorate and thiocyanate salts in acetonitrile showed a plateau region for changes in absorption at Ba^{2+}: valinomycin ratios of 1:1 indicating the formation of a 1:1 complex [75]. The magnitudes of the UV absorption changes are different for perchlorate and thiocyanate salts indicating an effect of the anion on complexation.

The CD spectra of the valinomycin-barium perchlorate system in acetonitrile are dependent on the absolute concentration of valinomycin [72]. Detailed analysis of the CD titration graphs led to the following conclusions: At low concentrations (~0.5 mM), valinomycin forms an equimolar complex with Ba^{2+} with a conformation different from that of the K^+ complex. At higher absolute concentrations of valinomycin, in addition to the 1:1 complex, 2:1 (valinomycin:Ba^{2+}) ion and 1:2 peptide sandwich complexes are formed. Further addition of barium salt resulted in the formation of a stable 1:2 complex, which is not a sandwich complex, but a "final complex" with a flat structure and with no internal hydrogen bonds. All these complexes are characterized by negative cotton effects at 217 nm and small positive cotton effects at 235 nm (Fig. 3). The CD spectral behavior for valinomycin-Sr^{2+} complex is similar to that of valinomycin-Ba^{2+} systems.

Marked changes were observed in the CD spectrum of free valinomycin in acetonitrile on gradual addition of calcium perchlorate salt [70]. The small CD molar ellipticity at 217 nm for free valinomycin, $[\theta]_{217} \simeq 4000$ deg·cm^2·d mol^{-1}, gradually changes into an intense negative band of molar ellipticity $[\theta]_{217} \simeq 12650$ deg·cm^2·d mol^{-1} at valinomycin-to-Ca^{2+} salt ratio of 1:1.4 (Fig. 3). A small CD band around 240 nm with positive molar ellipticity is also seen at high salt concentrations. The analysis of CD titration graph gave a predominantly 2:1 complex for the concentration ratios less than 1:0.5 and coexistence of 1:1 and 2:1 complexes for the concentration range 1:0.5 < valinomycin:Ca^{2+} < 1:1.5. The binding constants obtained for the species 1:1 and 2:1 are 0.5×10^3 M^{-1} and 1.0×10^3 M^{-1}, respectively [70]. The results obtained for the valinomycin-Mg^{2+} system were similar to those of the valinomycin-Ca^{2+} system [61,69].

IR studies on the complexes of valinomycin with Ca^{2+}, Mg^{2+}, Ba^{2+}, and Sr^{2+} in acetonitrile [61,69] indicated that in the 2:1 ion sandwich and 1:1 equimolar complexes, only part of the ester carbonyls are involved in complexation (Table 4). However, there is no evidence for the presence of a 1:2 peptide sandwich complex. In the final complex formed by Ba^{2+} and Sr^{2+}, the ester carbonyls are free and amide carbonyls coordinate to the cation.

Based on ^1H and ^{13}C NMR studies on the complexes of valino-mycin with Ca^{2+}, Mg^{2+}, Ba^{2+}, and Sr^{2+} [69-75], detailed conformational models have been proposed for these complexes. The NMR data on these complexes are summarized in Tables 5 and 6. The final

TABLE 4

IR Data of Valinomycin-Alkaline Earth Metal
Ion Complexes in Acetonitrile[a]

| Cation | Val/cation | Ester stretch (cm^{-1}) | | Amide I (cm^{-1}) |
		Free	Complexed	
–	–	1750	–	1650
Mg^{2+}	2:1	1750	1738	1670
Mg^{2+}	1:2	1750	1738	1670
Ca^{2+}	2:1	1750	1738	1670
Ca^{2+}	1:2	1750	1738	1670
Sr^{2+}	2:1	1750	1738	1670
Sr^{2+}	1:2	1750	–	1670, 1630
Ba^{2+}	2:1	1750	1738	1670
Ba^{2+}	1:2	1750	–	1670, 1630
Ba^{2+}	1:4	1750	–	1690, 1670, 1650-1620[b]

[a]Abstracted from Ref. 61.
[b]Broad.

TABLE 5

^{1}H NMR Chemical Shifts (in ppm) of Valinomycin-Alkaline Earth
Metal Ion Complexes in Acetonitrile

Cation	Val/cation	Amide protons		C^α protons				Ref.
		D-val	L-val	D-val	L-val	D-hyi	L-lac	
Mg^{2+}	2:1	7.50		4.17	4.17	4.89	5.14	69
Mg^{2+}	1:1	7.30	7.42	4.22	4.17	4.79	5.02	69
Ca^{2+}	2:1	7.57	7.65	4.36	4.18	4.80	5.11	70
Ca^{2+}	1:1	7.56	7.66	4.38	4.19	4.79	5.10	70
Sr^{2+}	1:2	7.54		4.20	4.24	4.76	5.10	69
Ba^{2+}	1:2	7.36	7.41	4.41	4.54	4.90	5.17	75

chemical shift positions of NH and C^αH protons for the Ba^{2+} complex
clearly show that the conformations of the Ba^{2+} complex are very
different from those of free valinomycin and its K$^+$ complex. Both
L-val and D-val C^α protons have moved downfield with Ba^{2+} complexation while they move upfield with K$^+$ complexation. The extent of

TABLE 6

Coupling Constant and dδ/dT Data of Valinomycin-Alkaline
Earth Cation Complexes in Acetonitrile

Cation	Val/cation	$^3J_{HNC^\alpha H}$ (Hz)		dδ/dT (ppb/°C)		Ref.
		D-val	L-val	D-val NH	L-val NH	
Mg^{2+}	1:1	6.62	6.62	0.86	1.36	69
Ca^{2+}	2:1	6.99	5.88	-	-	70
.Ca^{2+}	1:1	6.99	5.52	2.95		70
Sr^{2+}	1:2	6.60	6.60	2.70	2.90	69
Ba^{2+}	1:2	6.30	6.30	3.20		75

intramolecular hydrogen bonding involving the NH protons was studied for free valinomycin and its K^+ and Ba^{2+} complexes by the addition of 2,2,6,6-tetramethylpiperidin-1-oxyl, a nitroxide free radical [73]. Less than 0.1% addition of the free radical to the three systems, all containing the same amount of valinomycin dissolved in CD_3CN, caused broadening of NH signals in only the Ba^{2+} complex (Fig. 6) indicating the highly exposed nature of NH proton in this system when compared with the other two. The ^{13}C NMR spectral studies on the valinomycin-Ba^{2+} complex indicated that the L-lac and D-hyi carbonyl signals moved downfield (by ~1.2 ppm and ~1.3 ppm, respectively) as compared with small upfield shifts of L-val and D-val carbonyls [75]. It is clear from these results that amide carbonyls of L-lac and D-hyi residues are involved in metal binding in the valinomycin-Ba^{2+} system. The model proposed based on the NMR data (NH and $C^\alpha H$ chemical shift, $^3J_{HNC^\alpha H}$ of L-val and D-val residues, and ^{13}C carbonyl chemical shift) is consistent with a novel structure for the Ba^{2+} complex, which has a flat open structure without internal hydrogen bonds and with two Ba^{2+} ions per molecule [74,75] (Fig. 4). The presence of C_3-symmetric conformations in solution for the complex on the NMR time scale as indicated by the observation of only one set of ^1H and ^{13}C signals corresponding to one chemical repeating unit is due to

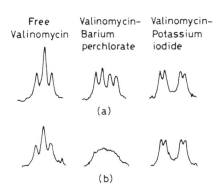

FIG. 6. Effect of free radical addition to valinomycin and its K^+ and Ba^{2+} complexes in CD_3CN: (a) NH region of ^1H NMR spectra and (b) the same region in the presence of 2,2,6,6-tetramethylpiperidin-1-oxyl. (Reproduced by permission from Ref. 73.)

many interconverting but closely related conformations for the
valinomycin-Ba^{2+} system. The weak binding of Ba^{2+} ion could allow
some flexibility in the complexed state for the molecule in solution.
Studies with barium thiocyanate salt gave similar results [74,75].
However, based on the low temperature NMR data, barium thiocyanate
was shown to form a 1:1 complex with the L-val carbonyls coordinating
to Ba^{2+}. It was suggested that the linear thiocyanate anion stabilizes this otherwise unstable complex by providing additional coordination to Ba^{2+} ion. The perchlorate anion with a low surface
charge density is a poor coordinator to Ba^{2+}

Detailed ^{1}H NMR studies on valinomycin-Ca^{2+} and valinomycin-
Mg^{2+} systems showed that the salt-induced chemical shift changes are
different for different protons. While L-val C^{α} proton signal re-
mained sharp and unshifted on addition of Ca^{2+} salt, the D-val C^{α}
proton signal became broad and moved continually downfield rather
rapidly till around a valinomycin/Ca^{2+} ratio of 1:0.5. The difference in the chemical shifts of the signals from valinomycin-Ca^{2+} and
free valinomycin beyond a molar ratio of 1:0.5 is very small and
reached a stabilized value around 1:1.5. The final chemical shift
and the $^3J_{HNC^{\alpha}H}$ coupling constant values for the complex are given
in Tables 5 and 6. The temperature coefficient of L-val, D-val, NH
chemical shifts, and free radical broadening experiments indicated
that the conformations of the Ca^{2+} and Mg^{2+} complexes have all six
intramolecular hydrogen bonds intact in a bracelet structure. D-val
ester carbonyls shifted downfield by ~1.7 ppm in the complex while
the L-val carbonyls shift upfield by 0.35 ppm with respect to free
valinomycin, indicating that only D-val carbonyls bind to Ca^{2+}.
Based on the ^{1}H NMR chemical shift, the coupling constant, information on intramolecular hydrogen bond, and ^{13}C carbonyl chemical shift
data, it was proposed that at low Ca^{2+} concentrations (1:0.5) valinomycin forms a 2:1 ion sandwich complex. As the salt concentration
is increased, in addition to the ion sandwich complex, an equimolar
(1:1) complex different from the K^+ complex is also observed. The
proposed model for the 2:1 valinomycin-Ca^{2+} complex is shown in
Fig. 7. The 1:1 complex is similar to the one proposed for the Li^+

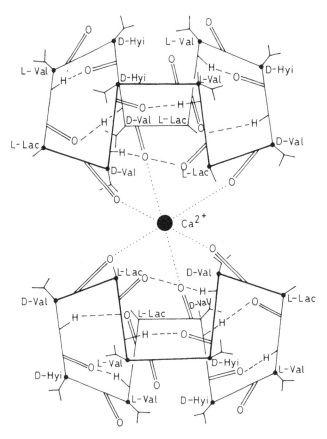

FIG. 7. Model for the 2:1 ion sandwich complex. (Reproduced from
Ref. 70 by permission of the American Chemical Society.)

complex (see Fig. 5). In both models, the NH's of L-val and D-val
residues are intramolecularly hydrogen-bonded and the Ca^{2+} ion is
coordinated on the "D-val—L-lac" side of the bracelet.

3.2.2. *Solid State Conformation*

Of the various valinomycin-divalent cation complexes the crystal
structures of only valinomycin-barium perchlorate [75,76] and
valinomycin-barium thiocyanate [75] have been studied so far.
These two complexes have almost identical conformations in their

respective crystal structures. The solid state conformation is
similar to the one proposed in solution, where the valinomycin
molecule in the complex stabilizes in a flat open structure without
any internal hydrogen bonds and with two Ba^{2+} ions per molecule
(Fig. 4). In the valinomycin-barium perchlorate crystal, each Ba^{2+}
is liganded with three consecutive amide carbonyls and is separated
from the other by 4.57 Å. The other atoms coordinating to the Ba^{2+}
ions belong to perchlorate groups and solvent molecules. The coordination number for one of the Ba^{2+} ions is 8 while for the other
it is 9 [73]. The crystal structure contains infinite layers of
valinomycin molecules interconnected by perchlorate groups. The
solvent molecules in the structure are sandwiched between successive
layers.

In the valinomycin-barium thiocyanate complex [75], in addition
to the three consecutive amide carbonyls, four water molecules and an
acetonitrile molecule coordinate to each Ba^{2+}. Of these, two water
molecules and an acetonitrile molecule coordinate to both Ba^{2+} ions.
It is interesting to note that, unlike the perchlorate groups in the
barium perchlorate complex, the thiocyanate groups are not involved
in metal coordination in the present structure.

The two valinomycin-Ba^{2+} complexes differ from the complexes
of valinomycin with monovalent ions in three important respects:
(1) the molecule has an open conformation with no intramolecular
hydrogen bonds, (2) amide carbonyls instead of ester carbonyls are
involved in metal coordination, and (3) the molecule binds two metal
ions instead of one. The valinomycin-Ba^{2+} complexes thus have a
unique structure fundamentally different from those observed in the
crystal structures of valinomycin complexes with alkali metal ions.

3.3. Summary

The results discussed in this section show that valinomycin forms
nonequimolar complexes with widely different conformations. Of the
five IA group cations, Li^+ and Na^+ form equimolar complexes with the

cation located on the periphery of the molecule held by the D-val ester carbonyls. The other side is sterically more crowded and is unapproachable by the cation. K^+, Rb^+, and Cs^+ form equimolar complexes with the well-known valinomycin-K^+ bracelet conformation.

The IIA group cations form widely different complexes with valinomycin and conformations ranging from compact, fully hydrogen-bonded ones to open, labile structures. Ca^{2+} and Mg^{2+} form predominantly 2:1 (ion sandwich) complexes with valinomycin at low salt concentration, while at high salt concentration both 1:1 D-val—L-lac side (different from the 1:1 valinomycin-K^+ complex) and 2:1 (ion sandwich) conformation coexist. Ba^{2+} and Sr^{2+} do form different types of complexes like the 2:1 (ion sandwich), 1:1 D-val—L-lac side and 1:2 (peptide sandwich) on gradual addition of salt. However, these conformations are only transient intermediates and are not stable. The final stable complex formed by these divalent cations is a novel one in which the valinomycin molecule assumes a flat structure devoid of internal hydrogen bonds and with two barium ions per molecule. The final complex in solution is an average of many interconverting but closely related structures having an average C_3 symmetry in NMR time scale. This solution conformation is identical to the solid state conformation obtained from x-ray crystallography.

The selectivity of valinomycin for K^+ is mainly due to the cavity size formed in the all-in A_2 conformation being ideal for K^+ (radius = 1.33 Å) to snugly fit in. For cations of radius lower than 1.33 Å the ligand-cation distances are not ideal in the all-in A_2 conformation [77]. Indeed, it has been shown that an unfavorable energy balance is obtained between the binding energies and the cation solvation energies for the complexing of Mg^{2+} and Ca^{2+} [78]. This leads to the formation of complexes with different conformations with one set of carbonyls opening out and complexing with the cation. The remaining coordination is provided by the solvent molecules in the 1:1 complexes and the same set of carbonyls from another valinomycin molecule in the 2:1 complexes. However, the arrangement of

the ligands cannot be adjusted corresponding to any desired coordination number due to the conformational restraints of the valinomycin
bracelet. When the cation has a larger coordination number, the conformations described for the 1:1 and 2:1 complexes become unfavorable
at higher concentrations of the cation and the valinomycin molecule
loses its bracelet structure to provide an optimum coordination for
the cation. This might be what is happening in the case of Sr^{2+} and
Ba^{2+}, which form 1:2 complexes with coordination numbers of 8 and 9
as observed in the crystal structure of the Ba^{2+} complexes.

4. VALINOMYCIN-METAL ION COMPLEXES IN TRANSMEMBRANE ION TRANSPORT

The observation of the various cation complexes of valinomycin helps
us to understand the mechanism of carrier-mediated transport at the
molecular level. The observation of the peripheral 1:1 D-val—L-lac
side complexes represents the complexation-decomplexation equilibrium
at the membrane-water interface. The cation could be complexed by
the D-val carbonyls at one of the interfaces and taken to the center
of the bracelet by a folding in of the ester carbonyls. The selectivity of the cation to the D-val carbonyls in preference to the
L-val carbonyls implies that at the other interface, the complexation-
decomplexation would be from the D-val side. This requires that the
valinomycin molecule turn around before it reaches the other side and
be ready to load/unload the cation unless the interface imparts stability to the L-val carbonyl-bound complex.

The importance of the heterogenous complexation reaction is
well-known in BLM electrical measurements [79]. In this reaction,
the complexation reaction takes place between the cation which is in
the aqueous layer and the carrier situated at the membrane-water
interface. It is very likely that the conformation of the complex
formed by this reaction is similar to that observed for the 1:1
peripheral complexes. Also, for valinomycin-mediated K^+ transport,
the presence of a 1:1 "loose complex" has been invoked to explain
the BLM conductance data [80]. This valinomycin-K^+ loose complex,

though not thermodynamically stable, might be having the conformation described for the 1:1 peripheral complexes.

The formation of the 2:1 ion sandwich complexes suggests a relay carrier mechanism [62]. In this mechanism, the cation is handed over from one carrier molecule to another in the lipid bilayer. The selectivity of the cation to the D-val side requires that the handing over is done when two valinomycin molecules come face to face. Transfer of a cation from one molecule to another can be visualized as an opening up of the D-val ester carbonyls resulting in a formation of a peripheral 1:1 complex from a valinomycin-K^+ type of complex, formation of a 2:1 ion sandwich complex, transfer of the cation from the first to the second molecule resulting in another peripheral 1:1 complex, and a folding in of the D-val carbonyls. When the rates of the complexation, decomplexation, and rearrangement processes are fast compared with the cation translocation constants, the relay-carrier and diffusive carrier mechanism cannot easily be distinguished by the widely employed steady-state conductance measurements on BLMs [81]. Typically, the effective rate constant for the handing over can be estimated to be about 1×10^{-8} sec [45] whereas the translocation rate constants for the carrier-cation complex are of the order of 1×10^{-5} sec [80,81]. Although these estimates are approximate because of the extrapolation of data in organic solvents and membrane-water interfaces to the environment in the lipid bilayer, it is very likely that at high membrane concentrations of valinomycin, the cations are transported by a relay-carrier mechanism.

ABBREVIATIONS AND DEFINITIONS

BLM	Black lipid membrane
CD	Circular dichroism
DMSO	Dimethyl sulfoxide
Glu	Glutamic acid
Hyi	Hydroxyisovaleric acid
Ion sandwich	Metal ion sandwiched between two valinomycins

IR	Infrared
Lac	Lactic acid
Lys	Lysine
NMR	Nuclear magnetic resonance
Peptide sandwich	Valinomycin sandwiched between two metal ions
UV	Ultraviolet
Val	Valine
δ	Chemical shift in ppm
$d\delta/dT$	Temperature coefficient of chemical shift

ACKNOWLEDGMENTS

The untiring efforts of the author's associates, Drs. Devarajan, Vishwanath, and Sankaram, who have contributed significantly to the studies reported in the text, are gratefully acknowledged. The author is also grateful to Dr. M. B. Sankaram for the discussions and help in preparation of this chapter.

REFERENCES

1. P. Läuger, *J. Membrane Biol., 57,* 163 (1980).

2. P. Läuger, R. Benz, G. R. Stark, E. Bamberg, P. C. Jordan, A. Fahr, and W. Brock, *Quart. Rev. Biophys., 14,* 513 (1981).

3. G. W. Feigenson and P. R. Meers, *Nature, 283,* 313 (1980).

4. G. Stark, in *Membrane Transport in Biology,* Springer-Verlag, Berlin, 1978, p. 447.

5. S. J. Singer, *J. Supramol. Struct., 6,* 313 (1977).

6. S. J. Singer, *Ann. Rev. Biochem., 43,* 805 (1974).

7. G. Giebisch, D. C. Tosteson, and H. H. Ussing, in *Membrane Transport in Biology,* Vol. III, Springer-Verlag, 1980.

8. B. C. Pressman, *Ann. Rev. Biochem., 45,* 501 (1976).

9. E. P. Bakker, in *Mechanisms of Action of Antibacterial Agents* (F. E. Hahn, ed.), Springer-Verlag, Berlin, 1979.

10. Yu. A. Ovchinnikov, V. T. Ivanov, and A. M. Shkrob, *Membrane Active Complexones*, Elsevier, Amsterdam, 1974.

11. F. M. Harold and J. R. Baarda, *J. Bacteriol., 94,* 53 (1967).

12. H. A. Lardy, S. N. Graven, and S. Estrada-O, *Fed. Proc., 26,* 1355 (1967). .

13. B. C. Pressman, E. J. Harris, W. S. Jagger, and J. H. Johnson, *Proc. Natl. Acad. Sci., 58,* 1949 (1967).

14. T. E. Andreoli, M. Tieffenberg, and D. C. Tosteson, *J. Gen. Physiol., 50,* 2527 (1967).

15. E. J. Harris, G. Catlin, and B. C. Pressman, *Biochemistry, 6,* 1360 (1967).

16. A. A. Lev and E. P. Bujinsky, *Tsitologiya, 9,* 102 (1967).

17. P. Mueller and D. O. Rudin, *Biochem. Biophys. Res. Commun., 26,* 398 (1967).

18. P. J. F. Henderson, J. D. McGivan, and J. B. Chappell, *Biochem. J., 111,* 52 (1969).

19. G. Szabo, G. Eisenman, and S. Ciani, *J. Membrane Biol. 1,* 346 (1969).

20. S. G. A. McLaughlin, G. Szabo, S. Ciani, and G. Eisenman, *J. Membrane Biol., 9,* 3 (1972).

21. W. C. McMurray and R. W. Begg, *Arch. Biochem. Biophys., 84,* 546 (1959).

22. H. Brockmann and G. Schmidt-Kestner, *Chem. Ber., 88,* 57 (1955).

23. C. Moore and B. C. Pressman, *Biochem. Biophys. Res. Commun., 15,* 562 (1964).

24. B. C. Pressman, *Proc. Natl. Acad. Sci., 53,* 1076 (1965).

25. G. R. Painter and B. C. Pressman, *Topics in Current Chemistry, 101,* 83 (1982).

26. M. M. Shemyakin, Yu. A. Ovchinnikov, V. T. Ivanov, and A. A. Kiryushkin, *Tetrahedron, 19,* 581 (1963).

27. B. C. Pressman, G. R. Painter and M. Fahin, in *Inorganic Chemistry in Biology and Medicine* (A. E. Martell, ed.), American Chemical Society, Washington, D.C., 1980, p. 1.

28. R. Brown, J. Brennan, and C. Kelley, *Antibiot. Chemother., 12,* 482 (1962).

29. H. Hishimura, M. Mayama, T. Kimura, A. Kimura, Y. Kawamura, K. Tawara, Y. Tanaka, S. Okamoto, and H. Kyotani, *J. Antibiot. Ser. A, 17,* 11 (1964).

30. Yu. A. Ovchinnikov, *Eur. J. Biochem., 94,* 321 (1979).

31. Yu. A. Ovchinnikov and V. T. Ivanov, *Tetrahedron, 30,* 1871 (1974).

32. H. R. Wyssbrod and W. A. Gibbons, *Survey of Proc. Chem., 6,* 209 (1973).

33. M. Pinkerton, L. K. Steinrauf, and P. Dawkins, *Biochem. Biophys. Res. Commun.*, *35*, 512 (1969).

34. D. H. Haynes, A. Kowalsky, and B. C. Pressman, *J. Biol. Chem.*, *244*, 502 (1969).

35. V. T. Ivanov, J. A. Laine, N. D. Abdulaev, L. S. Senyavina, E. M. Popov, Yu. A. Ovchinnikov, and M. M. Shemayakin, *Biochem. Biophys. Res. Commun.*, *34*, 803 (1969).

36. D. J. Patel and A. E. Tonelli, *Biochemistry*, *12*, 486 (1973).

37. D. J. Patel and A. E. Tonelli, *Biochemistry*, *12*, 496 (1973).

38. Yu. A. Ovchinnikov and V. T. Ivanov, *Tetrahedron*, *31*, 2177 (1975).

39. J. D. Glickson, S. L. Gordon, J. P. Pitner, D. G. Agresti, and R. Walter, *Biochemistry*, *15*, 5721 (1975).

40. D. B. Davis and Md. A. Khalid, *J. Chem. Soc. Perkin II*, 1327 (1976).

41. V. T. Ivanov, in *Proceedings oɪ the Fifth American Peptide Symposium* (M. Goodman and J. Meinhofer, eds.), John Wiley, New York, 1977.

42. V. F. Bystrov, Y. D. Gavrilov, V. T. Ivanov, and Yu. A. Ovchinnikov, *Eur. J. Biochem.*, *78*, 63 (1977).

43. E. Grell, T. Funck, and H. Sauter, *Eur. J. Biochem.*, *34*, 415 (1973).

44. V. T. Ivanov, A. I. Miroshnikov, N. B. Abdulaev, L. S. Senyavina, S. F. Arkhipova, N. N. Uvarov, and Yu. A. Ovchinnikov, *Biochem. Biophys. Res. Commun.*, *42*, 654 (1971).

45. E. Grell and T. Funck, *J. Supramol. Struct.*, *1*, 307 (1973).

46. K. J. Rothschild, I. M. Asher, H. E. Stanley, and E. Anastassakis, *J. Am. Chem. Soc.*, *99*, 2032 (1977).

47. D. F. Mayers and D. W. Urry, *J. Am. Chem. Soc.*, *94*, 77 (1972).

48. B. Maigret and B. Pullman, *Theor. Chim. Acta*, *37*, 17 (1975).

49. W. L. Duax, H. Hauptman, C. M. Weeks, and D. A. Norton, *Science*, *176*, 911 (1972).

50. G. D. Smith, W. L. Duax, D. A. Lang, G. T. de Titta, J. W. Edmonds, D. C. Rohrer, and C. W. Weeks, *J. Am. Chem. Soc.*, *97*, 7242 (1975).

51. I. L. Karle, *J. Am. Chem. Soc.*, *97*, 4379 (1975).

52. G. Stark, R. Benz, B. Kettner, and P. Läuger, in *Molecular Mechanisms of Antibiotic Action on Protein Biosynthesis and Membranes* (E. Munoz, F. Garcia-Fernandiz, and D. Vazquez, eds.), Elsevier, Amsterdam, 1972.

53. E. Grell, T. Funck, and F. Eggers, in *Molecular Mechanisms of Antibiotic Action on Protein Biosynthesis and Membranes* (E. Munoz, F. Garcia-Fernandiz, and D. Vazquez, eds.), Elsevier, Amsterdam, 1972.

54. E. Grell, T. Funck, and F. Eggers, in *Membranes: A Series of Advances* (G. Eisenman, ed.), Marcel Dekker, New York, 1974, Vol. III, Chap. 1.

55. M. C. Ross and R. W. Henkens, *Biochem. Biophys. Acta, 372*, 426 (1974).

56. I. M. Asher, K. J. Rothschild, and H. E. Stanley, *J. Mol. Biol., 89*, 205 (1974).

57. D. G. Davis and D. C. Tosteson, *Biochemistry, 14*, 3962 (1975).

58. M. Ohnishi and D. W. Urry, *Science, 168*, 1091 (1970).

59. N. R. Krishna, D. G. Agresti, J. D. Glickson, and R. Walter, *Biophys. J., 24*, 791 (1978).

60. M. B. Sankaram and K. R. K. Easwaran, *Biopolymers, 21*, 1557 (1982).

61. M. B. Sankaram, Ph.D. thesis, Indian Institute of Science, Bangalore, India, 1984.

62. V. T. Ivanov, *Ann. N.Y. Acad. Sci., 264*, 225 (1975).

63. Yu. A. Ovchinnikov and V. T. Ivanov, in *Biochemistry of Membrane Transport* (G. Semenza and E. Carafoli, eds.), Springer-Verlag, Berlin, 1977, pp. 123-146.

64. M. Pinkerton, L. K. Steinrauf, and P. D. Dawkins, *Biochem. Biophys. Res. Commun., 35*, 512 (1969).

65. K. Neupert-Laves and M. Dobler, *Helv. Chim. Acta, 58*, 432 (1975).

66. J. A. Hamilton, M. N. Sabesan, and L. K. Steinrauf, *J. Am. Chem. Soc., 103*, 5880 (1981).

67. H. W. Huang and C. R. Williams, *Biophys. J., 33*, 269 (1981).

68. L. K. Steinrauf, J. A. Hamilton, and M. N. Sabesan, *J. Am. Chem. Soc., 104*, 4085 (1982).

69. M. B. Sankaram and K. R. K. Easwaran, unpublished results.

70. C. K. Vishwanath and K. R. K. Easwaran, *Biochemistry, 21*, 2612 (1982).

71. C. K. Vishwanath, Ph.D. thesis, Indian Institute of Science, Bangalore, India, 1983.

72. S. Devarajan and K. R. K. Easwaran, *Biopolymers, 20*, 891 (1981).

73. S. Devarajan, M. Vijayan, and K. R. K. Easwaran, *Int. J. Prot. Pep. Res., 23*, 324 (1984).

74. S. Devarajan and K. R. K. Easwaran, *J. Biosci., 6*, 1 (1984).

75. S. Devarajan, Ph.D. thesis, Indian Institute of Science, Bangalore, India, 1983.

76. S. Devarajan, C. M. K. Nair, K. R. K. Easwaran, and M. Vijayan, *Nature, 286,* 640 (1980).

77. W. Simon and W. E. Morf, in *Membranes: A Series of Advances* (G. Eisenman, ed.), Marcel Dekker, New York, 1973, Vol. II, Chap. 4.

78. N. Gresh, C. Etchebest, O. de la Luz Rojas, and A. Pullman, *Int. J. Quant. Chem. Quant. Biol. Symp. No. 8,* 109 (1981).

79. P. Läuger and G. Stark, *Biochim. Biophys. Acta, 211,* 458 (1970).

80. S. B. Hladky, *Curr. Top. Membrane Trans., 12,* 1 (1979).

81. S. M. Ciani, G. Eisenman, R. Laprade, and G. Szabo, in *Membranes: A Series of Advances* (G. Eisenman, ed.), Marcel Dekker, New York, 1973, Vol. II, Chap. 2.

Chapter 6

BEAUVERICIN AND THE OTHER ENNIATINS

Larry K. Steinrauf
Department of Biochemistry
Indiana University School of Medicine
Indianapolis, Indiana

1. INTRODUCTION

1.1. General Introduction

Enniatin A, B, C and beauvericin are cyclic hexadepsipeptides having
an obvious relationship to valinomycin, which is a cyclic dodecadepsipeptide. All are produced by various strains of fungi. The
enniatins are frequently accompanied by smaller amounts of closely
related molecules. The sequence of beauvericin, which is cyclo(D-
Hiv-N-methyl-L-Phe)$_3$, is shown in Fig. 1. For enniatins A, B, and
C the N-methylated amino acid is L-isoleucine, L-valine, and L-
leucine, respectively, and valinomycin is cyclo(L-Val-D-Hiv-D-Val-
L-Lac)$_3$, in which Hiv is the residue of α-hydroxyisovaleric acid and
Lac is the residue of lactic acid. Each of these molecules has a
threefold axis of symmetry and an approximate inversion center.
These cyclodepsipeptides have moderate antimicrobial activity, are
of low solubility in water, have no free charged groups, bind cations
strongly, and in some circumstances bind anions. These cyclodepsi-

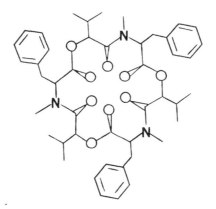

FIG. 1. Idealized drawing of the molecules of beauvericin. Notice
the threefold symmetry and notice that three of the carbonyl oxygen
atoms are pointing above the page and three pointing below the page.
For enniatins A, B, and C the N-methyl amino acids are isoleucine,
valine, and leucine, respectively.

peptides have the unusual property of catalyzing the translocation
of cations from one side of a lipid bilayer membrane to the other
according to the direction of the electrochemical potential gradient
of the cation. Most of the antimicrobial activity can probably be
expected to be caused by the translocation of sodium and potassium
ions through cell membranes so as to discharge the concentrations
created by the metabolism of the cell. Beauvericin and the enniatins
are probably free-moving carriers and do not form pores as do gramicidin or alamethicin. The availability of the homologous series has
been useful for the comparison of properties which result from the
differences between the members.

1.2. Discovery

The production of substances with antimicrobial activity by strains
of *Fusaria* had been observed since the 1920s [1]. The first efforts
at isolation and characterization were reported in 1947 by two groups.
Cook et al. [2] described the isolation of five crystalline compounds
from several strains of *Fusaria*, and Gaumann et al. [3] obtained a
compound from *F. orthoceras* var. *enniatinum* which they named enniatin.
Early estimates of the molecular size corresponded to what would be
a tetrapeptide instead of what is now known to be the hexapeptide.
The exact sequences of enniatins B and A were established by chemical
synthesis at Hoffman La Roche & Co. A.G., Basel [4] in 1963 and 1964.

Beauvericin was discovered by workers at the Eli Lilly Company
in Indianapolis. The first publication about beauvericin was concerned with its effects on the swelling of mitochondria in the presence of cations by Dorschner and Lardy [5] in 1968, and the isolation
and purification from *Beauveria bassiana* was published by Hamill et
al. [6] in 1969. Beauvericin was first synthesized by Ovchinnikov
et al. [7] in 1971 and by a slightly different method by Roeske et
al. [8] later that year.

Enniatin and the other ion transporters have been the subject
of extensive reviews during 1974-1977 by Ovchinnikov, Ivanov, and
their colleagues [9-13].

1.3. Biological Properties

The enniatins were first noticed due to their antimicrobial properties. They were found to be active [2,3,11,15,16] against *Mycobacterium phlei, M. tuberculosis, M. paratuberculosis, Staphylococcus
aureus, Sarcina lutea, Bacillus mycoides, Candida albicans,* and
Saccharomyces cerevisia. The enniatins and beauvericin cause swelling of mitochondria when incubated with a substrate and a cation.
In general, the biological activities of the enniatins and of beauvericin [6,21] are similar to but less than those of valinomycin.

1.4. Analogs

Many analogs of the enniatins have been synthesized by Russian
workers. Most of these are reviewed by Ovchinnikov et al. [13].
Among their many findings was that the n-methyl groups were abso-
lutely necessary for antimicrobial activity and for catalysis of
the transport of cations through lipid bilayer membranes, but not
necessary for the binding of cations in solution. Loss of the
N-methyl groups probably decreases the solubility in lipid and
nonpolar solvents, or introduces possibilities for hydrogen bonding
that are counterproductive for biological activity and membrane
transport. The changing of the chirality of either the amino acid
residue or the hydroxy acid residue would eliminate antimicrobial
activity and transport, and reduce cation binding. However, the
complete enantiomorphs of the enniatins behaved in exactly the
same manner as did the natural product in all respects as far as
antimicrobial activity, transport, and cation binding were concerned. This observation was strong evidence that the antibiotic
was not exerting its biological activity by interacting with a complex compound such as an enzyme or a binding site. Only biological
species not having chirality could be involved, such as sodium or
potassium.

2. ION BINDING

2.1. Solution Methods

The abilities of the ion-binding antibiotics to form complexes with
cations in aqueous solution are rather meager. If the hydrogen-
bonding abilities or the dielectric constant of the solution are
decreased, then the cation becomes much more interested in the antibiotic. Complexation is much stronger from ethanol or methanol
solutions in which the group I or group IIa cations form recogniz-
able complexes with the enniatins: participation of solvent molecules
is possible. In chloroform or hydrocarbon solutions the cations are
complexed exclusively by the antibiotic molecules, binding to the
carbonyl oxygen atoms (or alcohol or ether oxygen atoms in the case
of the monensin-type antibiotics [14]). Indeed, valinomycin or
enniatin/beauvericin in chloroform will quickly dissolve solid
potassium picrate or potassium thiocyanate. The conformation of
the antibiotic probably undergoes a transformation also, being open
in water solution, an equilibrium of several forms and species of
complexes with cations in alcohol solutions, and of a form closed
around the cation in chloroform solutions.

Studies on the structures of the antibiotics in solutions are
relevant to membrane transport insofar as the environment at the
lipid bilayer-water interface is probably similar to that of an
ethanol solution and that of the bilayer interior is similar to
that of a chloroform solution. Since most measurements giving
information about molecular structure must be made on a bulk larger
than that of a lipid bilayer, such studies have been made on homogenous solutions of the antibiotic with various cation-anion combinations.

Several techniques have been used to investigate the structures
and conformations of the antibiotics and their complexes in different
solvents. Shemyakin et al. [15] followed the complexation of enniatins A, B, and C with group I cations. They found that the electrical conductance of a solution of potassium chloride in absolute

alcohol would decrease as an enniatin was added due to the lower
mobility of the complex. They found that the optical rotatory dispersion (ORD), which is characteristic of the conformation of the
polypeptide chain, was much the same in polar and nonpolar solvents
for enniatins A and B, but not for C, and that the ORD of A and B
changed when potassium thiocyanate was added. They also isolated
the complexes of enniatins B and C with potassium thiocyanate and
found that the infrared (IR) spectra contained a significant batho-
chromic shift for the ester carbonyl bond after complexation. In a
further study of enniatin B, Shemyakin et al. [16] now found differences in the ORD spectra of the complexes with potassium in solvents
of different polarity. They decided that uncomplexed enniatin B has
a different conformation in polar solvents than in nonpolar solvents.
Enniatin B complexed with potassium has the same conformation in any
solvent; this conformation is the same as that of uncomplexed enniatin B in polar solvents. From nuclear magnetic resonance (NMR)
studies at different temperatures, they concluded that the nonpolar
form must have the three N-methyl groups in different environments.
The polar form, which retained the threefold symmetry, could exist
in only two possible conformations: that with the isopropyl groups
axial or that with the groups equatorial. They assumed that the
cation was situated inside the cage of the antibiotic and chose the
equatorial form on the basis of the decrease of the amide to α-carbon
coupling constant with increasing cation size for the (tri-N-desmethyl)
enniatin B. In this study they decided from the IR spectra, this time
in solution, that all six carbonyl oxygen atoms were participating in
the binding of the cation.

Ivanov et al. [17] and Ovchinnikov et al. [10] recognized the
ability of the enniatins to form complexes with cations of antibiotic/
cation ratios other than 1:1. Using the largest NMR chemical shifts
of certain hydrogen nuclei upon complexation, they carefully followed
these signals, such as from the hydrogens of the α-carbon atoms and
of the N-methyl groups, as an increasing amount of potassium thiocyanate was added (Fig. 2). They found clear evidence of a 2:1

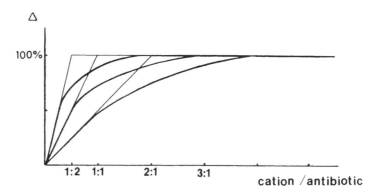

FIG. 2. Behavior of the spectral change (ordinate) of enniatin with
increasing cation concentration (abscissa) showing complexes of
antibiotic/cation ratios of 2:1 (left set), 1:1 (middle set), and
2:1 (right set). Light lines: complex of infinitely large binding
constant. Heavy lines: complex of moderate binding constant.

complex, which at higher salt concentrations became a 1:1 complex.
They found evidence for only the 1:1 complex with lithium; with
sodium they found a 1:1 and perhaps a 2:1; and with cesium they
found a 1:1 and probably a 3:2 complex. They also reported that the
circular dichroism (CD) spectrum for the 2:1 complex was very similar
to that of the 1:1 complex, which would suggest very similar conformation and bonding. They also observed a second order dependence of
the lipid bilayer conductance on enniatin concentration with potassium as the transporting cation, which would now suggest that the
2:1 complex was the dominant species in transport. They offered the
explanation that the 2:1 species would transport better in spite of
having a lower stability than the 1:1 complex because the cation
would be better shielded from the membrane interior than for the 1:1
complex. They also found higher than first order dependence of lipid
bilayer conductance on beauvericin concentration in potassium chloride.
The authors interpreted their observations by postulating the existence
of sandwich complexes in which the cation is bound by three carbonyl
oxygens of each of two different molecules. Although the authors do
not so state, their results would not require that the cation be
inside the cage of the antibiotic for the 1:1 complex.

The studies of bisenniatin B [12,17] provided new information
about the structure and function of the enniatins. This was synthesized by peptide condensation from enniatin B analogs with but a
single residue changed, (N-methyl-L-lysyl)enniatin and (N-methyl-L-
glutyl)enniatin (Fig. 3); this was indeed two enniatin molecules
permanently connected in approximately the correct manner to accommodate a cation between the enniatin molecules. Whereas the concentration dependence of the membrane conductance of enniatin B was the
second power of the antibiotic, the dependence of bisenniatin B was
first order, thus providing further evidence for the activity of the
sandwich type of complex such as (En·K$^+$·En). Moreover, solvent
extraction observations on enniatin gave one cation per macrocyclic
ring, while the connected rings of bisenniatin extracted but one
cation between them. Bisvalinomycin under the same conditions extracted and transported a total of two cations.

The behavior of the charged enniatin analogs, N-methyl-L-lysyl
and N-methyl-L-glutamyl, were described by Ovchinnikov et al. [12].
They found also that the negatively charged analog was more stable
with the cation than was the neutral which, in turn, was more stable
than the positively charged. They found that there was more influence of charge with the enniatin analogs than with the corresponding
valinomycin analogs. They interpreted these results to mean that
the cation is less shielded by enniatin than by valinomycin. They

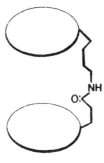

FIG. 3. Idealized drawing of the molecule of bisenniatin B. Imagine
a cation residing between the two cyclohexadepsipeptide rings.

also interpreted their results with spin-labeled enniatin analogs
as showing evidence of an interaction between the bound cation and
the labeled side chain. For the corresponding valinomycin analogs
the effects were much smaller. It might be noted that Tosteson [18]
and Steinrauf et al. [19,20] found evidence of anion influences on
valinomycin complexes in solution and in crystal structures, respectively. Therefore, anion influences on enniatin/beauvericin complexes would not be unexpected.

The first measurements of the ORD and CD spectra of beauvericin
by Ovchinnikov et al. [21] showed that the behavior in different solvents was very similar to that of the enniatins. The far-IR spectra
of beauvericin and valinomycin in chloroform were studied by Ivanov
et al. [22], who found one wide cation-oxygen stretching frequency
for beauvericin with sodium, two with both lithium and potassium,
and four with cesium. These observations were interpreted using a
1:1 internal complex model from which dihedral angles of the macrocyclic chain were estimated. In this study the authors apparently
did not attempt to determine analytically the antibiotic/cation
ratios in their preparations, and therefore the actual compositions
of the complexes are unknown.

Some indication that beauvericin-induced cation transport
could be influenced by anions was given by Estrada-O. et al. [23],
who reported that the rates of oxygen consumption of isolated, intact
liver mitochondria were accelerated by beauvericin in the presence
of external monovalent cations. With β-hydroxybutyrate as the oxy-
dizable substrate the cation preference sequence was Na^+ > (Rb^+,
Cs^+, K^+, Li^+), while with glutamate plus malate the sequence was
found to be $K^+ > Rb^+ > Cs^+ > Li^+ > Na^+$.

Roeske et al. [24] prepared crystals and gave the space groups
and unit cell parameters for the complexes of beauvericin with sodium,
potassium, rubidium, calcium, and barium picrate. They also carried
out solvent extraction and measured the initial rate of U-tube transport with these cations and with lithium, from which they found that
beauvericin has the highest activity with rubidium, potassium, and
barium. The subsequent crystal structure determination of beauvericin-

barium picrate will be described later. In this same group Prince
et al. [25] had compared beauvericin, enniatin, and valinomycin as
cation-transporting agents in liposomes and bacterial chromatophores.
They found that calcium was transported well by beauvericin but not
by enniatin or valinomycin.

Because of the marked specificity of the macrocyclic carriers
for certain cations, it has been natural to attempt to use these
compounds to construct ion-specific electrodes. Such electrodes,
largely due to the work of Simon and colleagues [26], are now available and widely used, particularly to follow cation changes in biological systems.

Grell et al. [27] reviewed ultrasonic absorption methods used
on enniatin B to investigate the relaxation kinetics of conformational transitions as a function of solvent polarity and of added
salt. These methods can detect conformational transitions if the
transition occurs within the frequency range used. For enniatin B
they observed a single transition at about 100 MHz in n-hexane, the
most nonpolar solvent used. As they added ethanol and then trifluoroethanol, they observed new transitions at about 20 MHz and
1 MHz. In pure trifluoroethanol all transitions in the range 1-
100 MHz disappeared. By this means a minimum of three transitions
was detected. Somewhat similar results were obtained for valinomycin. Grell et al. [27] also reported CD and absorption spectra
changes with solvent polarity and with added salt, which they interpreted as indicating orientation changes of the ester and peptide
groups and also solvent hydrogen bonding by solvent molecules. They
were able to calculate apparent stability constants for both monovalent and divalent cations. The same anion, however, was not used
throughout. Grell et al. also reported observations on the spectra
of valinomycin and enniatin B in lecithin bilayer vesicles (liposomes, Bangham [28]) which were suspended in aqueous solutions containing various salts. They observed that the CD spectra of cation-
free enniatin B was most similar to that from homogenous 9:1 methanol/
water and the potassium complex most similar to that from homogenous
pure methanol. The corresponding spectra of valinomycin corresponded

to that from homogenous nonpolar solvents. From these observations
they concluded that valinomycin is mostly in the interior of the
vesicle bilayer while enniatin B is mostly at the interface.

Other reports on the activity of beauvericin with calcium
include the NMR studies of Davies and Khaled [29], which gave some
indication that the alkali cations are bound in polar solvents predominantly by the peptide carbonyl oxygens. They also found that
complexes of lithium and cesium have much smaller proton shifts than
complexes of sodium and potassium. Ciani and Pajong [30] suggested
that beauvericin can transport calcium as ion pairs with $CaNO_3^+$ being
transported better than $CaCl^+$.

2.2. X-Ray Crystallography

X-ray crystallography has made significant contributions to the
understanding of the valinomycin type and the monensin type of ion-
transporting antibiotics. Therefore, it is not unexpected that the
first, and for some time the only, crystal structure of enniatin B
with potassium iodide should be highly influential on the thinking
about the site of cation binding. Dobler et al. [31] in 1969 were
able to produce from three-dimensional data only a two-dimensional
view of the complex, that looking down the threefold axis. By
assumption they placed the potassium ion inside the cage, although
based on the information available it could equally well have been
above and below the cage, bonded to three carbonyl oxygens each.
Thus the concept of the cation within the enniatin cage was firmly
established.

The second enniatin-type structure, that of the beauvericin-
barium picrate complex, was produced by Hamilton et al. [32,33] in
1975. This fully three-dimensional structure determination showed
for the first time clusters or sandwiches of carrier and ions of a
type somewhat similar to that suggested by the Russian workers [10,
17]. The structure contained one free picrate ion plus the cluster
$(Bv \cdot Ba \cdot Pic_3 \cdot Ba \cdot Bv)^+$ (Fig. 4), in which the three-fold symmetry was

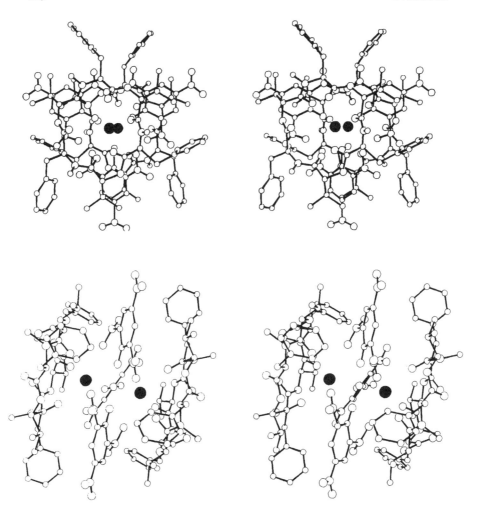

FIG. 4. Two views of the crystal structure determination of the
complex $(Bv \cdot Ba \cdot Pic_3 \cdot Ba \cdot Bv)^+$, Pic^-. The nonbinding picrate is not
shown.

well preserved and the cations were bound to the isovaleryl carbonyl
oxygens (away from the N-methyl groups). These same researchers also
produced a structure for beauvericin-potassium picrate (Bradford
Braden, Ph.D. thesis, Indiana University, 1976). This structure from
poorly resolved data showed the cluster $(Bv \cdot K \cdot Bv \cdot K \cdot Pic_3 \cdot K \cdot Bv \cdot K \cdot Bv)^+$

FIG. 5. Schematic arrangement of beauvericin-potassium picrate as deduced from the crystal structure results. [= beauvericin, K = potassium, ⅄ = picrates.]

in which both ends of the beauvericin must be involved in the binding of the cation (Fig. 5).

In 1976 Tishchenko et al. [34a] published preliminary details of uncomplexed enniatin B which they had obtained from acetonitrile also containing potassium thiocyanate. This structure had a pseudo-threefold axis and pseudoinversion center, and, they reported, was very similar to that which had been predicted from polar solutions. Later details of the crystal structure determination [34b,c] revealed that the crystal form was first thought to be triclinic with a single molecule of enniatin without molecular symmetry as the repeat unit. After refinement the authors noticed that the threefold symmetry of the molecule was exceedingly good. They were able to reindex the diffraction data into a rhombohedral form with one-third the amount of data and one-third as many atoms in the repeat unit. The single molecule of enniatin B now was required to have perfect threefold symmetry. The authors then found two peaks of electron density on the threefold axis, one nearly in the center of the cavity. Efforts to refine these features as potassium thiocyanate were unsuccessful. Their final conclusion was that the extra electron density represented water molecules originating probably from the atmosphere during the crystallization procedure. These water molecules had to be disordered since the molecule of water does not have threefold symmetry.

Also in 1976 Geddes and Akrigg [35] published full details of the structure of uncomplexed beauvericin crystallized from n-heptane. This structure had an approximate threefold symmetry, which in this case was influenced by three imperfectly ordered water molecules, one within the cage and the other two each forming hydrogen bonds to two of the carbonyl oxygen atoms at each end of the beauvericin.

Crystal structure determinations have also been produced by
Russian workers on biologically inactive and nontransporting analogs
of the enniatins.

In 1981 Zhukhlistova and Tishchenko [36] published the results
from a crystal believed to be a complex of enniatin B and potassium
thiocyanate, but which in fact had turned out to be enniatin B-
sodium nitrate-nickel nitrate. No explanation was offered as to
how this transmutation occurred. In this structure the antibiotic
had exact threefold symmetry and a pseudoinversion center. The
enniatin bonded to one of the partly occupied sodium positions by
the three valyl carbonyl oxygens and to water molecules by the isovaleryl carbonyl oxygens. The other sodium position was surrounded
by an octahedron of oxygen atoms from methanol molecules. The nickel
was surrounded by an octahedron of water molecules which then formed
hydrogen bonds to the enniatin.

Rotations of the dihedral angles of the polypeptide chain of
up to 50° were found between the free and complexed enniatin B.
Complexation had caused the carbonyl oxygen atoms of all six residues
to move inward together while the N-methyl groups moved outward. A
triangle is formed by the three carbonyl oxygen atoms of the Hiv
residues, as the oxygens of the Val (Phe) residues form a similar
triangle on the other side of the molecule. Going from one crystal
structure to another, the legs of these triangles change considerably. These distances and the legs of the triangles formed by the
N-methyl groups are given in Table 1.

TABLE 1

Structure	Hiv oxygen distances	Val (Phe) oxygen distances	N-methyl distances
BvBaPic	3.72-4.11 Å	3.72-3.87 Å	6.61-7.09 Å
Free Bv	4.33-4.84 Å	3.38-4.15 Å	6.12-6.78 Å
EnNaNi	3.93 Å	3.79 Å	6.41 Å
Free En	5.79 Å	5.61 Å	5.15 Å

Clearly, more x-ray crystal structure determinations are needed. The evidence thus far shows only external binding of the cation by either or both the valyl and the isovaleryl carbonyl oxygens. When the carbonyl oxygens rotate inward for binding, there is a corresponding outward rotation of the N-methyl groups. Thus the N-methyl groups provide little or no resistance to the binding of cations. From x-ray crystallography should come new insight about the geometry of cluster formation and suggestions as to the design of new and interesting analogs. The conformation when binding a cation has very close to threefold symmetry since either the three related isovaleryl residues or the three amino acid residues, or both will be involved in the binding.

2.3. Cation: Internal or External?

Only for valinomycin is there direct evidence from x-ray crystal structure determinations of the internal binding (valinomycin-potassium picrate) and the external binding (valinomycin-sodium picrate) of the cation [19,20]. The transition from internal to external binding is accomplished with little conformational change by valinomycin. In fact, as seen in Fig. 6, even when the cation is bound externally, the internal binding site is retained with a water molecule occupying the position and becoming a part of the new site for the sodium. The conformational change between complexed and uncomplexed enniatin B and between complexed and uncomplexed beauvericin in the crystal structures was far less than that between complexed and uncomplexed valinomycin. Therefore, since enniatin/beauvericin is less flexible than is valinomycin, it might be a reasonable exercise to place an artificial cation at the center of the six carbonyl oxygen atoms from the x-ray crystal structure determinations and thereby be able to calculate bond distances and angles. When this was done for the enniatin B-sodium, nickel nitrate [36], the artificial cation site was found to be at fractional coordinates (0.3093, -0.3093, 0.3093), the cation to

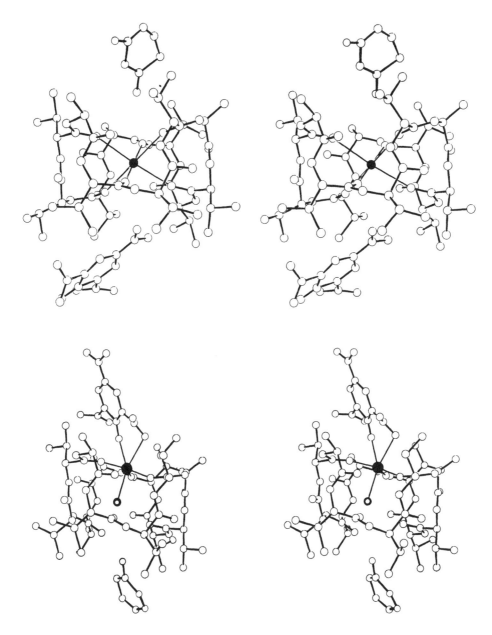

FIG. 6. Stereo view of the crystal structures of valinomycin-
potassium picrate and of valinomycin-sodium picrate as seen from
the same orientation. o water, ● cation.

carbonyl oxygen distances were all 2.62 Å, the cation to carbonyl
carbon distances were 2.94 Å, and the cation-oxygen-carbon bond
angles were 92°. Similar calculations based on the atomic coordinates from beauvericin-barium picrate [33] gave the artificial
cation to carbonyl oxygen distances of 2.41-2.77 Å, cation to carbon
distances of 2.75-3.11 Å, and cation-oxygen-carbon angles of 93-101°.
Uncomplexed beauvericin [35] is much less symmetric due to the absence of an external cation to force the depsipeptide chain into
something close to threefold symmetry. Here the artificial cation
site gave cation to oxygen distances of 2.41-3.22 Å, cation to carbon
distances of 2.81-3.21 Å, and cation-oxygen-carbon angles of 76-101°.
An atom placed at the center of the uncomplexed enniatin B [34b,c]
will find the carbonyl oxygen atoms farther away (3.68-3.75 Å) than
the carbonyl carbon atoms (2.93-2.94 Å). The bond angles would be
61-63°. From calculations such as these it may be seen that the
cation to oxygen distances are reasonable at least for the smaller
cations, but the real question is whether or not the cation to carbonyl carbon distances are too short, and consequently the cation-
oxygen-carbon angles too small. It should be noticed that the same
bond angles for the valinomycin structures were in the range 160-
170°.

In 1977 Steinrauf and Sabesan [37] presented semiempirical
calculation on the stability of cation sites based on monopole-
monopole electrostatic interactions for beauvericin as well as for
valinomycin, monensin, nigericin, and dianemycin. These calculations gave reasonable results for the internal binding of cations
by all the above molecules except for beauvericin. On the other
hand, Kostetsky et al. [38] found very small bond angles from their
quantum mechanical study of the enniatin system. Other theoretical
studies by Maigret and Pullman [39] and by Ramachandran and Chandrasekharan [40] did not specifically question the bond angles.

2.4. Anion Binding

There is no evidence thus far for the binding of anions in the
absence of some bound cation. The first reported involvement of
anions was the previously mentioned study by Estrada-O. et al. [23],
and Ciani and Pajong [30] may have observed anions being involved
with the transport of calcium by beauvericin. A similar situation
may have been observed by Prince et al. [25], who found that calcium
was transported with the same apparent charge as was sodium and
potassium by beauvericin.

The involvement of anions with complexation of cations by
valinomycin has been studied in detail by Davis and Tosteson [18].
Evidence of the ability of valinomycin to bind cations within the
cage and external to the cage has recently been reviewed by Steinrauf et al. [20] along with the binding of anions in solution and
in crystal structures (see Fig. 6). Crystal structures have now
been obtained in which the anion picrate assumes quite a variety of
relationships with the rest of the complex, from slight in valinomycin-potassium picrate, to moderate in valinomycin-ammonium picrate
and valinomycin-cesium picrate, to strong in valinomycin-sodium
picrate, to dominating in beauvericin-barium picrate and beauvericin-
potassium picrate.

The possibility certainly exists that anions can strongly
influence the binding and transport of cations by the enniatin-type
carriers even more strongly than has been found for valinomycin.
Obviously, more research is necessary along the lines which have
been applied to valinomycin by Kuo et al. [41], by Tosteson et al.
[42], or Ginsberg and Stark [43] as reviewed by Steinrauf et al.
[20].

2.5. Binding of Neutral Molecules

Evidence for the binding of molecular species other than of ions by
the ion-transporting antibiotics exists principally for valinomycin.

Such binding has not been the subject of much investigation. In
1970 Hinton and O'Brian [44] reported that the insecticide DDT would
inhibit the electrical conductance of potassium through lipid bilayer
membranes treated with valinomycin. Similarly Kuo et al. [41] observed that halogenated benzimidazoles would block the valinomycin-
mediated conductance of lipid bilayer membranes, although the blocking was better when the benzimidazoles were in the anionic form. In
1978 Levitt et al. [45] presented evidence that water molecules were
transported through lipid bilayer membranes by the cation complexes
of valinomycin, nonactin, and gramicidin.

The presence of a water molecule in the cage of uncomplexed
beauvericin, of uncomplexed enniatin B, and of the complex of valinomycin with sodium picrate may seem at first somewhat surprising. If
a water molecule can reside in the cage of beauvericin/enniatin, why
not a cation? The answer to this question must take into account
the differences between water and a metal cation. It is here most
significant that the metal cation is spherically symmetric and water
is a dipole, actually two dipoles. The interiors of valinomycin and
of enniatin/beauvericin are lined with different kinds of atoms at
different distances from the center of the cage. In the cage of
valinomycin the inner lining is composed of the carbonyl oxygen atoms
at a radius of 2.7 Å. The next layer is composed of the carbonyl
carbon atoms much more distant at 3.9 Å. The next layer is composed
of several other types at greater than 4 Å. The cage of enniatin/
beauvericin is lined with several kinds of atoms at a distance of
2.7-3.0 Å. These include the carbonyl oxygen and carbon atoms. As
a consequence of this the center of the cage of valinomycin is relatively uniform because only the carbonyl oxygen atoms are close.
The center of the cage of enniatin/beauvericin must be changing very
much with distance because both electronegative and electropositive
atoms are about the same distance away. Therefore, it might well be
easier to place the dipole in the enniatin/beauvericin cage, which
has both negative and positive regions in its inner layer. The
valinomycin cage has only negative atoms in its inner layer, except

when there is a cation bound to one pocket as is the case for valinomycin sodium picrate.   Here the water has the carbonyl oxygen atoms
to bind the hydrogen atoms and the sodium to bind the oxygen atom.

3. MEMBRANE TRANSPORT

In 1967 Mueller and Rudin [46] published their observations that
valinomycin, the actins, and the enniatins would greatly decrease
the electrical conductance of lipid bilayer membranes as obtained
from a variety of sources in the presence of monovalent cations.
They found that divalent cations had little effect.   They proposed
the hydrated cation to be the species that was being carried by the
ring of the antibiotic through the membrane.

   A detailed study of the effects of enniatins A and B and beauvericin on the conductance of lipid bilayer membranes was made by
Benz [47] in 1978.   He formed his membranes from a solution of synthetic lipid in n-decane by brushing the decane solution over a
small hole in a septum separating two aqueous solutions.   He used
dioleoyl phosphatidylcholine (diO-PC) and monoglycerides of various
unsaturated fatty acids (GMOs).   He observed that the conductance
was first order with respect to the concentration of enniatin A or
B over nearly three orders of magnitude and first order with respect
to cation concentration over about two orders of magnitude for the
group I cation chlorides.   The conductance was approximately third
order with respect to beauvericin concentration.   For both enniatin
A and B he found that the cation preference sequence was $K^+ > Rb^+ >$
$Cs^+ > Na^+ > Li^+$.   From the current-voltage relationships at steady
state he deduced that the migration at the membrane interface is
fast and that the rate-limiting step is the passage through the
middle of the membrane.   He further suggested that the complexation
steps occur at the interface rather than in the aqueous solution.
When using membranes made of diO-PC he obtained K^+/Na^+ preference
ratios of 30:1, 15:1, and 5:1 for enniatin A, B, and beauvericin,
respectively.   He obtained relaxation times due to the membrane

transport from charge pulse experiments. The membrane was charged
from a second set of electrodes by a brief current pulse. The resulting voltage transient was recorded on a storage oscilloscope and
analyzed [48,49] according to the association constant, dissociation
constant, translocation rate constant of the complex, and translocation rate complex of the free carrier (Fig. 7). It was possible to
calculate these parameters for valinomycin and for enniatins A and
B. For beauvericin the expected three relaxation times were not
resolved. Benz et al. [48,49] stated that they found no indications
of the 2:1 or 3:2 complexes found by Ivanov et al. [10,17], although
their results on behavior of beauvericin were similar to that observed by the Russian workers. Benz et al. suggested that it would
be possible, but not highly probable, that the third order dependence
by beauvericin could be caused by a concentration-dependent partition
coefficient. They also noted that the cation preferences are much
smaller for the enniatins than for valinomycin or the actins. To
that note it is worth adding that only for valinomycin and the actins
has it been proven by x-ray crystallography that the cation actually
resides entirely within the cage of the antibiotic.

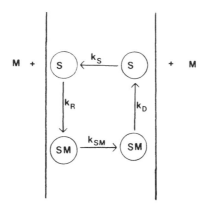

FIG. 7. Steps in the translocation process of a free carrier moving
from left to right. M is the cation being transported by the free
carrier S, k_{SM} is the rate of transport of the complex, k_S is the
rate of transport of the free carrier, k_R is the rate of association
between the carrier and the cation, and k_D is the rate of disassoci-
ation.

In 1973 [50] we extended our own investigations of the enniatins to include lipid bilayer transport, from which we reported
that the conductance was approximately second order for beauvericin
with calcium, barium, and other cations. In 1980 the author initiated a study on the enniatins as a guest in the laboratory of Dr.
Daniel C. Tosteson, Harvard Medical School, Boston. The study was
to investigate the possibility of the existence of clusters of ions
and antibiotics such as had been observed by us in the x-ray crystal
structures of beauvericin-barium picrate and potassium picrate. In
this study steady-state voltage and current measurements were taken
on solvent-free membranes composed of diphantoyl phosphatidylcholine
(diP-PC), bacterial phosphatidylethanolamine (b-PE), and glycerol
monoleate (GMO). Membranes were constructed by the method of Alverez
and Latorre [51] in which the lipid material, dissolved in highly
volatile pentane, was spread on the surface of two solutions separated by a thin partition. The two solutions can be connected only
by a small hole (about 1 mm in diameter punched through a thin plas-
tic film stuck to the partition by stopcock grease) well above the
initial surface of the solutions. After the evaporation of the
pentane the lipid will form a monolayer at all surfaces. As the
level of both sides was carefully raised the monolayer would coat
all fresh surfaces covered by the solutions, including the small
hole in the partition, which was now covered by a monolayer from
both sides if the levels of the two solutions were raised at the
same time. The resulting bilayer had thus been constructed from a
small, measured amount of lipid and had a reproducible surface area
and consequently has a reproducible conductance. Antibiotic added
in ethanol to the water solutions was quickly taken up by the lipid;
the concentration of the antibiotic in the lipid is thus fixed. A
diagram of the apparatus is given in Fig. 8. The log conductance
vs. log concentration antibiotic curves so obtained are given in
Fig. 9 for enniatin A and beauvericin through diP-PC, b-PE, and GMO
membranes using K^+, Na^+, and Ca^+ as the transported cations. A
common feature was a slope of 1 at lower antibiotic concentrations,

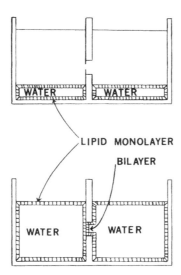

FIG. 8. Lipid bilayer transport apparatus. About 10 ml of aqueous
solution is contained in each side. Electrodes are placed in each
side to measure the current flow. More aqueous solution has been
added in the lower figure, raising the level of both sides above the
small hole, which now contains a lipid bilayer.

a slope of 2 at intermediate enniatin concentrations, or a slope of
2-4 at intermediate beauvericin concentrations, and a limiting conductance at highest concentrations. The K^+/Na^+ preferences were
dependent on lipid composition, being 20:1 to 40:1 for enniatin B
and 0.3:1 to 10:1 for beauvericin. In all cases the preference for
Ca^{2+} was slightly lower over the concentration range of antibiotic
observed, which was about 100 in b-PE, 200 in diP-PC, and 3,000 in
GMO. Note that in Fig. 9 the concentration of calcium is 0.1 M while
that of sodium and potassium is 1.0 M. The effects of membrane composition were most apparent on the transport by beauvericin with K^+
and Na^+ which produce a slope of 4 in diP-PC and b-PE but a slope of
2 through GMO. The slope of transport by beauvericin in diP-PC was
not constant but depended on both antibiotic and cation concentrations. At different cation concentrations (Fig. 10) the maximum
slope changed from 4 at 1.0 M KCl to 3 at 0.1 M KCl to 2 at 0.1 M
KCl. These results are best interpreted as the transport of clusters,

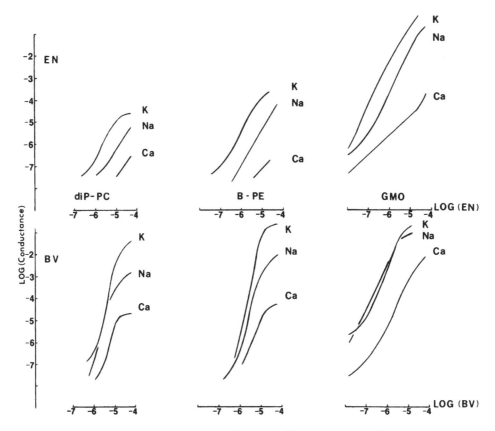

FIG. 9. Membrane transport results. Ordinate: log conductance in mhos/cm, abscissa: log antibiotic concentration in mol/liter. Curves for beauvericin and for enniatin B with 1.0 M potassium chloride, 1.0 M sodium chloride, and 0.1 M calcium chloride, through diphantoyl phosphatidylcholine, bacterial phosphatidylethanolamine, and glycerol monooleate (diP-PC, B-PE, and GMO, respectively). Notice particularly the maximum slope displayed by each curve.

or multiple sandwiches, of antibiotics and cations, such as $En \cdot K^+ \cdot En$ with perhaps some $En \cdot K^+$ at lower concentrations, and with $Bv \cdot K^+ \cdot Bv \cdot K^+ \cdot Bv \cdot K^+ \cdot Bv$ with smaller clusters at lower concentrations. With calcium the clusters $Bv \cdot Ca^{2+} \cdot Bv$ and $En \cdot Ca^{2+} \cdot En$ seem to dominate with perhaps $En \cdot Ca^{2+}$ being the major species through GMO. The dependence of conductance on changing cation concentration is more difficult to follow because of the changing ionic strength. (With valinomycin it

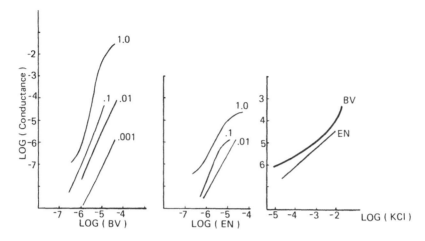

FIG. 10. Membrane transport results as in Fig. 9, showing the results of varying the concentration of potassium chloride for beauvericin and of enniatin B through diphantoyl phosphatidylcholine. Notice that the slope for both antibiotics approaches 2 as the concentration is lowered.

is possible to keep the ionic strength constant by the addition of the nontransporting lithium chloride.) However, the results clearly show a curve of increasing slope with increasing potassium concentration which is in agreement with the cluster of $Bv_4K_3^{3+}$ plus smaller clusters. From these observations and the tentative conclusions drawn therefrom two questions arise: (1) Why are clusters necessary? (2) Why is beauvericin better at clustering than enniatin B?

In order to attempt answers to these questions some speculations are necessary. Let it be assumed that the cation is bound to one or two external sites of beauvericin and the enniatins. For brevity, let us call such a site a pocket (Table 2). Obviously, a cation bound to a single hexadepsipeptide would be only about half shielded from the external environment. Such a "half sandwich" ($En \cdot K^+$) would be less soluble in nonpolar solvents than the "full sandwich" ($En \cdot K^+ \cdot En$) in which a second antibiotic molecule provided a pocket for the other side of the cation. The half sandwich would be expected to be found only in the more polar solvents, through

TABLE 2

Antibiotic	Center binding	Pocket binding	Clustering ability	Membrane solubility	Cation specificity
Actins	Strong	Weak?	Poor	High	Good
Valinomycin	Strong	Medium	Poor	High	Very good
Prolinomycin	Very strong	Weak	Poor	Medium	Very good
Enniatin	Weak	Strong	Fair	High	Poor
Beauvericin	Weak	Strong	Good	High	Poor

more fluid membranes, or at lower concentrations of antibiotic. Beauvericin would be able to form double or triple sandwiches by adding one or two layers of cation and antibiotic. The more complete shielding of the cation would be necessary in nonpolar solvents, through membranes of high potential barrier, and at high concentration of antibiotic. Furthermore, it might be suggested that the double and triple sandwiches might be formed because of high concentrations of cations and antibiotic molecules at the water-membrane interface caused in part by a dipole layer at the interface. Membranes of GMO do not have the dipole layer as do those made of diP-PC or b-PE. Another suggestion could be the need to have a large electrical charge on the transported species in order to penetrate the hydrophobic center of the membrane. If such be the case, the divalent cations would not require clusters as large as would the monovalent cations. Some observations that could provide useful additional information would be the effects of different anions and of phloretin or other surface-active agents on membrane transport and the effects of membranes differing only in the fluidity of the interior.

The cation preferences observed for beauvericin and the enniatins are not as great as those to be expected from a cation sitting at the center of an inflexible cage. However, the staggered trigonal bipyramid of oxygen ligands provided by the pockets of two antibiotic molecules at opposite sides of a cation will adapt to provide reason-

able coordination for cations of any size. The triangle of the
pocket can open slightly to provide a larger size, and of course
the two antibiotic molecules can move further apart to provide the
needed room. These accommodations are accomplished smoothly and
without any great expenditure of energy, in which case the major
competition for the cation would be the hydration shell of water
molecules displaced by the antibiotic molecules.

The superior clustering abilities of beauvericin obviously
must be a consequence of the three benzene rings. As seen in Fig. 4,
these benzene rings contribute rather neatly to the compact nature of
the cluster, which would be considerably more exposed without them.
These benzene rings also cause beauvericin to be more soluble in nonpolar solvents than enniatin B. Therefore for beauvericin both the
transported complex and the returning free carrier would be more
stable in the membrane.

4. ANALOGS

4.1. Previous Work

Researchers at the Shemyakin Institute in Moscow have synthesized
many analogs of the enniatins [13]. Studies on the macro ring size
have shown that any deviation from six residues results in almost
complete loss of biological activity and ability to bind cations.
Likewise any deviation from the natural chirality of the residues
gives almost complete loss of activity. The sole exception to this
was the synthetic mirror image of enniatin B, cyclo(N-Methyl-D-Val-
L-Hiv)$_3$, which was found to have much the same activity as natural
enniatin B. The replacement of a single amino acid residue of
enniatin B with a lysyl or glutamyl residue has been discussed. It
would have been interesting to have determined the potassium/calcium
preference for particularly the glutamyl analog and for bisenniatin B.

Indeed, a major objective in the enniatin/beauvericin research
should be to find an analog that will transport calcium and not potassium or sodium.

4.2. Analogs of Possible Interest

(The following discussion is based on conversations with the late
Dr. Baltz F. Gisin.)

It appears that there are three loci on which changes can be
made which may usefully alter the activity of the cyclohexadepsipeptide and not abolish it altogether.

1. One or more of the N-methyl groups may be changed to an
    ethyl or larger alkyl group.
2. One or more of the N-methyl amino acid residues may be
    changed.
3. One or more of the hydroxy acid residues may be changed.

If all three of the N-methyl groups were converted to ethyl or
isopropyl, the pocket lined by these groups would be so deep as to
prevent the approach of another cyclohexadepsipeptide molecule to
form another layer of sandwich. Thus ethyl beauvericin should be
strictly limited to the single sandwich (EBv·M·EBv) for both mono
and divalent cations. Since most of the transport of sodium and
potassium is carried by supersandwiches of beauvericin, ethyl beauvericin may well have reduced preference for sodium and potassium.

By analogy to the proline valinomycin (or prolinomycin) of
Gisin and Merrifield [52], two analogs of, for example, beauvericin
may have interesting properties. One analog would be cyclo(D-Pro-N-
Methyl-L-Phe)$_3$. A proline-containing analog would be expected to
have altered dimensions of the internal cavity and much reduced
flexibility of the backbone. Prolinomycin [52] was found to have
even better cation binding than valinomycin but poorer membrane
transport of cations [53]. Another cyclohexadepsipeptide would be
cyclo(D-Hiv-L-Pro)$_3$, which would have no need for the protective
methyl groups.

An increase in the lipophilic nature of the side chains may
result in an increase in the rate of cation transport through membranes such as is observed going from enniatin B to beauvericin.
Such an analog of beauvericin might be cyclo(D-Hpa-N-Methyl-L-Phe)$_3$

where Hpa is the residue α-hydroxyphenyl acetic acid. A different
means of accomplishing the same objective might be the tryptophan
analog of beauvericin cyclo(D-Hiv-N-Methyl-L-Try)$_3$.

Finally, the N-methyl groups make the organic synthesis more
difficult but are necessary to remove the peptide hydrogen. From
looking at molecular models it would seem possible to block the
hydrogen by intramolecular hydrogen bonding such as might be obtained from a hydrogen bond receptor some two or three atoms along
the side chain. An ether, carbonyl, or ester at the end of the side
chain might accomplish this. An example might be cyclo(D-Hiv-O-
Methyl-L-Glu)$_3$ in which the carbonyl oxygen of the side chain would
form a hydrogen bond with the peptide hydrogen. The ester, in this
case methyl, would regulate the hydrophobicity of the molecule.

Such considerations do not take into account difficulties in
the syntheses. Some of these problems have recently been discussed
in a review by Roeske and Kennedy [54].

5. CONCLUSIONS

The enniatin/beauvericin series has provided an interesting oppor-
tunity to explore some of the aspects of the membrane transport
behavior of free carriers. Some potentially useful possibilities
remain yet to be investigated. Analogs of the series may yet be
developed which have high preferences for calcium. The behavior
of beauvericin in particular seems to be sensitive to the properties
of the membrane and thus this molecule may find use as a means of
characterizing membranes.

ABBREVIATIONS

Bv   beauvericin
CD   circular dichroism
En   enniatin

Glu   glutamic acid residue

GMO   glycerol monooleate lipid bilayer

Hiv   α-hydroxyisovaleric acid residue

IR   infrared

Lac   lactic acid residue

Lys   lysine residue

NMR   nuclear magnetic resonance

ORD   optical rotatory dispersion

PC   phosphatidylcholine lipid bilayer

PE   phosphatidylethanolamine lipid bilayer

Phe   phenylalanine residue

Pro   proline residue

Val   valine residue

ACKNOWLEDGMENTS

The author would like to thank Dr. Daniel C. Tosteson for his hospitality, Dr. Ramon Latorre for helpful discussions, and the late Dr. Balthazar F. Gisin for his help and interest. The author would also like to acknowledge grant support from the National Science Foundation (NSF PCM 82-07908).

REFERENCES

1. (a) H. W. Wilkins and C. G. M. Harris, *Brit. J. Exp. Path.*, *24*, 141 (1943); (b) S. A. Waksman and E. Horning, *Mycologia, 35,* 47 (1943); (c) J. Vonkennel, J. Kimmig, and A. Lembke, *Klin. Wschr.*, *22*, 321 (1943); (d) C. W. Carpenter, *Hawaii Plant Rec.*, *49*, 41 (1945); (e) S. A. Waksman, *Microbial Antagonism and Antibiotic Substances*, New York, 1945, p. 264.

2. A. H. Cook, S. F. Cox, T. H. Farmer, and M. S. Lacey, *Nature, 160,* 31 (1947).

3. (a) E. Gaumann, S. Roth, L. Ettlinger, P. A. Plattner, and U. Nager, *Experientia, 3,* 202 (1947); (b) P. A. Plattner and U. Nager, *Experientia, 3,* 325 (1947).

4. (a) P. A. Plattner, K. Vogler, R. O. Studer, P. Quitt, and W. Keller-Schierlein, *Helv. Chim. Acta, 46,* 927 (1963); (b) P. Quitt, R. O. Studer, and K. Vogler, *Helv. Chim. Acta, 46,* 1715 (1964), *47,* 166 (1964).

5. E. Dorschner and H. Lardy, *Antimicrob. Agents Chemother.,* 1968, 11.

6. R. L. Hamill, C. E. Higgins, H. E. Boaz, and M. Gorman, *Tetrahedron Lett.,* 1969, 4255.

7. Yu. Ovchinnikov, V. T. Ivanov, and I. I. Mikhaleva, *Tetrahedron Lett.,* 1971, 159.

8. R. W. Roeske, S. Isaac, L. K. Steinrauf, and T. E. King, *Fed. Proc., 30,* abs 1282 (1971).

9. Yu. A. Ovchinnikov, V. T. Ivanov, and A. M. Shkrob, *Membrane Active Complexones,* Elsevier, Amsterdam, 1974.

10. Yu. A. Ovchinnikov, V. T. Ivanov, A. V. Evstratov, I. I. Mikhaleva, V. F. Bystrov, S. L. Portnova, T. A. Balashova, E. N. Meshcheryakova, and V. M. Tulchinsky, *Int. J. Peptide Protein Res., 6,* 465 (1974).

11. Yu. A. Ovchinnikov, *FEBS Lett., 44,* 1 (1974).

12. Yu. A. Ovchinnikov and V. T. Ivanov, *Tetrahedron, 31,* 2177 (1975).

13. Yu. A. Ovchinnikov and V. T. Ivanov, *Biochemistry of Membrane Transport* (G. Semenza and E. Carafoli, eds.), Springer-Verlag, Berlin, 1977, p. 123 ff.

14. A. Agtarap, J. W. Chamberlin, M. Pinkerton, and L. K. Steinrauf, *J. Am. Chem. Soc., 89,* 5737 (1967).

15. M. M. Shemyakin, Yu. A. Ovchinnikov, V. T. Ivanov, V. K. Antonov, A. M. Shkrob, I. I. Mikhaleva, A. V. Evstratov, and G. G. Malenkov, *Biochem. Biophys. Res. Commun., 29,* 834 (1967).

16. M. M. Shemyakin, Yu. A. Ovchinnikov, V. T. Ivanov, V. K. Antonov, E. I. Vinogradova, A. M. Shkrob, G. G. Malenkov, A. V. Evstratov, I. A. Laine, E. I. Melnik, and I. D. Ryabova, *J. Memb. Biol., 1,* 402 (1969).

17. V. T. Ivanov, A. V. Evstratov, L. V. Sumskaya, E. I. Melnik, T. S. Chumburidze, S. L. Portnova, T. A. Balashova, and Yu. A. Ovchinnikov, *FEBS Lett., 36,* 65 (1973).

18. D. G. Davis and D. C. Tosteson, *Biochemistry, 14,* 3962 (1975).

19. J. A. Hamilton, M. N. Sabesan, and L. K. Steinrauf, *J. Am. Chem. Soc., 103,* 5880 (1981).

20. L. K. Steinrauf, J. A. Hamilton, and M. N. Sabesan, *J. Am. Chem. Soc., 104,* 4085 (1982).

21. Yu. A. Ovchinnikov, V. T. Ivanov, and I. I. Mikhaleva, *Tetrahedron Lett.,* 1971, 159.

22. V. T. Ivanov, G. A. Kogan, V. M. Tulchinsky, A. V. Miroshinikov, I. I. Mikhalyova, A. V. Evstratov, A. A. Zenkin, P. V. Kostetsky, and Yu. A. Ovchinnikov, *FEBS Lett.*, *30*, 199 (1973).

23. S. Estrado-0., C. Gomez-Lojero, and M. Montal, *Bioenergetics*, *3*, 417 (1972).

24. R. W. Roeske, S. Isaac, T. E. King, and L. K. Steinrauf, *Biochem. Biophys. Res. Commun.*, *57*, 554 (1974).

25. R. C. Prince, A. R. Crofts, and L. K. Steinrauf, *Biochem. Biophys. Res. Commun.*, *59*, 697 (1974).

26. Z. Stefanac and W. Simon, *Microchem. J.*, *12*, 125 (1967); W. Simon and E. Carafoli, *Methods in Enzymol.*, *LVI*, 439 (1979).

27. E. Grell, Th. Funck, and F. Eggers, in *Molecular Mechanisms of Antibiotic Action on Protein Biosynthesis and Membranes* (E. Munoz, F. Garcia-Ferrandiz, and D. Vazquez, eds.), Elsevier, Amsterdam, 1972, p. 646.

28. A. D. Bangham, *Advan. Lipid Res.*, *1*, 65 (1963).

29. M. A. Khaled and D. B. Davis, *J. Chem. Soc. Perkin Trans.*, *2*, 1327 (1976).

30. S. Ciani and S. Pajong, *Biophys. J.*, *25* (part 2), 8A (1979).

31. M. Dobler, J. D. Dunitz, and J. Krajewski, *J. Mol. Biol.*, *42*, 603 (1969).

32. J. A. Hamilton, L. K. Steinrauf, and B. Braden, *Biochem. Biophys. Res. Commun.*, *64*, 151 (1975).

33. B. Braden, J. A. Hamilton, N. Sabesan, and L. K. Steinrauf, *J. Am. Chem. Soc.*, *102*, 2704 (1980).

34. (a) G. N. Tishchenko, Z. Karimov, B. K. Vainshtein, A. V. Evstratov, V. T. Ivanov, and Yu. A. Ovchinnikov, *FEBS Lett.*, *65*, 315 (1976); (b) G. N. Tishchenko, A. I. Karaulov, and Z. Karimov, *Cryst. Struct. Comm.*, *11*, 45 (1981); (c) G. N. Tishchenko, A. I. Karaulov, and Z. Karimov, *Sov. Phys. Crystallogr.*, *26*(5), 559 (1981).

35. A. J. Geddes and D. Akrigg, *Acta Cryst.*, *B32*, 3164 (1976).

36. N. E. Zhukhlistova and G. N. Tishchenko, *Kristallografiya*, *26*, 1232 (1981).

37. L. K. Steinrauf and M. N. Sabesan, in *Metal-Ligand Interactions in Organic Chemistry and Biochemistry,* Part 2 (B. Pullman and N. Goldblum, eds.), Reidel, Dordrecht, Holland, 1977, pp. 43-57.

38. P. V. Kostetsky, V. T. Ivanov, Yu. A. Ovchinnikov, and G. Shchembelov, *FEBS Lett.*, *30*, 205 (1973).

39. B. Maigret and B. Pullman, *Biochem. Biophys. Res. Commun.*, *50*, 908 (1973).

40. G. N. Ramachandran and R. Chandrasekharan, *Prog. Pept. Res.*
     (Proc. Am. Pept. Symp.), *2*, 195 (1972).

41. K.-H. Kuo, T. R. Fukuto, T. A. Miller, and L. J. Bruner, *Bio-phys. J.*, *16*, 143 (1978).

42. H. Ginsberg, M. T. Tosteson, and D. C. Tosteson, *J. Mol. Biol.*,
     42, 153 (1978).

43. H. Ginsberg and G. Stark, *Biochim. Biophys. Acta*, *455*, 685
     (1976).

44. B. D. Hinton and R. D. O'Brian, *Science*, *168*, 841 (1970).

45. D. G. Levitt, S. R. Elias, and J. M. Hautman, *Biochim. Biophys.
     Acta*, *512*, 436 (1978).

46. P. Mueller and D. O. Rudin, *Biochem. Biophys. Res. Commun.*, *26*,
     398 (1967).

47. R. Benz, *J. Memb. Biol.*, *43*, 376 (1978).

48. R. Benz, O. Frohlick, and P. Läuger, *Biochim. Biophys. Acta*,
     464, 465 (1977).

49. R. Benz and P. Läuger, *J. Memb. Biol.*, *27*, 171 (1976).

50. M. Yafuso, A. R. Freeman, L. K. Steinrauf, and R. W. Roeske,
     Fed. Proc., *33*, abs 1258 (1974).

51. G. Alverez and R. Latorre, *Biophys. J.*, *21*, 1 (1978).

52. B. F. Gisin and R. B. Merrifield, *J. Am. Chem. Soc.*, *94*, 6165
     (1972).

53. B. F. Gisin, H. P. Ting-Beall, D. G. Davis, E. Grell, and D. C.
     Tosteson, *Biochim. Biophys. Acta*, *509*, 201 (1978).

54. R. W. Roeske and S. J. Kennedy, in *Chemistry and Biochemistry
     of Amino Acids, Peptides, and Proteins*, Vol. 7 (B. Weinstein,
     ed.), Marcel Dekker, New York, 1983, pp. 205-256.

Chapter 7

COMPLEXING PROPERTIES OF GRAMICIDINS

James F. Hinton and Roger E. Koeppe II
Department of Chemistry
University of Arkansas
Fayetteville, Arkansas

1. INTRODUCTION

The transfer of ions across lipid bilayer membranes requires trace
amounts of either "carrier" or "pore-forming" molecules which facilitate the passage of ions through the hydrophobic interior of the membrane. In the absence of specific agents which can mediate such
transport of ions, lipid bilayers are extremely impermeable to inorganic ions. Thus, the conductance of a pure phosphatidylcholine/
decane membrane is less than 10^{-9} pS cm^{-2} in 0.1 M NaCl [1]. This
conductance can be increased to nearly that of the aqueous phase by
the addition of certain antibiotics, such as valinomycin or nonactin,
which function as ion carriers, or the linear gramicidins, which
function as pore formers or ion channels [1]. This chapter summarizes aspects of the structural, ion-binding, and ion transport
properties of the linear gramicidins, particularly gramicidin A.

That gramicidin A functions as a channel was demonstrated by
Hladky and Haydon [2,3], who investigated the unit conductance channel. In membranes treated with very small amounts of gramicidin,
the conductance was observed to vary in discrete steps of uniform
size. The unit conductance of about 2 x 10^7 ions/sec is too large
for an ion carrier. Furthermore, in membranes of different thick-
nesses, gramicidin channels exhibit identical conductances but
different stabilities, the stability decreasing with increasing
membrane thickness [3]. The conductance due to a carrier would
depend on membrane thickness.

Gramicidin channels are ideally selective for monovalent
cations and the single-channel conductances for the alkali cations
are ranked in the same order as the aqueous mobilities of these ions
[4]. Divalent cations such as Ca^{2+} or Ba^{2+} block the channel by
binding near the mouth of the channel [5,6]. Monovalent cations
move through the channel in single file, as shown by tracer-flux
measurements [7,8] and the mole-fraction dependence of channel conductance in the presence of symmetric mixtures of K^+X^- and either
Tl^+X^- [9-12] or Ag^+X^- [13]. The channel is filled with about six
water molecules, almost all of which must be displaced when an ion

is transported [14-17]. Protons are transported much more rapidly than other ions by means of a proton-hopping mechanism with no displacement of water [4,16].

The gramicidin channel is partially selective for potassium over sodium. For 10 mM Na^+ and K^+ at 25°C, the biionic potential is 27.7 mV [18]. This corresponds to a permeability ratio of 2.9, a value which agrees with the measured conductance ratio G_K/G_{Na} of 2.9 at 27 mV [19]. The agreement of these permeability and conductance ratios implies that most of the gramicidin channels in a membrane are not occupied by Na^+ or K^+ at 10 mM concentrations [19].

Investigations of the ion-binding and ion transport properties of gramicidin A have been aided by the availability of either naturally occurring or chemically synthesized analogs which differ from the parent molecule by only a single amino acid substitution [20-23]. Because these analogs provide subtle and well-defined perturbations of the chemical constitution of gramicidin A, they represent powerful tools for deciphering the chemical mechanisms which govern ion-peptide interactions.

2. STRUCTURAL PROPERTIES

2.1. Amino Acid Sequence

The linear gramicidins A, B, and C are synthesized by the aerobic spore-forming bacterium *Bacillus brevis* (ATCC 8185) [24]. [Gramicidin S (Soviet), a cyclic decapeptide ion carrier from a different strain of *B. brevis* [25] will not be discussed here.] These peptides contain D-amino acids and are made in vivo by total enzymatic synthesis, without the use of ribosomes or a template RNA [26]. Exclusively L-amino acids are activated by ATP to enzyme-bound aminoacyladenylates, and then transferred to enzymic sulfhydryl groups and finally to peptide intermediates that are also covalently thioesterlinked to enzymes. In cases where D-amino acids appear in the final peptide, the configuration of the corresponding L-amino acid precursor is inverted at some step in the process [26,27].

A mixture of six pentadecapeptides, all having N-terminal
formyl and C-terminal ethanolamine blocking groups, are synthesized
by *B. brevis*. The major component is valine gramicidin A [28]:

Formyl-L-Val-Gly-L-Ala-D-Leu-L-Ala-D-Val-L-Val-D-Val-L-Trp-
            1   2   3   4   5   6   7   8   9

D-Leu-L-Trp-D-Leu-L-Trp-D-Leu-L-Trp-ethanolamine
  10   11   12   13   14   15

Other components in the mixture have different amino acids at position one (Ile instead of Val) and/or at position eleven (Phe [grami-
cidin B], or Tyr [gramicidin C] instead of Trp). The relative abun-
dances of the various amino acids in the growth medium can affect
the relative amounts of the six linear gramicidins which are produced by *B. brevis* [29], and so the relative proportions of gramicidins A, B, and C in different commercial lots can vary. An A/B/C
ratio of about 7:1:2 is typically observed (Fig. 1). Further variations in amino acid sequence can be produced by chemical synthesis
or semisynthesis; specific substitutions at positions one [21,23]
and two [22] have been reported. These gramicidin analogs often
exhibit altered membrane channel properties and offer a powerful
means of investigating structure-function relationships (see Sec. 4).

Gramicidins and other peptides such as tyrocidine are synthesized when a culture of *B. brevis* begins to sporulate. Gramicidin-
negative mutants fail to form heat-resistant spores [30]. Through
effects which are apparently unrelated to membrane channel activities, gramicidins inhibit RNA synthesis by interfering with the
formation of stable complexes between RNA-polymerase and DNA in
vitro [31-33]. These effects will not be discussed in this chapter.

2.2. Three-Dimensional Structure

If Gly is counted as a D-amino acid, the gramicidin sequence consists
of a strictly alternating series of L- and D-amino acids. Early x-ray
diffraction patterns [34] indicated predominant helical components to
the structure, and yet none of the α family of helices would be stable

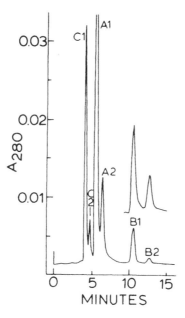

FIG. 1. HPLC resolution of six linear gramicidins from a commercial
mixture on a Dupont Zorbax C-8 column (4.6 x 250 mm), eluted with
83% methanol at 1.5 ml/min. The peaks are coded as follows: A1,
valine gramicidin A; A2, isoleucine gramicidin A; B1, valine gramicidin B; B2, isoleucine gramicidin B; C1, valine gramicidin C; C2,
isoleucine gramicidin C. *Inset*: Elution of a purified gramicidin B
sample, enriched in the isoleucine component. Note that ε_{280} for
gramicidin B is only 0.75 that of gramicidin A [94].

for an alternating L,D sequence of amino acids. In 1971, Urry [35]
proposed that helical variations of peptide β-sheet structures could
provide stable environments for the side chains of an alternating
L,D sequence. In such single-stranded π-(L,D)- or β helices the
intramolecular hydrogen-bonding pattern resembles that of parallel
β-pleated sheet. In 1974, Veatch et al. [36] pointed out that
various double-stranded β helices could be formed by intertwining
two (L,D) peptides in either parallel or antiparallel fashion. In
all β-type helices, as in β-sheet structures, the basic repeating
unit is a dipeptide. The evidence summarized in the succeeding
paragraphs indicates that gramicidin adopts both single- and double-
stranded β-helical conformations depending on its environment.
Figure 2 shows examples of both types of helix.

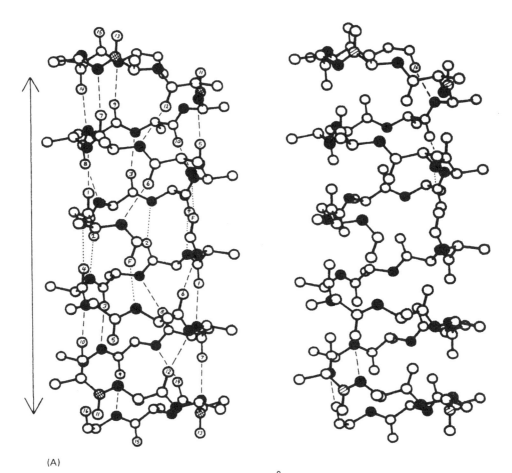

(A)

FIG. 2. Stereo ORTEP drawings of 26-Å-long β-helical polypeptide
backbone models of gramicidin dimers. (A) Antiparallel dimer of
$\beta^{6.3}$-single-stranded helices, the generally accepted structure for
the gramicidin channel in membranes. The coordinates are from Ref.
58. (B) Antiparallel $\beta^{7.2}$-double-stranded helix. The coordinates
were generated as in Ref. 58, using the helical parameters of Ref.

Circular dichroism and infrared spectroscopy indicate that
gramicidin adopts different conformations in different solvents.
Four conformers can be separated by thin layer chromatography in
either dioxane or ethyl acetate [36]. These conformers intercon-
vert in ethanol, but not in nonpolar solvents. Changes in fluorescence intensity, fluorescence polarization, and circular dichroic

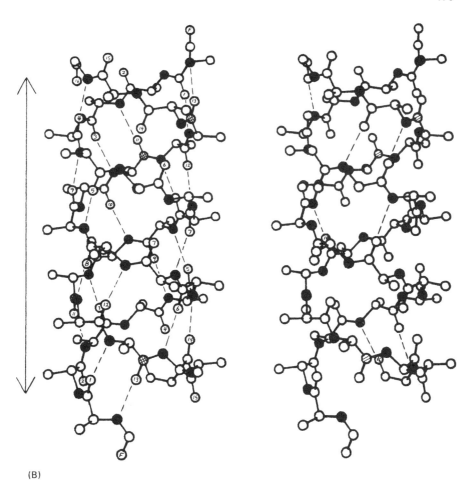

(B)

38. Models in both (A) and (B) contain a central 3.8-Å pore. Atom codes:
F, formyl oxygen; numbered atoms, other oxygens; white, carbons; black,
nitrogens. The carbonyl carbons of residues 11 and 13 are hatched to
indicate nuclei whose ^{13}C-resonance frequencies are strongly shifted
when cations are bound. The 21-Å-long arrow in each figure represents
the distance between the tight cation-binding sites.

ellipticity upon dilution of any of the pure conformers to 10^{-6} M

gramicidin concentration suggest that all of the conformers are dimers

which dissociate to monomers when diluted [37]. The half-times for

the spectral changes vary from minutes in dimethyl sulfoxide, to hours

in ethanol, to weeks in dioxane. The estimated equilibrium constants

for dimerization range from about 3 M^{-1} in dimethyl sulfoxide to about

10^5 M^{-1} in dioxane. An infrared amide I band at 1680 cm^{-1} suggests
that one of the conformers is an intertwined antiparallel double-β-
helical dimer. Parameters for such double helices having 5.6, 7.2,
9.0, or 10.8 residues per turn have been described for fibers of
polybenzyl-(D,L)-glutamate at various temperatures and pressures
[38]. The very slow rates of dissociation of the other conformers
to monomers suggest they may also be intertwined double-helical
dimers [37]. A variety of parallel and antiparallel, left-handed
and right-handed forms of different diameters are possible. Each
conformer exhibits a characteristic circular dichroism spectrum.
However, "because of the possibility of strong interactions among
the tryptophan transitions and those of the peptide" [36], the circular dichroism spectrum has not been used to specify a specific helix
or handedness for any of the conformers.

The circular dichroism spectrum for gramicidin incorporated
into phospholipid vesicles or micelles is quite different from that
of any of the conformers observed at high concentrations in nonpolar
organic solvents [39-41], indicating that the conformation in membranes is different from those that are present in solution. A first
step in defining the conformation of the membrane-bound channel-
forming species of gramicidin A was the unambiguous demonstration
that a dimer constitutes the transmembrane channel. Early experiments indicated that the membrane conductance depends on the second
power of the gramicidin concentration [42]. Furthermore, the channel
activity of a covalent dimer, malonylbis(desformylgramicidin) [43];
relaxation kinetic analyses of the trans bilayer conductance responses
to voltage jumps [44]; autocorrelation analyses of spontaneous fluctuations in the number of channels in a bilayer [45,46]; simultaneous
conductance and fluorescence studies of a fluorescent analog of gramicidin [47]; and finally, conductance and fluorescence energy transfer
measurements of hybrid channels formed from two different kinds of
gramicidin analogs [48] have all now firmly established that a dimer
is the active transmembrane channel-forming species. The possibility
of tetramer or higher multimer models has been excluded. In addition,

more than 95% of the dimers present in a bilayer are conducting
channels, there being no major inactive dimer states [47].

The model originally proposed by Urry in 1971 [35], in which
a gramicidin dimer is formed by the formyl-to-formyl association of
two single-stranded $\beta^{6.3}$-helical monomers, is now firmly established
as the conformation of conducting channels in membranes. A number
of lines of evidence require that the amino terminal formyl groups
of the two monomers which constitute a conducting dimer be near
each other, and buried near the center of the bilayer away from the
membrane surface:

(1) Channel activity is retained when the formyl groups are
chemically crosslinked to give a malonylbis(desformylgramicidin)
dimer [43,49].

(2) Replacements of the formyls by acetyl groups leads to a
twofold decrease in the single-channel conductance and a 50-fold
decrease in the channel lifetime, presumably due to steric crowding
of the added methyl groups at the interface between two associated
β-helical monomers at the center of a bilayer [50]. The substitution of a negatively charged and bulkier N-pyromellityl [51] or
N-succinyl [52] derivative for the N-formyl group causes a more
drastic reduction in activity.

(3) By contrast, O-pyromellityl gramicidin is nearly as active
as the native gramicidin, if added to both sides of a membrane, presumably because the ethanolamine moieties are situated toward the
outside of the bilayer in antiparallel fashion [53].

(4) Weinstein et al. [54,55] directly demonstrated that the
C terminals of the gramicidin channel are located near the membrane
surfaces whereas the N terminals are buried within the bilayer.
Using gramicidin in which either the N or C terminal had been chemically labeled with ^{13}C or ^{19}F, it was shown that paramagnetic ions
in aqueous solution alter the chemical shifts and spin lattice
relaxation times of nuclei located on the C terminal, whereas paramagnetic nitroxide groups on carbon-16 of stearoyl/myristoylphos-
phatidylcholine influence nuclear spins located at the N terminal.

(5) The thallium ion-induced shifts in the ^{13}C-resonance frequencies of all of the L-residue peptide carbonyl carbons are consistent only with an antiparallel head-to-head single-stranded β-
helical dimer [56]. The chemical shifts of carbonyls 11 and 13 are
most strongly influenced by Tl$^+$; those of 9 and 15 are somewhat
affected and the others are unchanged. Thus there is general agreement regarding the structure of the gramicidin channel in membranes.

Among single-stranded β helices, both left-handed and right-
handed models of various diameters are possible [43,57]. The peptide
(φ,ψ) torsion angles for a number of minimum energy structures have
been presented [38,57], as has a scheme for generating the coordinates of a β helix having a specified pitch and number of residues
per turn [58]. The ion selectivity sequence (H^+ > NH_4^+ > Cs^+ > Rb^+ >
K^+ > Na^+ > Li^+ > $(CH_3)_4N^+$ in 0.1 M salt [4]) is consistent with a
pore diameter of about 4 Å, and conductance measurements in membranes
of different thickness [3,47,59] suggest that a 26-Å-long gramicidin
channel spans the hydrocarbon region of a lipid bilayer. These
parameters are consistent with the $β^{6.3}$ helix, a single-stranded
structure having 6.3 residues per turn. A gramicidin β helix would
be right-handed if the formyl group made an intramonomer hydrogen
bond or would be left-handed if the formyl group pointed away from
the rest of the monomer so as to be available to form an intermolecular hydrogen bond to another monomer upon dimerization [35,58].
Early evidence that the formyl group is required for channel formation [35] led to the idea that the formyl group would be likely to
be involved in intermolecular hydrogen bonding, a situation which
would demand a left-handed helix. Furthermore, as the L- and D-side
chains differ, the left-handed helix was perceived to produce less
crowding of the side chains [35,43]. The left-handed sense of the
helix has been recently confirmed by measuring the thallium ion-
induced shifts in the ^{13}C-resonance frequencies of carbonyl carbons
8, 11, 13, and 14 of gramicidin [60]; the results are inconsistent
with a left-handed $β^{6.3}$ helix.

Gramicidin has been crystallized from methanol and ethanol
[61,62] and in complexes with CsSCN and KSCN [63,64]. In all of

these crystal forms, the basic repeating unit is a dimer of gramicidin. For the ion-free crystals, the asymmetric unit is a dimer;
for the Cs^+ and K^+ crystals the asymmetric unit is two dimers which
are related by approximate crystallographic symmetry [63]. In the
ion-free crystals, the dimer is arranged as a 32-Å-long helical
cylinder aligned parallel to a crystallographic axis and having a
pore size of about 2 Å, the backbone atoms on opposite sides of the
cylinder being separated by about 5 Å [63]. The amide I band at
1680 cm^{-1} in the infrared spectrum of redissolved ion-free crystals
of gramicidin has been attributed to an antiparallel double-stranded
β helix [36], and infrared and Raman spectra of the crystals themselves have led to the same conclusion [65]. The 32-Å length for a
dimer and the strong 5.6-Å helical repeat evident from the diffraction pattern [63] serve to select the $\beta^{5.6}$-double helix from among
the family of antiparallel double helices [38]. Although no isomorphous heavy atom derivatives are available for an x-ray phase analysis of the ion-free crystal structure, deuterium-hydrogen difference
neutron diffraction has provided phases at low resolution and shows
promise of revealing the high-resolution structure [66]. The 2-Å
pore size of the $\beta^{5.6}$-double helix is too small to pass many of the
ions which are transported through gramicidin channels. The mechanisms and folding intermediates which may be involved in converting
double-helical structures to the active channel structure, and vice
versa, have not been investigated. The mechanisms and structures
which are involved as the peptide inserts into a membrane are also
unknown.

When complexed with either Cs^+ or K^+ cations and crystallized
from methanol, the gramicidin dimer assumes a 26-Å-long structure
[64], the same length as has been inferred for the membrane channel
structure. A 3.8-Å pore is observed through the gramicidin cylinder
in these crystals, the centers of the atoms lining the pore being
separated by a 6.8-Å diameter [63]. The K^+- and Cs^+-complexed gramicidin crystal structures are isomorphous [64]. A length of 26 Å
would be consistent with either a head-to-head dimer of $\beta^{6.3}$ single-
stranded Urry-type helices [57,58,67], or a $\beta^{7.2}$ double helix [38].

The lengths and diameters of these two models are almost identical.
In both cases the pore size of about 3.8 Å is consistent with the
diameters of cations which are known to pass through the gramicidin
channel. For the $\beta^{6.3}$ single-stranded helix to precisely fit the
crystallographic dimensions, the pitch of the energy-minimized model
[57,69] would have to be shortened from 5.0 to 4.85 Å/turn, resulting
in a shortening of the intrastrand hydrogen bonds from 2.93 to 2.81 Å
[58]. This slightly shortened model is depicted in Fig. 2(A). However, recent infrared and Raman data on Cs^+-gramicidin crystals suggest that the dimer is a double helix in these crystals as well as
in the ion-free crystals [65]. In addition, the circular dichroism
spectrum of Cs^+-gramicidin in methanol is different from the spectrum
of gramicidin in phospholipid vesicles [41,70]. Thus the gramicidin
dimer may be a $\beta^{7.2}$-double helix [Fig. 2(B)] in cation-complexed
crystals as opposed to two associated $\beta^{6.3}$-single helices in membranes, although these models have similar overall dimensions.
Nevertheless, the locations of the tight ion-binding sites found
for K^+ and Cs^+ in gramicidin crystals are in complete agreement with
those subsequently determined for Na^+ and Tl^+ in phospholipid-
packaged gramicidin (see Sec. 3.2).

To summarize structural aspects, gramicidin exists predominantly as a dimer in its membrane channel state, in various crystal
forms, and in nonpolar organic solvents. The alternating (L,D)
amino acid sequence of gramicidin dictates a β-helical structural
motif, although several different families of β helices are possible
and are in fact observed in different environments. The ion-conduct-
ing gramicidin channel in membranes is a 26-Å-long dimer of $\beta^{6.3}$
single-stranded helices hydrogen-bonded at their formylamino terminal ends in the center of a lipid bilayer. The polypeptide backbone
of this channel structure [Fig. 2(A)] is coiled around a 3.8-Å pore,
through which monovalent cations can pass. The conformation of
gramicidin in ion-free crystals is quite different, the dimer being
32-Å long and having a pore of only 2 Å; spectroscopic evidence

favors an antiparallel $\beta^{5.6}$ double-helical assignment for this
structure. The molecular dimensions deduced for the cation-bound
crystal structure of gramicidin [63,64] are consistent with either
the $\beta^{6.3}$ single-helical membrane channel structure depicted in Fig.
2(A) or the $\beta^{7.2}$ double-helical model depicted in Fig. 2(B); once
again, spectroscopic data seem to favor the double-helical choice
of model. Nevertheless, the ion-binding sites which are observed
in crystalline complexes [64] agree with those which are found in
gramicidin membrane channels, near carbonyls 11 and 13 at each end
of the dimer [71]. This means that if Fig. 2(A) represents the
structure in membrane channels and Fig. 2(B) represents the structure in crystals, then these two conformations bind ions similarly
near each end of the dimer.

It should be borne in mind that physical studies of gramicidin
conformations in membranes use much higher concentrations of the
polypeptide than do single-channel studies of ion transport. Therefore, one must verify that the most prevalent conformation observed
by physical measurement is the same as the conformation which is
responsible for the single-channel events. This question seems to
have been resolved by the simultaneous conductance and fluorescence
measurements of Veatch et al. [47], who showed that all of the channels in a membrane are dimers and nearly all of the dimers are channels; thus, there does not exist any significant population of inactive dimers in a membrane.

The $\beta^{6.3}$ single-helical channel structure places the amino
acid side chains at some considerable distance (5-10 Å) from the
pathway for translocation of cations. Nevertheless, the side chains
are important for channel activity as well as for the solubility of
gramicidin in membranes, and relatively minor chemical changes in
individual side chains can rather dramatically alter the conductances
and lifetimes of single channels (see Sec. 4). Figure 3 shows an end
view of the channel lumen, with side chains attached to the $\beta^{6.3}$-
single helix in sterically favorable orientations.

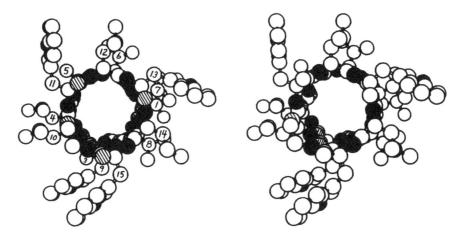

FIG. 3. End view of a monomer from the model in Fig. 2A, pseudo-
space filling, with side chains attached in sterically allowed
orientations. Oxygens are hatched, nitrogens black, and carbons
white. Selected β carbons are numbered. The side chain torsion
angles [72] have been arbitrarily chosen for the illustration:
$\chi^1_{Val} = 300°$, $\chi^1_{Leu} = 180°$, $\chi^2_{Leu} = 60°$, $\chi^1_{Trp} = 180°$, $\chi^2_{Trp} = 90°$.

3. ION-COMPLEXING PROPERTIES

3.1. Ion Selectivity

A detailed understanding of the transport properties of the linear
gramicidins requires a knowledge of ion selectivity, equilibrium
binding constants, thermodynamics of the binding process, the locations of the binding sites, and the chemical structure and conformational energetics of the gramicidin channel. These parameters and
their relationships to one another provide a valuable picture of the
mechanism of transport for this unique model channel system.

The gramicidin channel is known to be essentially impermeable
to anions [4]. However, there are conditions under which some anion
permeation may be observed. Investigations with thallous acetate and
thallous chloride have indicated that at high thallium occupancy of
the channel, the acetate and chloride anions may be about 0.1 as
permeant as the thallous ion [73]. Since anion binding is not

possible with the carbonyl groups that are responsible for cation binding, Urry suggests that L-residue NH groups midway through the channel may be responsible for the anion binding [74]. However, the evidence regarding anion permeability has been disputed [18,75].

The permeability of the gramicidin channel to divalent cations is extremely low [3,5]. Conductance studies of the blocking of the gramicidin channel by the divalent cations Ca^{2+}, Ba^{2+}, Mg^{2+}, and Zn^{2+} suggest that although these cations are bound at the channel entrance, they do not move through the channel [3]. These ions interfere with the transport of monovalent cations.

Monovalent cations are transported through the gramicidin channel; however, there is a large range in the selectivity among these ions. For gramicidin channels in erythrocyte lipid/n-decane planar bilayers, the relative cation permeability was found to be: Li^+ (0.33) < Na^+ (1.0) < K^+ (3.9) < Rb^+ (5.5) < Cs^+ (5.8) [4]. This is the same ordering as observed for a number of cation properties such as relative ionic mobility: Li^+ (0.77) < Na^+ (1.0) < K^+ (1.48) < Rb^+ (1.55) < Cs^+ (1.56) [74]; the relative ionic radius of the cations: Li^+ (0.74) < Na^+ (1.0) < K^+ (1.4) < Rb^+ (1.55) < Cs^+ (1.83) [76]; and the relative hydration energy: Li^+ (0.85) < Na^+ (1.0) < K^+ (1.11) < Rb^+ (1.14) < Cs^+ (1.19) [77].

3.2. Molecular Details of Binding

Having established that the gramicidin channel is cation-selective, several important problems must then be addressed in order to understand this selectivity: (1) the location of the binding site or sites must be determined and (2) the binding constants and associated thermodynamic parameters must be determined for each cation.

Conductivity, Na-23 NMR, and Tl-205 NMR studies indicate that there are two types of cation-binding sites for the channel, one being a tight binding site and the other a weak binding site [3,11, 68,73,75,78-88]. A Cs^+-K^+ x-ray difference analysis [64] showed two binding sites per dimer, the sites located 2.5 Å from the ends of the channel (an alternate assignment of sites symmetrically located 2.5 Å on either side of the center of the channel could not be strictly excluded). Using a series of ^{13}C-enriched gramicidins made by total chemical synthesis, Urry and co-workers [60,71,89,90] elegantly showed that both Na^+ and Tl^+ bind near the carbonyls of residues 11 and 13, a conclusion based on the cation-induced shifts in the ^{13}C resonance frequencies of these carbonyl carbons and on the much lesser or undetectable changes in the resonance frequencies of other carbonyls. In these experiments, the gramicidin was packaged in lysophosphatidylcholine micelles, in which the channel is an antiparallel formyl-to-formyl dimer of $\beta^{6.3}$-single helices. In this model the carbonyls of residues 11 and 13 are about 2.5 Å from either end of the channel [71, and Fig. 2(A)], in agreement with the x-ray results. Similar ^{13}C NMR studies with the divalent cations Ca^{2+} and Ba^{2+} indicate that the divalent cations also bind to the gramicidin A channel; however, they do so at a site that is displaced outward toward the aqueous phase about 1.5 Å from the Na^+-binding site [6, 71,91].

Since the binding site is essentially the same for all monovalent cations tested, but slightly different for the divalent cations, it is essential to determine the binding constants for both types of cations to further understand the cation selectivity of the gramicidin channel.

The binding constant of gramicidin A for the Tl^+ cation in trifluoroethanol has been determined using a Tl-205 NMR chemical shift technique [88]. In trifluoroethanol gramicidin exists as a helical dimer similar to that found in model membranes [36,37,39]. The Tl-205 data were found to fit a model in which there are two binding sites with equal binding constants accessible to the Tl^+ cation near the ends of the channel dimer. The value of the binding constant was found to be about 500 M^{-1}. In DMSO where the dominant species of gramicidin A is the monomer, the complexation of the Tl^+ cation was also studied using Tl-205 NMR spectroscopy [92]. A binding constant of about 450 M^{-1} was obtained based on a model of

gramicidin A in which only one binding site of the molecule was used
for binding. This model was confirmed by C-13 NMR studies of gramicidin A in DMSO in the absence and the presence of the Tl^+ cation,
in which carbonyl C-13 chemical shifts were observed for the tryptophan end of the molecule but not for the formyl, or opposite, end.

In an ethanol-water mixture of 90:10, where gramicidin A exists
as a dimer, a binding constant of 4 M^{-1} at 36°C was obtained for the
Na^+ cation [81]. In this study the linewidth of the Na-23 NMR resonance signal as a function of the gramicidin/Na^+ mole ratio was used
as a probe for obtaining the data necessary for determining the binding constant. The binding site was assumed to be at the channel
entrance.

Equilibrium binding constants of gramicidin A incorporated
into micelles or bilayers have been determined for several cations
using a number of different techniques. Some of the techniques, such
as equilibrium dialysis and NMR, make direct measurements of the equilibrium binding constant while conductance experiments generate binding constants indirectly as parameters used for fitting the conductance data to various kinetic models describing ion movement through
the gramicidin channel. Table 1 contains some of the equilibrium
constants for the binding of the Li^+, Na^+, K^+, Rb^+, Cs^+, Tl^+, and
Ca^{2+} cations to gramicidin channels in lipid environments. Although
there are rather large variations among the various determinations
of the equilibrium binding constant for a given cation, it is tempt-
ing to correlate the constants with cation properties such as hydration energy, ionic radius, or ionic mobility. For example, it appears
that the magnitude of the equilibrium binding constant increases with
increasing cation polarizability. One must be very cautious in making
such generalizations with the data since the binding constants can be
model- and system-dependent. Using Na-23 NMR relaxation times as a
function of Na^+ concentration in the presence of gramicidin A or covalently dimerized malonylbis(desformyl gramicidin A) incorporated
into micelles, two ion-binding sites (i.e., one tight and one weak)
with equilibrium binding constants of about 100 and 1 M^{-1}, respec-

TABLE 1

Equilibrium Binding Constants (M^{-1}) for Gramicidin A
in Bilayers and Micelles

Method	Ions Li^+	Na^+	K^+	Rb^+	Cs^+	Tl^+	Ca^{2+}
Li-7 NMR [102]	14 (0.5)						
Conductance [73]	9 (1.2)						
Na-23 NMR [6]		100 (1)					
Na-23 NMR [78,68,93]		70 (1.4)					
Dialysis [84]		<30					
Conductance [73]		20 (2)					
Conductance [86]		1.43 (0.18)					
Conductance [8]		5					
Conductance [73]			40.4 (1.5)				
Conductance [86]			1.37 (0.68)				
K-39-NMR [93]			(8)				
Dialysis [84]				<30			
Conductance [73]				57.3 (0.9)			
Cs-133 NMR [93]					55 (4)		
Conductance [73]					100 (0.8)		
Dialysis [84]						500-1000	
Tl-205 NMR [103]						900	
Conductance [73]						950	
Conductance [86]						400-500	
C-13 NMR [6]							1

Note: Numbers in parentheses represent equilibrium binding constants
for weak binding sites.

tively, have been reported at 30°C [6,78,93]. However, the equilibrium dialysis measurements of Veatch and Durkin [84] performed with gramicidin A incorporated into vesicles gave a different result: an equilibrium binding constant for Na^+ of <30 M^{-1} at 23°C for a singly occupied channel.

Single-channel conductance studies have shown that the transport properties of gramicidin are significantly affected by changes in amino acid residues (see Sec. 4) [20]. The extent to which ion-binding properties are affected by changes in amino acid sequence are only beginning to be studied. Single-channel conductance studies of modified gramicidins show that the binding constant for Na^+ is different for the various derivatives used [23]. The values of the binding constants varied from 28.6 to 11.1 M^{-1}. These are within the range of the Na^+-gramicidin A binding constants found in Table 1. Tl-205 NMR chemical shift studies of the Tl^+ complex of gramicidin A and gramicidin B in trifluoroethanol suggest that the binding constant of gramicidin B is different to that for gramicidin A [94]. The different affinity of gramicidin B may contribute significantly to its lower conductance. This promises to be an interesting area for future investigations.

In order to more fully understand the cation selectivity of gramicidin A, it is important to determine the thermodynamic parameters for the binding of cations. An activation energy of 7.4 kcal/mol has been determined for the exchange process of the Na^+ cation with the gramicidin A channel incorporated into micelles using Na-23 NMR linewidths as a function of temperature [95,96]. This activation energy is associated with the exchange on and off the binding site at the entrance to the channel. The activation energy for the transport of Na^+ cations through the gramicidin A channel is 7.3 kcal/mol [97]. Since the activation energy for the transport process is similar to that for the exchange process, the inference to be drawn is that the dominant energy barrier is at the channel entrance rather than near the center of the membrane (where the effect of the low dielectric

constant of the lipid would be most pronounced). This would be consistent with having the ionic hydration energy being satisfied by
the lateral peptide carbonyl coordination and the bound water that
follows and preceeds the ion through the channel [43]. The activation energy for transport would thus be primarily associated with
replacing some of the water of hydration around an ion with peptide
carbonyl groups. A study has been made of the thermodynamics of the
binding of the thallous ion by gramicidin A incorporated into micelles
using Tl-205 NMR chemical shifts to obtain equilibrium binding constants as a function of temperature [98]. A value of 11.3 kcal/mol
was obtained for the enthalpy of the exchange process. Activation
energy for transport of Tl^+ cations through the gramicidin A channel
is 6.7 kcal/mol [99].

Pullman and Etchebest [100] calculated the energy profile of
the Na^+ cation in the gramicidin A dimer channel. An energy minimum
of -31.4 kcal/mol was found external to the channel with the Na^+
cation bound to the carbonyl of Trp 15 at 2.02 Å from it. This binding energy is 5 kcal/mol more negative than the corresponding energy
of the single binding of Na^+ to water.

A theoretical study of the relation between binding affinities
and selectivity of a pore was made [101]. The influence of energy
barriers and sites on the selectivity of a single-file pore was
investigated. The calculations show that if the pore is only weakly
occupied, the permeability ratio, which is used as a measure of
selectivity, depends entirely on the energies of the barriers. Therefore, the selectivity of the pore is independent of the binding affinities of these sites. The affinities, however, become important in
the saturation range provided that the pores have a sufficiently high
number of sites. In this sense, two sites are sufficient if the pore
is at an internal equilibrium. In this case the permeation process
then favors the component with the higher binding affinity.

There are two other factors that can play an important role in
cationic selectivity: (1) the repulsive image force of the lipid and
(2) the peptide libration that orients the coordinating carbonyls
with respect to the particular cation. As previously discussed,

divalent cations are not transported through the channel but are
known to bind at the channel entrance and undergo rapid exchange at
the binding site. The binding site for divalent cations is farther
out toward the aqueous phase than it is for monovalent cations [93].
Urry concludes that this demands that a center barrier be rate limiting in the transport process and, more specifically, it is the part
of the barrier that comes from the proximity to the lipid surface that
produces a repulsive image [93]. This repulsive image force is proportional to the square of the cation charge. Consequently, the
repulsive image force for monovalent cations is much less than that
for divalent cations; hence, a cation selectivity is produced.

The second factor, the energetics of peptide libration, involves the energy necessary to bring the peptide carbonyl into the
channel to coordinate with the cation [43]. The size of the channel
in its lowest energy configuration is about that necessary for the
coordination of Rb^+. For Cs^+ a slight outward libration may be
necessary while inward libration may be required for smaller ions.
This type of effect could account for some of the correlations
between ionic radius and binding constant.

4. TRANSPORT PROPERTIES

4.1. Current-Voltage Characteristics

The shape of the current-voltage curve for gramicidin channels
depends on the aqueous concentration of the permeant ion and changes
from sublinear at low ion concentrations (~0.1 M) to superlinear at
high ion concentrations (~4 M) [3,80,104,105]. At low permeant ion
concentrations, the single-channel current becomes almost completely
voltage-independent for V > 300 mV [105], and the magnitude of the
current is so high that ion entry into the gramicidin channels is
diffusion-controlled, at least for the most permeant ions such as
K^+, Rb^+, Cs^+, and NH_4^+ [105]. Because the single-channel currents
at high potentials decrease when the aqueous diffusion coefficients
of the permeant ion are lowered (by a substitution of D_2O, a change

of temperature, or an addition of sucrose or glycerol), the limiting currents are "primarily determined by the diffusion of ions through the aqueous phase up to the channel entrance" [106].

The change in the shape of the current-voltage curve at higher permeant ion concentrations reflects a change in the rate-determining step for ion translocation through the channel [3,105]. At high ion concentrations, the aqueous convergence conductance is no longer rate limiting, but is replaced by some other step, such as perhaps translocation through the channel or dissociation from the channel. The change of rate-determining step is accompanied by a large increase in the single-channel conductance.

4.2. Concentration Dependence

The gramicidin single-channel conductance is a saturating function of the permeant ion activity, reaching a plateau at 1-2 M salt and decreasing slightly thereafter [3,18,85,107], thus suggesting that ion binding to the channel is a saturable process. The conductance decrease at high ion activities, which is not observed for sodium [17], can be taken as evidence for multiple ion occupancy of the channel. Tracer-flux data [7,8,108] and biionic potential measurements [4,18,73] indicate that at least two K^+, Rb^+, Cs^+, NH_4^+, or Tl^+ can simultaneously occupy the channel. For the case of Na^+, however, the flux-ratio exponent [109] for gramicidin channels in diphytanoyl phosphatidylcholine membranes is identically 1.0 at all NaCl concentrations [8,17], thus indicating single-ion occupancy for Na^+. A different result has been obtained using mixed brain lipid membranes [108], and the Na^+ occupancy in glycerol monooleate membranes is still unresolved [106].

4.3. Ion Selectivity

Gramicidin channels are selectively permeable to alkali cations as noted in Sec. 3.1. The differences among ions are less pronounced

at high potentials indicating that the relative current ratios for
various pairs of ions are voltage-dependent. For example, i_{Cs}/i_{Na}
decreases from about 3.7 at 0 mV to about 1.9 at 500 mV applied
potential [105]. At low potentials, the conductances for different
ions vary more than do their aqueous diffusion coefficients; at high
potentials, the conductance ratios decrease and approach the diffusion coefficient ratios. This observation is consistent with an
aqueous diffusion limitation on the conductance at high transmembrane potentials.

Tl^+ and Ag^+ behave differently from the alkali cations in
transport experiments with gramicidin channels. These two ions are
bound more tightly than the alkali cations [84], their currents do
not reach limiting values at high potentials [104], and the currents
at 500 mV are more than twice those predicted from alkali metal ion
experiments [12]. Furthermore, either Tl^+ [10] or Ag^+ [13] at low
concentration partially blocks the conductance due to alkali metal
cations. Thus in symmetric mixtures of K^+ and Tl^+ at constant ionic
strength and a variable K^+/Tl^+ ratio, the conductance passes through
a minimum at about 0.05 mol fraction Tl^+; indicating that K^+ transport is blocked by strongly binding Tl^+ under these conditions [10-
12]. This phenomenon has been termed mole-fraction-dependent behavior
[110]. The permeability ratio P_{Tl}/P_K is 25, whereas the conductance
ratio is 0.8 in 1 M salt [12]. This behavior indicates multiple occupancy by one or both of the ions in the mixture. A molecular explanation for the effectively larger capture radius for Tl^+ or Ag^+ has not
yet emerged.

4.4. Ion-Water Interactions

Gramicidin channels are permeable to water. For a single-file channel
(see below), the average number of water molecules in the channel is
determined by the ratio of the osmotic permeability coefficient to
the diffusion permeability coefficient [17,111]. This ratio gives a
value of about five or six water molecules in the gramicidin channel

[15]. In an alternate determination of water occupancy, streaming
potentials of 3 mV per osmol of osmotic pressure across membranes
containing gramicidin [14] indicate that about 6.5 water molecules
are transported per ion crossing the membrane. Thus there is general
agreement from water-water and from ion-water interactions that there
are about six water molecules in a gramicidin channel [17].

4.5. Single-File Transport

That water and ions move through gramicidin channels in single file
is supported by the following observations: (1) the ~4-Å channel
diameter of the generally accepted $\beta^{6.3}$-helical model of gramicidin
is consistent with single-file transport. (2) Na^+ and H_2O cannot
pass each other in the channel [14-16], although water apparently
can pass a Na^+ ion situated at one of the binding sites [17]. (3)
The impermeability of the channel to large ions such as $(CH_3)_4N^+$ and
to nonelectrolytes such as urea, even in the presence of 1 M Na^+,
Tl^+, or Ag^+, which are transported [104], suggests that the channel
is narrow. (4) "Mole-fraction-dependent" conductances in mixtures
of Tl^+ or Ag^+ [9-13] with one of the alkali cations are consistent
not only with multiple ion-binding sites and ion-ion interactions
within the channel, but also with single-file behavior. (5) Tracer-
flux experiments demonstrate interactions between bidirectional
fluxes of Cs^+ or Rb^+ in gramicidin channels [7,8,108].

4.6. Kinetic Models

Kinetic descriptions of ion movement through the gramicidin channel
are as yet incomplete. Proposed models must account for the current-
voltage characteristics, the concentration dependence of conductance
and of the ion selectivity ratios, the mole-fraction-dependent conductances observed with mixtures of two permeable ions, the equilibrium binding constants for various cations, and the positions of the
tight binding sites. The simplest model that accounts qualitatively

for the above properties is the three-barrier two-site model with double occupancy [17,68,107]. In order to improve the quantitative agreement with experimental data, Eisenman and co-workers extended the model to include up to four ion sites [11,85,112,113]. Some researchers are beginning to discuss the limitations of Eyring rate theory and suggest alternatives to discrete site-and-barrier models [113-115]. At present, experimental limitations preclude an unambiguous choice of one particular model. Future experimental and theoretical developments toward this goal promise to be exciting.

4.7. Side Chain Effects

The dependence of single-channel properties on the side chain structure of gramicidin is emerging as an important technique for elucidating the molecular mechanism of ion transport through the channel. In the $\beta^{6.3}$-helical dimer, the side chains do not directly contact the passing ions, but are situated some 5-10 Å away from the path taken by the ions. Nevertheless, the side chains are important regulators of both channel conductance and channel lifetime. For example, changing Trp-11 of gramicidin A to Tyr-11 (gramicidin C) produces little change in conductance, while a change to Phe-11 (gramicidin B) lowers the conductance by about one-third [20]. Similarly, dansylation of Tyr-11 produces a molecule whose conductance is about the same as gramicidin B [48]. Hybrid dimers formed from dansyl gramicidin C and gramicidin A exhibit single-channel conductances intermediate between those of the respective homogenous dimers [48], suggesting that the structures of the channels formed by the various species are similar. If all of the four Trp's of gramicidin A are replaced by Phe [116], both the channel lifetime and the conductance are reduced by about a factor of 5. This latter compound has been made by total chemical synthesis.

Further side chain variations can be introduced either by total synthesis [117,118] or by semisynthetic modifications at or near the N terminal [21-23,119]. Using these techniques, it is now possible

to substitute any of the side chains of gramicidin and thus produce
families of closely related gramicidin analogs whose single-channel
properties can be related to subtle differences in chemical structure. The semisynthetic approach, using a diphenylphosphorazidate
coupling reaction and double-passage HPLC purification [119], provides a relatively facile method for preparing a large number of
highly pure gramicidins having different amino acids near their N
terminals, i.e., near the center of the lipid bilayer in which the
dimeric transmembrane channel is situated.

Morrow et al. [21] showed that a hydrophobic L residue is
required as the (formylated) N-terminal amino acid of gramicidin.
The strictly alternating (L,D) sequence must be preserved, the gramicidin being inactive if D-Val is substituted for L-Val at position
one. Both the polar, bulky p-I-Phe$_1$ and Gly$_1$ gramicidins exhibit
greatly reduced single-channel conductances [21]. Curiously, Gly
is the naturally occurring residue at position two, but is not
favored at position one. Substitution of D-Leu in place of Gly at
position two reduces the channel lifetime without affecting the
single-channel conductance [22].

The polarity of the N-terminal residue is more important than
its bulk, as Phe$_1$ gramicidin behaves similarly to Val$_1$ gramicidin
[23]. In contrast, the polar trifluoro-Val$_1$ gramicidin has a sixfold reduced single-channel conductance [120,121] (see Table 2).
Similarly, Tyr$_1$ gramicidin exhibits a lower conductance than Phe$_1$ or
Val$_1$ gramicidin [23]. In experiments with closely related gramicidin
analogs, gross changes in the structure of the conducting dimer are
not expected. Possible major structural alterations can be assayed
and excluded through the observation of hybrid channels formed by
the analog with a molecule of Val$_1$ gramicidin A. In most cases the
hybrid channels have conductances intermediate between those of the
analog and the Val$_1$ gramicidin A [21,23,48,121]. A notable exception
is Tyr$_1$ gramicidin C [23], for which hybrid channels of unusually low
conductance are observed. Detection and quantitation of hybrid channel conductances and frequencies require extremely pure samples of

TABLE 2

Single-Channel Conductances of Some Position One
Analogs of Gramicidin A

Side chain at position 1	Conductance in 1 M NaCl[a] (pS)	Ref.
Valine	12.4	105, 120, 121
Trifluorovaline	1.89	120, 121
Phenylalanine	10.2	23
Tyrosine	3.9	23
Norvaline	14.6	120, 121
S-methyl cysteine	9.1	120, 121

[a]Diphytanoyl phosphatidylcholine membranes; 25-50 mV applied potential.

gramicidin analogs, e.g., samples for which 95-99% of the single-channel transitions fall in a narrow peak on a current histogram [23].

The effect of a particular amino acid substitution depends on its position in the gramicidin sequence. A striking example is a comparison of the single-channel conductances of Phe_1-Tyr_{11}-gramicidin and Tyr_1-Phe_{11}-gramicidin [23, Fig. 4]. More polar substituents at the N terminal, near the center of the bilayer, tend to decrease conductance, while less polar substituents at position 11, near the tight ion-binding site, tend to decrease conductance.

Provided there is no structural change, polar side chains at position 1 could modulate the conductance by means of electrostatic ion-dipole interactions ("through space") and/or electron withdrawing or donating effects ("through bonds"). Experiments with a series of substituted phenylalanine side chains suggest that dipolar interactions are more important than inductive effects [122]. Further characterization of these and other analogs, with respect to cation-binding constants and rate constants for the elementary steps of transport, will help to further clarify the chemical and kinetic mechanisms of ion passage through gramicidin channels.

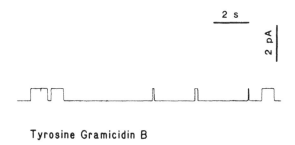

Tyrosine Gramicidin B

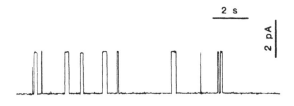

Phenylalanine Gramicidin C

FIG. 4. Comparison of single-channel events observed with Tyr_1-Phe_{11} gramicidin (upper) and Phe_1-Tyr_{11} gramicidin (lower). Note that the amplification and chart speed are the same in both traces. The figure illustrates that both the nature and position of a particular side chain influence the properties of gramicidin channels. 1.0 M NaCl, 200 mV, 25°C. (Reprinted from Ref. 23 with permission.)

ACKNOWLEDGMENTS

Research in the authors' laboratories has been supported in part by grants from the Research Corporation, the National Science Foundation (PCM-8300065), the U.S. Public Health Service (NIH-NS-00648 and NIH-NS-16449), and the NSF-Arkansas EPSCOR program.

REFERENCES

1. D. A. Haydon and S. B. Hladky, *Q. Rev. Biophys.*, *5*, 187 (1972).

2. S. B. Hladky and D. A. Haydon, *Nature (London)*, *225*, 451 (1970).

3. S. B. Hladky and D. A. Haydon, *Biochim. Biophys. Acta, 274,* 294 (1972).

4. V. B. Myers and D. A. Haydon, *Biochim. Biophys. Acta, 274,* 313 (1972).

5. E. Bamberg and P. Läuger, *J. Membr. Biol., 35,* 351 (1977).

6. D. W. Urry, T. L. Trapane, J. T. Walker, and K. U. Prasad, *J. Biol. Chem., 257,* 6659 (1982).

7. L. V. Schagina, A. E. Grinfeldt, and A. A. Lev, *Nature (London), 273,* 243 (1978).

8. J. Procopio and O. S. Andersen, *Biophys. J., 25*(2), 8a (abstr.), (1979).

9. O. S. Andersen, 50th Int. Biophys. Congr., Copenhagen, 1975, p. 112 (abstr.).

10. E. Neher, *Biochim. Biophys. Acta, 401,* 540 (1975); errata ibid., *469,* 359 (1977).

11. J. Sandblom, G. Eisenman, and E. Neher, *J. Membr. Biol., 31,* 383 (1977).

12. O. S. Andersen, in *Renal Function* (G. H. Giebisch and E. F. Purcell, eds.), Josiah Macy, Jr. Foundation, New York, 1978, p. 71 ff.

13. D. McBride and G. Szabo, *Biophys. J., 21*(2), 25a (abstr.) (1978).

14. P. A. Rosenberg and A. Finkelstein, *J. Gen. Physiol., 72,* 327 (1978a).

15. P. A. Rosenberg and A. Finkelstein, *J. Gen. Physiol., 72,* 341 (1978b).

16. D. G. Levitt, S. R. Elias, and J. M. Hautman, *Biochim. Biophys. Acta, 512,* 436 (1978).

17. A. Finkelstein and O. S. Andersen, *J. Membr. Biol., 59,* 155 (1981).

18. B. W. Urban, S. B. Hladky, and D. A. Haydon, *Biochim. Biophys. Acta, 602,* 331 (1980).

19. E. R. Decker and D. G. Levitt, *Biochim. Biophys. Acta, 730,* 178 (1983).

20. E. Bamberg, K. Noda, E. Gross, and P. Läuger, *Biochim. Biophys. Acta, 419,* 223 (1976).

21. J. S. Morrow, W. R. Veatch, and L. Stryer, *J. Mol. Biol., 132,* 733 (1979).

22. R. J. Bradley, K. U. Prasad, and D. W. Urry, *Biochim. Biophys. Acta, 649,* 281 (1981).

23. J.-L. Mazet, O. S. Andersen, and R. E. Koeppe II, *Biophys. J., 45,* 263 (1984).

24. R. D. Hotchkiss and R. J. Dubos, *J. Biol. Chem.*, *141*, 155 (1941).

25. G. F. Gause and M. G. Brazhnikova, *Nature*, *154*, 703 (1944).

26. K. Bauer, R. Roskoski Jr., H. Kleinkauf, and F. Lipmann, *Biochemistry*, *11*, 3266 (1972).

27. H. A. Akers, S. G. Lee, and F. Lipmann, *Biochemistry*, *16*, 5722 (1977).

28. R. Sarges and B. Witkop, *J. Am. Chem. Soc.*, *87*, 2011 (1965).

29. L. K. Ramachandran, *Biochemical Reviews*, *46*, 1 (1975).

30. P. Mukherjee and H. Paulus, *Proc. Natl. Acad. Sci. USA*, *71*, 780 (1977).

31. N. Sarkar, D. Langley, and H. Paulus, *Proc. Natl. Acad. Sci. USA*, *74*, 1478 (1977).

32. H. Paulus, N. Sarkar, P. Mukherjee, D. Langley, V. T. Ivanov, E. N. Shepel, and W. R. Veatch, *Biochemistry*, *18*, 4532 (1979).

33. J. Hansen, W. Pschorn, and H. Ristow, *Eur. J. Biochem.*, *126*, 279 (1982).

34. P. M. Cowan and D. C. Hodgkin, *Proc. Roy. Soc. Ser. B*, *141*, 89 (1953).

35. D. W. Urry, *Proc. Natl. Acad. Sci. USA*, *68*, 672 (1971).

36. W. R. Veatch, E. T. Fossel, and E. R. Blout, *Biochemistry*, *13*, 5249 (1974).

37. W. R. Veatch and E. R. Blout, *Biochemistry*, *13*, 5257 (1974).

38. B. Lotz, F. Colonna-Cesari, F. Heitz, and G. Spach, *J. Mol. Biol.*, *106*, 915-942 (1976).

39. D. W. Urry, M. M. Long, M. Jacobs, and R. D. Harris, *Ann. N.Y. Acad. Sci.*, *264*, 203 (1975).

40. L. Masotti, A. Spisni, and D. W. Urry, *Cell. Biophys.*, *2*, 241 (1980).

41. B. A. Wallace, W. R. Veatch, and E. R. Blout, *Biochemistry*, *20*, 5754 (1981).

42. D. C. Tosteson, T. E. Andreoli, M. Tieffenberg, and P. Cook, *J. Gen. Physiol.*, *51*, 373s (1968).

43. D. W. Urry, M. C. Goodall, J. D. Glickson, and D. F. Mayers, *Proc. Natl. Acad. Sci. USA*, *68*, 1907 (1971).

44. E. Bamberg and P. Läuger, *J. Membr. Biol.*, *11*, 177 (1973).

45. H. P. Zingshein and E. Neher, *Biophys. Chem.*, *2*, 197 (1974).

46. H. A. Kolb, P. Läuger, and E. Bamberg, *J. Membr. Biol.*, *20*, 133 (1975).

47. W. R. Veatch, R. Mathies, M. Eisenberg, and L. Stryer, *J. Mol. Biol.*, *99*, 75 (1975).

48. W. R. Veatch and L. Stryer, *J. Mol. Biol., 113,* 89 (1977).

49. E. Bamberg and K. Janko, *Biochim. Biophys. Acta, 465,* 486 (1977).

50. G. Szabo and D. W. Urry, *Science, 203,* 55 (1979).

51. E. Bamberg, H.-J. Apell, and H. Alpes, *Proc. Natl. Acad. Sci. USA, 74,* 2402 (1977).

52. R. J. Bradley, D. W. Urry, K. Okamoto, and R. Rapaka, *Science, 200,* 435 (1978).

53. H.-J. Apell, E. Bamberg, H. Alpes, and P. Läuger, *J. Membr. Biol., 31,* 171 (1977).

54. S. Weinstein, B. A. Wallace, E. R. Blout, J. S. Morrow, and W. R. Veatch, *Proc. Natl. Acad. Sci. USA, 76,* 4230 (1979).

55. S. Weinstein, B. A. Wallace, J. S. Morrow, and W. R. Veatch, *J. Mol. Biol., 143,* 1 (1980).

56. D. W. Urry, T. L. Trapane, and K. U. Prasad, *Science, 221,* 1064 (1983).

57. G. N. Ramachandran and R. Chandrasekharan, *Indian J. Chem. Biophys., 9,* 1 (1972).

58. R. E. Koeppe II and M. Kimura, *Biopolymers, 23,* 23 (1984).

59. J. R. Elliott, D. Needham, J. P. Dilger, and D. A. Haydon, *Biochim. Biophys. Acta, 735,* 95 (1983).

60. D. W. Urry, J. T. Walker, and T. L. Trapane, *J. Membr. Biol., 69,* 225 (1982).

61. R. L. M. Synge, *Cold Spring Harbor Symp. Quant. Biol., 14,* 191 (1949).

62. W. R. Veatch, Ph.D. thesis, Harvard University (1973).

63. R. E. Koeppe II, K. O. Hodgson, and L. Stryer, *J. Mol. Biol., 121,* 41 (1978).

64. R. E. Koeppe II, J. M. Berg, K. O. Hodgson, and L. Stryer, *Nature, 279,* 723 (1979).

65. V. M. Naik and S. Krimm, *Biophys. J., 45,* 109 (1984).

66. R. E. Koeppe II and B. P. Schoenborn, *Biophys. J., 45,* 503 (1984).

67. D. W. Urry, in *Conformation of Biological Molecules and Polymers, Jerusalem Symposium on Quantum Chemistry and Biochemistry* (E. D. Bergman and B. Pullman, eds.), Israel Academy of Sciences, Jerusalem, Vol. 5, 1973, p. 723 ff.

68. D. W. Urry, C. M. Venkatachalam, A. Spisni, R. J. Bradley, T. L. Trapane, and K. U. Prasad, *J. Membr. Biol., 55,* 29 (1980).

69. C. M. Venkatachalam and D. W. Urry, *J. Comp. Chem., 4,* 461 (1983).

70. B. A. Wallace, *Biopolymers, 22,* 397 (1983).

71. D. W. Urry, K. U. Prasad, and T. L. Trapane, *Proc. Natl. Acad. Sci. USA, 79,* 390 (1982).

72. IUPAC-IUB Commission on Biochemical Nomenclature, *J. Mol. Biol., 52,* 1 (1970).

73. G. Eisenman, J. Sandblom, and E. Neher, *Biophys. J., 22,* 307 (1978).

74. D. W. Urry in *Topics in Current Chemistry* (F. L. Boschke, ed.), Springer-Verlag, Heidelberg, Germany, 1984.

75. B. W. Urban, S. B. Hladky, and D. A. Haydon, *Fed. Proc. Fed. Am. Soc. Exp. Biol., 37,* 2628 (1978).

76. J. Kleinberg, W. J. Argersinger, and E. Griswold, *Inorganic Chemistry,* D. C. Heath, Boston, 1960, p. 74.

77. L. Benjamin and V. Gould, *Trans. Faraday Soc., 50,* 797 (1959).

78. C. M. Venkatachalam and D. W. Urry, *J. Magn. Reson., 41,* 313 (1980).

79. J. Hagglund, B. Enos, and G. Eisenmann, *Brain Res. Bull., 4,* 154 (1979).

80. S. B. Hladky, B. W. Urban, and D. A. Haydon, in *Membrane Transport Processes* (C. F. Stevens and R. W. Tsien, eds.), Vol. 3, Raven Press, New York, 1979, p. 89.

81. A. Cornelius and P. Laszlo, *Biochemistry, 18,* 2004 (1979).

82. H. Monoi and H. Vedaira, *J. Biophys., 25,* 535 (1979).

83. D. W. Urry, C. M. Venkatachalam, A. Spisni, P. Läuger, and M. A. Khaled, *Proc. Nat. Acad. Sci. USA, 77,* 2028 (1980).

84. W. R. Veatch and J. T. Durkin, *J. Mol. Biol., 143,* 411 (1980).

85. E. Neher, J. Sandblom, and G. Eisenman, *J. Membr. Biol., 40,* 97 (1978).

86. D. G. Levitt, *Biophys. J., 22,* 221 (1978).

87. J. F. Hinton, G. L. Turner, and F. S. Millett, *J. Magn. Reson., 45,* 42 (1981).

88. J. F. Hinton, G. L. Turner, and F. S. Millett, *Biochemistry, 21,* 646 (1982).

89. D. W. Urry, A. Spisni, and M. A. Khaled, *Biochem. Biophys. Res. Commun., 88,* 940 (1979).

90. A. Spisni, M. A. Khaled, and D. W. Urry, *FEBS Lett., 102,* 321 (1979).

91. D. W. Urry, T. L. Trapane, and K. V. Prasad, *Int. J. Quantum Chem., Quantum Biol. Symp., 9,* 31 (1982).

92. G. L. Turner, J. F. Hinton, and F. S. Millett, *J. Magn. Reson., 51,* 205 (1983).

93. D. W. Urry, Ionic mechanism of the gramicidin transmembrane channel. In *Proceedings of the NATO ASI on Spectroscopy of Biological Molecules,* D. Reidel, Dordrecht, Holland, 1984.

94. G. L. Turner, J. F. Hinton, R. E. Koeppe, J. A. Parli, and F. S. Millett, *Biochem. Biophys. Acta, 756,* 133 (1983).

95. D. W. Urry, A. Spisni, and M. A. Khaled, *Biochem. Biophys. Res. Commun., 88,* 940 (1979).

96. D. W. Urry, A. Spisni, M. A. Khaled, M. M. Long, and L. Masotti, *Int. J. Quantum Chem. Quantum Biol. Symp., 6,* 289 (1979).

97. E. Bamberg and P. Läuger, *Biochim. Biophys. Acta, 367,* 127 (1974).

98. J. F. Hinton, G. Young, and F. S. Millett, *Biochem. Biophys. Acta, 727,* 217 (1983).

99. R. Henze, E. Neher, T. L. Trapane, and D. W. Urry, *J. Membr. Biol., 64,* 233 (1982).

100. A. Pullman and C. Etchebest, *FEBS Lett., 163,* 199 (1983).

101. H. H. Kohler and K. Heckmann, *J. Membr. Sci., 6,* 45 (1980).

102. D. W. Urry, T. L. Trapane, C. M. Venkatachalam, and K. U. Prasad, *J. Phys. Chem., 87,* 2918 (1983).

103. J. F. Hinton, G. Young, and F. S. Millett, *Biochemistry, 21,* 651 (1982).

104. O. S. Andersen and J. Procopio, *Acta Physiol. Scand. Suppl, 481,* 27 (1980).

105. O. S. Andersen, *Biophys. J., 41,* 119 (1983).

106. O. S. Andersen, *Biophys. J., 41,* 147 (1983).

107. B. W. Urban and S. B. Hladky, *Biochim. Biophys. Acta, 554,* 410 (1979).

108. L. V. Schagina, A. E. Grinfeldt, and A. A. Lev, *J. Membr. Biol., 73,* 203 (1983).

109. H. H. Ussing, *Acta Physiol. Scand., 19,* 43 (1949).

110. G. Eisenman, J. P. Sandblom, and F. L. Walker, Jr., *Science, 155,* 965 (1967).

111. D. G. Levitt, *Biochim. Biophys. Acta, 373,* 115 (1974).

112. J. Sandblom, G. Eisenman, and J. Hagglund, Jr., *Membr. Biol., 71,* 61 (1983).

113. G. Eisenman and J. Sandblom, *Biophys. J., 45,* 88 (1984).

114. P. C. Jordan, *Biophys. J., 39,* 157 (1982).

115. O. S. Andersen, *Ann. Rev. Physiol., 46,* 531 (1984).

116. F. Heitz, G. Spach, and Y. Trudelle, *Biophys. J., 40,* 87 (1982).

117. K. U. Prasad, T. L. Trapane, D. Busath, G. Szabo, and D. W.
      Urry, *Int. J. Peptide Protein Res.*, *19*, 162 (1982).

118. K. U. Prasad, T. L. Trapane, D. Busath, G. Szabo, and D. W.
      Urry, *J. Protein Chem.*, *1*, 191 (1982).

119. L. B. Weiss and R. E. Koeppe II, submitted.

120. E. W. Barrett, L. B. Weiss, O. S. Andersen and L. Stryer,
      Biophys. J., *33* (Vol. 2, pt. 2), 63a (1981) (abstr.).

121. E. W. B. Russell, L. B. Weiss, F. I. Navetta, R. E. Koeppe II,
      and O. S. Andersen, submitted.

122. R. E. Koeppe II, O. S. Andersen, and J.-L. Mazet, 8th Inter-
      national Biophysics Congress, Bristol, United Kingdom (1984)
      (Abstr.).

Chapter 8

NACTINS: THEIR COMPLEXES AND BIOLOGICAL PROPERTIES

Yoshiharu Nawata and Kunio Ando
Research Laboratories
Chugai Pharmaceutical Co. Ltd.
Takada, Toshima, Tokyo, Japan

Yoichi Iitaka
Faculty of Pharmaceutical Sciences
University of Tokyo
Hongo, Bunkyo, Tokyo, Japan

1. INTRODUCTION

Nactins were isolated from fermented browth of *Streptomyces* in 1955 [1]. They are macrotetrolide antibiotics (cyclodepsid) composed of four subunits of ω-hydroxycarboxylic acids, consisting of (+)- and (-)-nonactic acid, (+)- and (-)-homononactic acid, (+)- and (-)-bishomononactic acid (Fig. 1).

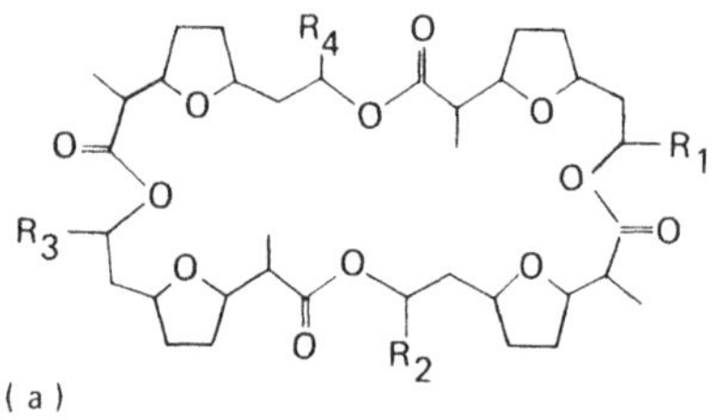

(a)

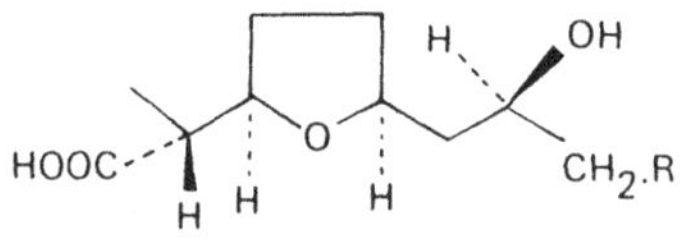

(b)

FIG. 1.   (a) Chemical structural formula of nactins: Nonactin (R_1, R_2, R_3, R_4 = methyl), monactin (R_1 = ethyl; R_2, R_3, R_4 = methyl), dinactin (R_1, R_3 = ethyl; R_2, R_4 = methyl), trinactin (R_1, R_2, R_3 = ethyl; R_4 = methyl), tetranactin (R_1, R_2, R_3, R_4 = ethyl), substance G (R_i = methyl, methyl, ethyl, isopropyl), substance D (R_i = methyl, ethyl, ethyl, isopropyl), substance C (R_i = methyl, ethyl, isopropyl, isopropyl), and substance B (R_i = ethyl, ethyl, isopropyl, isopropyl).   (b) Chemical structural formula of constituent acids: (-)nonactic acid (R = methyl), (-)homononactic acid (R = ethyl), and (-)bishomononactic acid (R = isopropyl).

These constituent acids are connected to each other through ester linkages to form a 32-membered ring. The isolated nactins include all the homologous compounds composed of these acids. Nonactin, monactin, dinactin, trinactin, tetranactin, and substances G, D, C, and B were thus obtained and characterized [2-4].

Nactins are neutral ionophoric substances like valinomycin (macrocyclic depsipeptide), they are able to uncouple the oxidative phosphorylation of mitochondria of rat liver in a low concentration [5], and they can also carry cations across biological and artificial membranes [6]. Although nactins exhibit high cation selectivity for K^+ and Rb^+ among the alkali metal cations [7], they exhibit the highest selectivity for NH_4^+ [8] and Tl^+ [9]. Nactins are used as NH_4^+ sensor in membrane electrodes [10] and also as pesticides [39].

2. MOLECULAR STRUCTURE OF NACTINS

In order to clarify the mechanism of the ionophoric activity of
nactins, molecular structures and physicochemical properties of
nactins have been investigated. In aqueous solution, four subunits
of nonactin were shown to be magnetically equivalent [11], and this
is also true for tetranactin [12]. Accordingly, the molecular symmetry of nactins in solution is considered to be S4. In the crystalline state, the nonactin molecule is flat and looks like a torus
which has crystallographic C2 and approximate S4 symmetries [13]
[Fig. 2(a)]. This form is considered to be consistent with the
molecular symmetry deduced from NMR spectra observed in solution.
On the other hand, dinactin possesses overall the shape of a lobster
and is asymmetric, [14] [Fig. 2(b) and (d)], and tetranactin resembles a propeller with C2 symmetry [15] [Fig. 2(c)].

To compare the molecular conformation among these three structures, torsional angles (conformational angles) along the 32-membered
ring were plotted for each subunit [14]. As is shown in Fig. 3, the
conformations of the four subunits in dinactin are different from
each other, and the conformational types of these four are designated
as D1, D2, D3, and D4, respectively. They are partly similar to those
observed in nonactin (conformational types of the subunits are N and
-N, where the "-" indicates the enantiomer of the corresponding conformation) or tetranactin (T1, -T1, T2, and -T2). Major differences
in conformation among these subunits are whether they possess trans
or gauche form at the bonds adjacent to the tetrahydrofurane ring.
Except for these bonds, one may conclude that the molecular conformation of dinactin is constructed by a hybridization of the subunits
observed in nonactin and tetranactin.

For these three types of molecular conformation, steric energy
(conformational energy) was estimated and minimized [16] by the
method of molecular mechanics using the program MMPI [17]. The
lowest steric energy was derived from a dinactin-type structure and
the highest energy was obtained from a tetranactin-type structure.

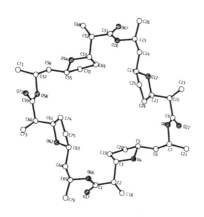

(a)

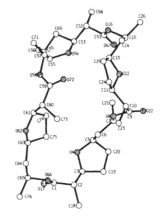

(b)

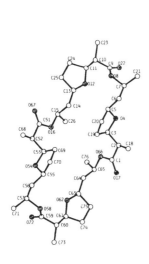

(c)

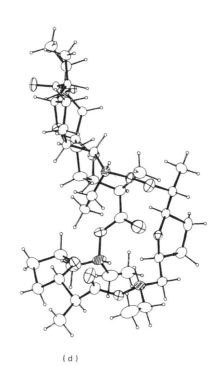

(d)

However, none of these structures can afford the vicinal proton-proton coupling constant ($^3J_{H,H'}$) consistent with those observed in ^1H NMR spectra. Especially the differences between the observed and the calculated $^3J_{H,H'}$ values for a nonactin-type structure are serious. It is obvious that none of these three structures observed in the solid state can explain the coupling constant $^3J_{H,H'}$. It is also noteworthy that a remarkably large dipole moment was calculated for dinactin (4.3 D by MMPI program). We therefore calculated the electrostatic potential curves around the dinactin molecule. Figure 4 shows the lines of electric force emitted from the molecule. It is clear that the majority of electric lines are concentrated to one part of the molecule showing that a dinactin-type structure is suitable to attract cations by ion-dipole interactions and to begin the complex formation.

The IR spectra of dinactin in the *crystalline state* are similar to those of nactins *in solution*, and all these spectra are rather simple and broad. It is considered that the extent of the deformations of molecular conformation of nactins in solution may be of the same order of magnitude as those observed in the four subunits of the dinactin molecule in the crystal [14].

FIG. 2. Molecular conformation of nactins. (a) Nonactin (projection of nonhydrogen atoms along the crystallographic c axis; carbon (open circle) and oxygen (shaded circle) atoms with atomic numbering are depicted). (b) Dinactin (projected along the crystallographic c axis as described in (a); ethyl groups are replaced by methyl groups). (c) Tetranactin (projected along the crystallographic b axis as described in (a); ethyl groups are replaced by methyl groups). (d) Dinactin (projected along the crystallographic a axis; two ethyl groups are distributed at the four C7 sites (shaded atoms), and nonhydrogen atoms are represented by the thermal ellipsoids of 20% probability. The principal ellipses are added to the oxygen atoms). (Parts (a), (b), and (c) are from Ref. 16, with permission of the publisher. Part (d) is from Ref. 14 with permission of the publisher.)

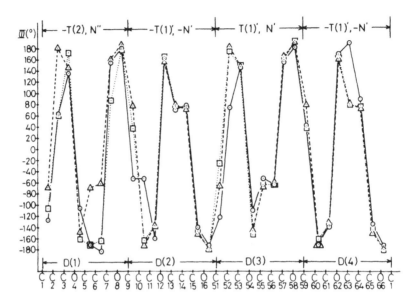

FIG. 3. A comparison of the backbone torsional angles (Φ) along
the 32-membered ring: dinactin ($\odot$), nonactin ($\triangle$), and tetranactin
($\square$). The conformations of the subunits in dinactin are grouped in
the four types, D(1), D(2), D(3), and D(4), and they are compared
with those of tetranactin [T(1), T(2)] and of nonactin (N, corresponding Φ values in the four subunits are averaged assuming the
approximate S4 symmetry), where a minus sign indicates the enantiomer of the subunit and the primes denote that one or two Φ values
differ from dinactin more than 90° (double primes denote more differences). (Reproduced from Ref. 14 with permission of the publisher.)

3. COMPLEXES

Nactins can complex with various cations. Alkali metal complexes,
with Na^+, K^+, Rb^+, and Cs^+, were obtained in the crystalline state
[18-23]. In addition, NH_4^+, Tl^+, Ag^+, Ca^{2+}, Cu^{2+}, Ba^{2+}, and Pr^{3+}
complexes were obtained also in the crystalline state [24-27]. The
IR spectra of the last two crystals are apparently different from
those of the others, especially at the stretching vibration absorption band of the carbonyl group [46], namely, strong singlet bands

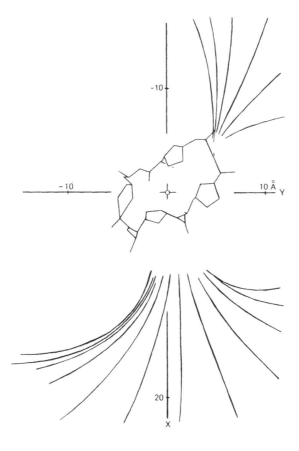

FIG. 4. Lines of electric force in the dinactin-type conformer.
The lines and the nonhydrogen atoms of dinactin, in which ethyl
groups are replaced by methyl groups, are projected along the direction perpendicular to X (the direction of maximum ion-dipole interaction) and parallel to the plane defined by X and the crystallographic [100]-plane. The lines are drawn by calculating the Coulomb
force acting on a positive unit charge from the partial electric
charges on atoms of dinactin (for details, see Ref. 21) at every
1-Å step along the electric vectors. The lines start at 18 points
(every 20° of longitude) on the equator of the sphere with a radius
of 40 Å and centered at the center of gravity of the molecule. The
north pole is taken to be the projection direction.

are observed in the alkali-metal complexes, and also in the NH_4^+, Tl^+, and Ag^+ complexes, while those of Ba^{2+} and Pr^{3+} show a doublet and multiplet, respectively. Consequently, the coordination geometries of the last two complexes may be extremely deformed from those of the others.

3.1. Complex Formation

The complex formation equilibrium between nactins and cations is under the strong influence of counterions and solvents. The Ca^{2+}-nonactin complex was obtained in the crystalline state from an acetone solution containing equimolar amounts of nonactin and $Ca(ClO_4)_2$ [27]. But the Ca^{2+}-tetranactin complex could not be obtained as crystals from the same solvent containing equimolar amounts of tetranactin and $Ca(SCN)_2 \cdot 2H_2O$, probably because of the coexistence of water [26]. The complex formation constant (K) of nonactin with $KClO_4$ in dry acetone is almost the same as that in wet acetone. On the contrary, the K value of nonactin with $NaClO_4$ in dry acetone is approximately in the same order as that of non-actin-$KClO_4$, but it decreases markedly in wet acetone. This shows that the hydration of the cation may be one of the most important factors of cation selectivity of nactins in biological systems [44]. The K values increase with increasing degree of methylation of the nactins. The values increase in the order of nonactin, monactin, dinactin, trinactin, and tetranactin. These variations may be due to the Karkwood-Westheimer field effect or Taft induction effect of methyl groups [36]. Thermodynamic studies on the interaction of cations with nactins have been extensively carried out by Simon and his colleagues [7,45].

3.2. Structure of Alkali-Metal Complexes

Nactins can take a cation into the central cavity of the molecule forming stoichiometric complexes of distorted cubic coordination

[23]. In these complexes, 32-membered rings are folded into the
form of the seam of a tennis ball, and the four oxygen atoms of the
carbonyl groups and the four ether oxygen atoms are arranged in an
approximate symmetry of S4 to form a complex of 8-coordination
(Fig. 5).

Among the alkali-metal ion-nactin complexes, the distortion of
the coordination geometry is remarkable in the Na^+ complexes [18,20].
The fact that the coordination distances of carbonyl oxygen atoms
$[M^+$ . . . O (carbonyl) distance] are approximately equal to the sum
of the ionic radius of the cation and the van der Waals radius of
the oxygen atom (M^+ + O distance), while the M^+ . . . O (ether) distances are longer than the sum for small ionic radii and shorter for
larger ionic radii as shown in Fig. 6, may indicate that the deforma-
tion of nactin molecules can easily occur about the ester groups,
while conformational changes around the tetrahydrofurane ring are
restricted by steric hindrance. In the Cs^+-tetranactin complex,
the M^+ . . . O (ether) distances are shorter than the M^+ + O distance, and the repulsive forces between Cs^+ and O (ether) atoms are
obvious [22]. On the contrary, in the Cs^+-nonactin complex, the
central cavity of the nonactin molecule is expanded by lengthening
the M^+ . . . O (ether) distances to diminish the repulsive forces
between M^+ and O (ether) atoms. The conformational changes of the
nonactin molecule, which possesses no ethyl groups, are more easily
performed than of tetranactin, which possesses four ethyl groups.
Actually, the coordination polyhedra of Na^+-nactin complexes are
remarkably deviated from the typical cubic coordination structure
due to the smaller ionic radius of Na^+. It seems to be extremely
difficult to form a stable complex with monovalent cations smaller
than Na^+. In this connection, it is interesting to see that the
cation selectivity of nonactin for Na^+ and Cs^+ ($Na^+ < Cs^+$) is reversed
to that of tetranactin ($Na^+ > Cs^+$) [36], which shows that the selectivities are sensitive to the flexibility of the molecular structure.

In order to compare the conformational stability of alkali
metal-nactin complexes, steric energies were estimated in terms of
nonbonded interaction (U_{nb}), torsional energy (U_{tor}), dipole-dipole

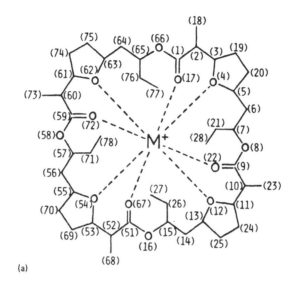

(a)

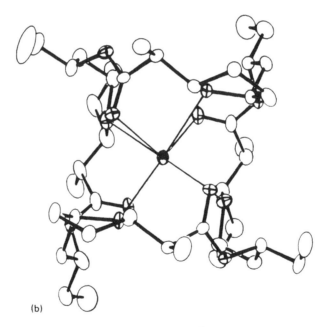

(b)

FIG. 5.   Coordination geometry of the Rb[+]-tetranactin complex.
(a) Structural formula with atomic numberings and (b) molecular
structure.

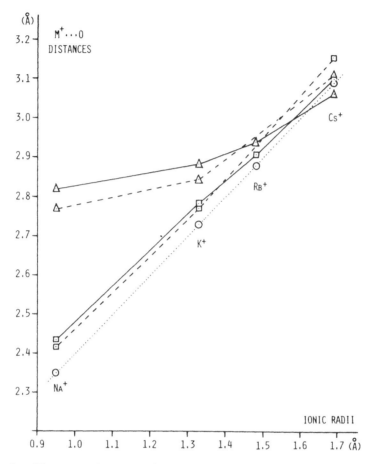

FIG. 6. Distances between the central cation and the ligand oxygen
atoms. Observed values for the tetranactin complexes (solid lines)
and those for the nonactin complexes (broken lines) are plotted
against the ionic radii of the cation (abscissa). (△) Distance
between the cation and the ether oxygen atoms. (□) Distance between
the cation and the carbonyl oxygen atoms. (⊙) Sum of the ionic
radius of the central cation and the van der Waals radius of the
oxygen atom (M^+ + O distance). (Reproduced from Ref. 22 by courtesy
of the publisher.)

interaction (U_{el}), and ion-dipole interaction (U_{id}) [21]. The U_{id} is strongly dependent on the ionic size. In the Na^+ complexes, ligand oxygen atoms contract to the center of the molecular cavity and produce the largest U_{id} energy, whereas in the Cs^+ complexes, the ligand oxygen atoms are away from the molecular center and the weakest U_{id} interaction is estimated. In tetranactin, the steric energy of the Cs^+ complexes is further decreased by the repulsive forces between Cs^+ and ligand oxygen atoms due to the close approach caused by the less flexible structure. The U_{nb} interaction energy is also proportional to the ionic size of the central cation. The U_{nb} of Na^+ complexes, in which all the atoms are shrinked to the center of the molecule, is larger than in other complexes with alkali metal cations. However, the U_{el} interaction energy is positive (= repulsive) in all of the alkali metal complexes, and it is especially remarkable in those with the smaller cations. The U_{tor} interaction energies are also positive in all complexes but are not in correlation with the ionic size. The steric energies of these complexes are proportional to the ionic radius of the cation, i.e., the conformation of the complex with the smaller cation is more stable than that with the larger cation.

In aqueous solution or in alcohol, the K values for K^+ and Rb^+ are larger than those for Li^+, Na^+, and Cs^+. The K values do not correlate directly to the steric energies because these values express the thermodynamic equilibrium in solution and the steric energies depend on the three-dimensional arrangements of atoms. Still, it is evident that the 32-membered ring of the nactins is easily folded to form the conformation of the seam of a tennis ball (S4 symmetry), and the size of the central cavity is suitable to accommodate K^+ and Rb^+ but is too small for Cs^+ and too large for Na^+. It is also clear that there are some differences in the flexibility among the nactin homologues, which cause a difference in the cation selectivity for Na^+ and Cs^+ as observed in nonactin and tetra-nactin.

3.3. NH$_4^+$ Complex

Nactins exhibit the highest cation selectivity for NH$_4^+$, although the interactions between NH$_4^+$ and carbonyl oxygen atoms are weaker than those observed in K$^+$ and Rb$^+$ complexes [37]. As is shown in Fig. 7, the molecular conformation of the NH$_4^+$ complex is isostructural to those of the alkali metal-nactin complexes, and the conformational differences among these complexes are subtle [24,25]. But the co-ordination geometry around the cation in the NH$_4^+$ complex brought a specific selectivity for NH$_4^+$. Although the ionic radius of NH$_4^+$ is

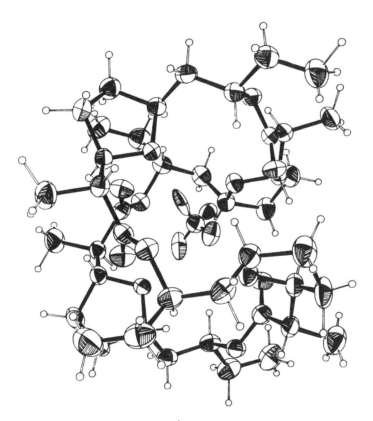

FIG. 7. ORTEP drawing of the NH$_4^+$-tetranactin complex. (From Ref. 24 by courtesy of the publisher.)

almost the same as that of Rb$^+$ (1.48 Å), the M$^+$ . . . O (ether) distances are shorter than the M$^+$ . . . O (carbonyl) distances in the NH$_4^+$ complex and vice versa in the Rb$^+$ complex.

These facts suggest that in the NH$_4^+$ complex the interaction between NH$_4^+$ and ether oxygen atoms is stronger than that between NH$_4^+$ and carbonyl oxygen atoms. In fact, hydrogen bonds between NH$_4^+$ and ether oxygen atoms are observed in the NH$_4^+$ complex [30], namely, the four N-H bonds are directed to the four lone pair orbitals of the ether oxygen atoms which are arranged in the approximate symmetry of S4 and form a central cavity of suitable size to accommodate NH$_4^+$ (Fig. 8). In this coordination geometry, none of the

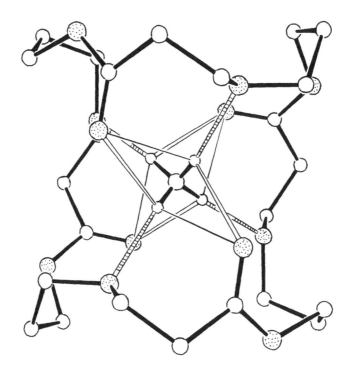

FIG. 8. The 32-membered ring and the coordination geometry of the NH$_4^+$-tetranactin complex viewed down the b axis. Hydrogen bonds are shown as shaded bonds. Weak and strong electrostatic interactions are indicated by interatomic single and double lines, respectively. Oxygen atoms are shown as stippled circles. (Reproduced from Ref. 24 by courtesy of the publisher.)

carbonyl oxygen atoms can form a hydrogen bond with NH_4^+ because the
lone pair orbitals of carbonyl oxygen atoms are directed perpendicular to the cation. Consequently, the interaction between the NH_4^+
and the carbonyl oxygen atoms is electrostatic and its energy estimated by the molecular orbital method is about half the value of
that between NH_4^+ and the ether oxygen atoms [31]. For a model
structure, in which NH_4^+ is best fitted to the carbonyl groups, the
interaction energy is raised about 9.8 kcal/mol above that of the
observed structure [31].

3.4. Ternary Complexes

Nactins can transfer both cation and its counterion from aqueous
layers into organic solvents by forming ternary complexes. In these
complexes, a stoichiometric relation is preserved between nactins,
cation, and counterion. Accordingly, colored [28] or fluorescent
anions [29] are selected as reagents for quantitative analysis, and
in fact the Rb^+-tetranactin-picrate complex was obtained in the
crystalline state [47]. In the crystal structure, the Rb^+-tetra-
nactin complex cation and the picrate anion are piled up alternatingly and equidistantly to form columns, in which the nearest position of one of the o-nitro groups in the picrate anion to Rb^+ is
5.79-6.15 Å [Fig. 9(a)]. Although the crystal structure is regarded
as a distorted NaCl-type structure, the electrostatic interaction
between the complexed cation and the picrate anion is weaker than a
nonbonded interaction.

SF6847 (3,5-di-tert-butyl-4-hydroxybenzylidenemalononitrile;
hereafter abbreviated as SFH) is a strong uncoupler and a protonophore
(proton carrier) [33]. In the presence of valinomycin, SFH promotes
a leakage of K^+ from the liposomes, in which potassium phosphate is
enclosed, and it also promotes proton uptake into liposomes to com-
pensate a potential difference generated at the liposomal membranes
as K^+ goes out [34]. In this process of flux and reflux of ions,
K^+ is considered to be translocated out from the liposomes in the

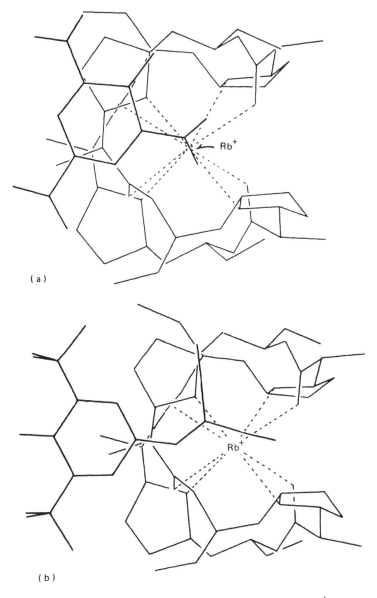

(a)

(b)

FIG. 9. Ternary complexes of tetranactin: (a) The Rb$^+$-tetranactin-picrate complex and (b) the Rb$^+$-tetranactin-SF$^-$ complex. (Reproduced from Ref. 49 with permission of the publisher.)

form of a ternary complex, i.e., of K⁺-valinomycin-SF⁻ [38], where
SF⁻ stands for the anionic form of SFH.

The ternary complexes were obtained in the crystalline state
from a methanol solution containing equimolar amounts of RbSF (rubidium salt of SFH) and valinomycin (or tetranactin), and the crystal
structures were determined [35]. In the latter case, the Rb⁺-tetra-
nactin complexes and the SF⁻ anions are piled up alternatingly to
form columns in which each cationic complex is in a proximity of a
specific SF⁻ anion to form the ternary complex [32] [Fig. 9(b)].

In these columns, the malononitrilo group is in contact with
the surface of the cationic complex, and both the phenoxide (-O⁻)
and the two tert-butyl groups are away from the surface of the complex: -O⁻ . . . Rb⁺ = 9.95 Å. Although the phenoxide group is near
to the methyl group of tetranactin in the neighboring ternary complex,
it is far away from Rb⁺ (6.94 Å) (Fig. 10). A negative electronic

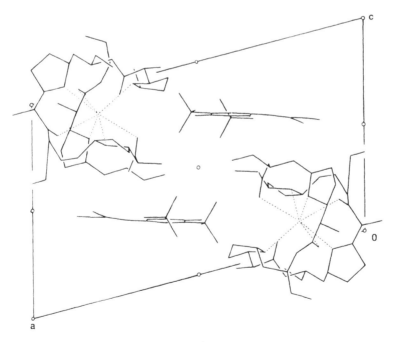

FIG. 10. Crystal structure of Rb⁺-tetranactin-SF⁻ complex viewed
down the b* axis.

charge (e^-) of the SF$^-$ anion, estimated by the molecular orbital method (CNDO/2), is slightly concentrated on the phenoxide group but considerably concentrated on the malononitrilo group. Therefore, the latter group plays an important role in the formation of the ternary complex, in which nonbonded interactions between SF$^-$ and the cationic complex are predominant over electrostatic interactions.

4. BIOLOGICAL PROPERTIES

Nonactin was named after its biological property of showing no bacteriocidal activity [1]. Nactins uncouple oxidative phosphorylation at a concentration of 10^{-7} M and induce adenosine triphosphatase in the presence of alkali metal cations (except Li$^+$) [48]. Nactins are of low toxicity to warmblooded animals but exert potent pesticidal activity against the carmine spider mite (*Tetranychus cinnabarinus*), for which the lethal concentration (LC$_{50}$) is 4.8 μg/ml with the spray method [40]. Although other insects, such as Azuki-bean weevil (*Callosobruchus chinensis*) and larvae of mosquito (*Culex pipiens molestus*) are moderately sensitive to nactins, house fly and cockroach are insensitive to them. However, nactins are only weakly ovocidal against mites. It is almost impossible for mites to acquire resistance to nactins [39]. Although the LC$_{50}$ of the sensitive strain of carmine spider mites doubled during the exposure to sublethal concentration for 6 years, the LC$_{50}$ of the resistant strain was only twice as high as that of the sensitive strain.

Nactins exhibit synergistic action with other insecticides, e.g., 2-sec-butylphenyl-N-methylcarbamate. Synergistic combination increased not only the lethal effects in imagoes but improved the residual ovocidal effect of nactins. The synergistic effects of valinomycin and uncoupler SFH were also considered to be important in the process of alkali metal cation and proton translocation through the membrane and in the action of the uncoupler [41,43].

Nactins are highly susceptible to sunlight, although they are resistant to weathering. When nactins were sprayed on tea leaves, they rapidly decomposed on exposure to sunlight. Degradation was more than 50% after 2 days exposure. On the other hand, degradation of nactins in soil is primarily due to soil bacteria such as *Bacillus* and *Micrococcus* sp, which readily hydrolyze nactins at the ester linkage yielding homononactic and/or nonactic acids. These ω-hydroxycarboxylic acids are decomposed by soil fungi and other bacteria into water and carbon dioxide.

The mode of miticidal action of nactins is unusual [42]. When mites are placed on leaves covered with a dry film of nactins, they are quite safe under various humidities, but all of the mites died soon after spraying water. Water is an essential medium for the miticidal activity of nactins. It is plausible that in the presence of water, nactins exhibit an ionophoric property of transporting K^+ from the intercellular medium through the cell membrane to the outer side, resulting in a destruction or a decrease of the cellular function.

REFERENCES

1. R. Corbaz, L. Ettlinger, E. Gäumann, W. Keller-Schierlein, F. Kradolfer, L. Neipp, V. Prelog and H. Zähner, *Helv. Chim. Acta, 38,* 1445 (1955).

2. J. Beck, H. Gerlach, V. Prelog and W. Voser, *Helv. Chim. Acta, 45,* 620 (1962).

3. H. Oishi, T. Hosokawa, T. Okutomi, K. Suzuki and K. Ando, *Agr. Biol. Chem., 33,* 1790 (1969).

4. W. Keller-Schierlein, H. Gerlach and J. Seibl, *Antimicrob. Agents Chemother.,* 644 (1966).

5. S. N. Graven, H. A. Lardy and A. Rutter, *Biochemistry, 5,* 1735 (1966).

6. G. Szabo, G. Eisenman and S. Ciani, *J. Memb. Biol., 1,* 346 (1969).

7. Ch. U. Züst, P. U. Früh and W. Simon, *Helv. Chim. Acta, 56,* 495 (1973).

8. R. P. Scholer and W. Simon, *Chimia, 24,* 372 (1970).

9. G. Eisenman and S. J. Krasne, in *MTP International Review of Science,* Biochemistry Series, Vol. 2 (C. F. Fox, ed.), Butterworths, London, 1973.

10. H. Degawa, N. Shinozuka, and S. Hayano, *Nippon Kagaku Kaishi,* 1462 (1980).

11. J. H. Prestegard and S. I. Chan, *Biochemistry, 8,* 3921 (1969).

12. Y. Kyogoku, M. Ueno, H. Akutsu, and Y. Nawata, *Biopolymers, 14,* 1049 (1975).

13. M. Dobler, *Helv. Chim. Acta, 55,* 1371 (1972).

14. Y. Nawata, T. Hayashi, and Y. Iitaka, *Chem. Lett.,* 315 (1980).

15. Y. Nawata, T. Sakamaki, and Y. Iitaka, *Acta Crystallogr., Sect. B, 30,* 1047 (1974).

16. Y. Nawata and Y. Iitaka, *Tetrahedron, 39,* 1133 (1983).

17. N. L. Allinger and Y. H. Yuh, program *MMPI,* QCPE No. 318 (1975).

18. M. Dobler and R. P. Phizackerley, *Helv. Chim. Acta, 57,* 664 (1974).

19. M. Dobler, J. Dunitz, and B. T. Kilbourn, *Helv. Chim. Acta, 52,* 2573 (1969).

20. Y. Iitaka, T. Sakamaki, and Y. Nawata, *Chem. Lett.,* 1225 (1972).

21. T. Sakamaki, Y. Iitaka, and Y. Nawata, *Acta Crystallogr., Sect. B, 32,* 768 (1976).

22. T. Sakamaki, Y. Iitaka, and Y. Nawata, *Acta Crystallogr., Sect. B, 33,* 52 (1977).

23. B. T. Kilbourn, J. Dunitz, L. A. R. Pioda, and W. Simon, *J. Mol. Biol., 30,* 559 (1967).

24. Y. Nawata, T. Sakamaki, and Y. Iitaka, *Acta Crystallogr., Sect. B, 33,* 1201 (1977).

25. K. Neupert-Laves and M. Dobler, *Helv. Chim. Acta, 59,* 614 (1976).

26. Y. Nawata, K. Ando, and Y. Iitaka, *Acta Crystallogr., Sect. B, 27,* 1680 (1971).

27. C. K. Vishwanath, N. Shamala, K. R. K. Easwaran, and M. Vijayan, *Acta Crystallogr., Sect. C, 39,* 1640 (1983).

28. K. Suzuki, Y. Nawata, and K. Ando, *J. Antibiot., 24,* 675 (1971).

29. K. Kina, K. Shiroishi, and N. Ishibashi, *Bunseki Kagaku, 27,* 291 (1978).

30. Y. Nawata, T. Sakamaki, and Y. Iitaka, *Chem. Lett.,* 151 (1975).

31. H. Umeyama, S. Nakagawa, T. Nomoto, and I. Moriguchi, *Chem. Pharm. Bull., 28,* 745 (1980).

32. Y. Iitaka, Y. Nishibata, and Y. Nawata, in *Collected Abstracts, 38th National Meeting of the Chemical Society of Japan*, Chemical Society of Japan, Tokyo, 1978, p. 967 ff.

33. H. Terada, *Biochem. Biophys. Acta, 387*, 519 (1975).

34. A. Yamaguchi, Y. Anraku, and S. Ikegami, *Biochem. Biophys. Acta, 501*, 150 (1978).

35. Y. Iitaka, Y. Nishibata, A. Itai, and Y. Nawata, *Acta Crystallogr., Sect. A, 37*, C-75 (1981).

36. G. Eisenman, S. Krasne, and S. Ciani, *Ann. N.Y. Acad. Sci., 264*, 34 (1975).

37. E. Pretsch, M. Vasak, and W. Simon, *Helv. Chim. Acta, 55*, 1098 (1972).

38. K. Yoshikawa and H. Terada, *J. Am. Chem. Soc., 103*, 7788 (1981).

39. K. Ando, in *Human Welfare and the Environment*, Vol. 2: *Natural Products* (J. Miyamoto and P. C. Kearny, eds.), Pergamon Press, Oxford, 1983, p. 253 ff.

40. T. Sagawa, S. Hirano, H. Takahashi, N. Tanaka, H. Oishi, K. Ando, and K. Togashi, *J. Econ. Entomol., 65*, 372 (1972).

41. A. Yamaguchi and Y. Anraku, *Biochem. Biophys. Acta, 501*, 136 (1978).

42. S. Hirano, T. Sagawa, H. Takahashi, N. Tanaka, H. Oishi, K. Ando, and K. Togashi, *J. Econ. Entomol., 66*, 349 (1973).

43. M. C. Blok, J. De Gier, and L. L. M. van Deenen, *Biochem. Biophys. Acta, 367*, 202 (1974).

44. J. H. Prestegard and S. I. Chan, *J. Am. Chem. Soc., 92*, 4440 (1970).

45. W. E. Morf and W. Simon, *Helv. Chim. Acta, 54*, 2683 (1971).

46. Y. Nawata, unpublished observations (1974).

47. Y. Nawata and Y. Iitaka, *Acta Crystallogr., Sect. A*, Part S4, S78 (1978).

48. S. N. Graven, H. A. Lardy, D. Johnson, and A. Rutter, *Biochemistry, 5*, 1729 (1966).

49. Y. Nawata and Y. Iitaka, *Biophysics (Seibutsu Butsuri), 19*, 295 (1979).

Chapter 9

CATION COMPLEXES OF THE MONOVALENT AND POLYVALENT CARBOXYLIC IONOPHORES: LASALOCID (X-537A), MONENSIN, A23187 (CALCIMYCIN), AND RELATED ANTIBIOTICS

George R. Painter
Wellcome Research Laboratories
Research Triangle Park, North Carolina

Berton C. Pressman
Department of Pharmacology
University of Miami Medical School
Miami, Florida

1. INTRODUCTION

Ionophores are small molecules with the ability to carry ions across hydrophobic phase barriers [1]. The transport process depends on the formation of lipid-soluble ion inclusion complexes which are stable within a low polarity environment and readily dissociate at an aqueous interface. By moving ions across the low polarity lipid interior of biological membranes, ionophores are able to alter membrane permeability and perturb transmembrane ion gradients. Consequently, ionophores have become a valuable and extensively used experimental tool for studying the effects of cations on the electrical and metabolic properties of cells [2].

Although the ionophores comprise a group of structurally diverse compounds, they may be divided into two classes depending on the mode of ion transport they promote [2]. The *neutral ionophores*, which lack ionizable functionality, form charged complexes with cations and consequently catalyze potential-sensitive or *electrophoretic* transport across membranes. Neutral ionophore complexes have already been discussed in the opening chapter of this book and are the main concern of Chaps. 5, 6, 8, and 12. The *carboxylic ionophores*, which are the subject of this chapter, contain an ionizable terminal carboxyl group. They form electrically neutral zwitterionic complexes which catalyze electrically silent *exchange-*

diffusion transport. This distinction is fundamental for explaining the profound differences in the biological behavior of the ionophore subclasses.

Structures of representative naturally occurring carboxylic ionophores are presented in Fig. 1. All of these compounds are produced by organisms of the order *Actinomycetales*. The majority are produced by the genus *Streptomyces*, a few by the genera *Actinomydura* and *Dactylopsorangium* [3]. On the basis of structural similarities and extensive studies of their biosynthetic pathways, it appears that all of the naturally occurring carboxylic ionophores, with the exception of A23187 and X-14547A, which contain pyrrole rings, arise from related ketide precursors [3]. Each ionophore contains from 11 to 18 residues of acetate, propionate, and sometimes butyrate. The initiating subunit for polyketide assembly is usually acetyl CoA, less frequently propionyl or butyryl CoA. The most common chain-terminating unit is propionate, less frequently acetate or butyrate. A lack of total selectivity for each acyl group in the sequence often results in one or more homologues of a basic structure being produced simultaneously. The characteristics of the synthetase enzymes and details of the steps involved in conversion of the polyketide into specific carboxylic ionophores has been the subject of a recent review [4].

This chapter will review and critically evaluate the techniques by which the cation affinities, selectivities, and transport turnover numbers of the carboxylic ionophores have been determined experimentally and how these parameters relate to ionophore structure. Accordingly, we will deal primarily with the chemical and physical properties of these compounds in single- and multiphase solvent systems. The activity of the carboxylic ionophores in biological test systems (e.g., mitochondria, erythrocytes) will also be described and related to their physiochemical properties. Finally, we will consider how ionophore-mediated transport affects the behavior of more highly integrated biological systems.

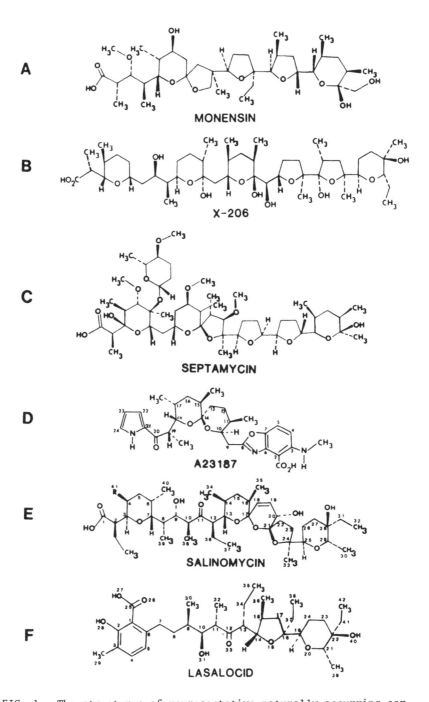

A   MONENSIN

B   X-206

C   SEPTAMYCIN

D   A23187

E   SALINOMYCIN

F   LASALOCID

FIG. 1. The structures of representative naturally occurring carboxylic ionophores: (A) monensin; (B) X-206; (C) septamycin; (D) A23187; (E) R = H for salinomycin, R = CH$_3$ for narasin; and (F) lasalocid. A23187, salinomycin, narasin, and lasalocid are numbered for the discussion of structural studies.

2. STRUCTURE OF IONOPHORE-CATION INCLUSION COMPLEXES

2.1. Mechanism of Ion Complexation

The basic chemical property underlying the cation transport properties of carboxylic ionophores is their ability to form lipophilic,
zwitterionic cation inclusion complexes. In more polar solvents,
such as water or methanol, the coordination process consists of the
ionophore garlanding itself around the cation and replacing the solvent molecules in its inner coordination sphere with the liganding
heteroatoms of the ionophore backbone [5]. The overall complexation
reaction presumably consists of a series of consecutive substitution
steps as each solvent molecule in the primary solvation sphere is
replaced. This can be inferred from the large magnitude of the
energy of total desolvation of cations, e.g., 92 and 75 kcal for Na^+
and K^+, respectively, in methanol [6]. If the interchange between
the solvent of the inner sphere and the ionophore backbone were not
smoothly coordinated, i.e., the complexation reaction depended on
extensive spontaneous cation desolvation, the energy of activation
of such a step would render its rate extremely slow and its temperature coefficient extremely large.

The limited data available indicate that the ion complexation-
decomplexation reactions of carboxylic ionophores in water and methanol probably proceed by *dissociative interchange* [7]. In this mechanism a solvent molecule leaves the inner coordination sphere just
prior to cation-ligand bond formation. The forward rate constants
for Mn^{2+}, Mg^{2+}, and Ni^{2+} complexation by lasalocid and A23187 in
water and methanol are consistent with the known rates of exchange
of these solvents into and out of the cation coordination sphere.
Since the rate constants approach the diffusional limit, the conformational changes accompanying complex formation must be extremely
rapid [8-11].

The ultimate stability of the inclusion complex is a function
of the ability of the ionophore to prevent the cation from reacting
with the bulk solvent. For the reaction:

$$M^+ + I^- \rightleftharpoons M^+I^- \tag{1}$$

where M^+ is a monovalent cation and I^- is an ionized carboxylic ionophore, complex stability can be expressed thermodynamically as the
sum of several free energy terms:

$$\Delta G_{complex} = \Delta G_{M^+ desolv} + \Delta G_{I^- desolv} + \Delta G_{I^- conf} - \Delta G_{lig} \qquad (2)$$

where $\Delta G_{M^+ desolv}$ is the net energy required to transfer the cation
from the bulk solvent to the liganding cavity of the ionophore,
$\Delta G_{I^- desolv}$ is the energy necessary to desolvate the liganding heteroatoms of the ionophore backbone in order to make way for the cation
[12,13], $\Delta G_{I^- conf}$ is the energy required to convert the uncomplexed
conformation of the ionophore to its complexed conformation, and
ΔG_{lig} is the energy of interaction between the desolvated cation and
the conformationally primed ionophore [2,14,15]. The magnitudes of
these terms for a given ionophore in a given environment determine
its ability to bind various cationic species and to discriminate
between them.

The conformational options available to the carbon skeleton
supporting the liganding heteroatoms determine the geometry of the
ionophore inclusion cavity which, in turn, determines the effectiveness of ligand focusing about the cation, i.e., the magnitude of
ΔG_{lig}. In the case of the macrocyclic neutral ionophores, the
covalently locked, cyclic conformation intrinsic to the molecule is
predominant in determining ligand focusing [16]. Consequently,
cation complexation is accompanied by relatively small changes in
$\Delta G_{I^- conf}$ [17]. However, the absence of covalent head-to-tail linkage
in the carboxylic ionophores permits greater backbone flexibility and
more profound conformational changes accompanying ion complexation
[18,19].

2.2. Architecture of the Ionophore-Ion Inclusion Complex

For monovalent ions the binding cavity is generally formed from the
liganding atoms of a single ionophore in a 1:1 complex and for divalent cations the liganding cavity is formed by two ionophore molecules in a 1:2 complex. In both cases the complexes are electrically

neutral zwitterions because of the ionized carboxyl group. An exception to the 1:2 rule for divalent ion complexes is ionomycin which
contains an additional ionizable functionality, a β-diketone, and
thus is able to form electrically neutral 1:1 complexes with divalent
ions [20,21].

The Ag^+ complex of monensin serves to illustrate the basic
architectural attributes of a 1:1 ionophore-ion complex (Fig. 2).
The monensin anion is wrapped around the Ag^+ ion which is situated
in the middle of an approximately spherical cavity coordinated to
six oxygen atoms; two hydroxyl groups, O(4)H and O(10)H, and four
ether oxygens, O(6), O(7), O(8), and O(9) [22-24]. The terminal
carboxylate is not a part of the liganding cavity because the shortest Ag^+-O^- (carboxyl) distance is 3.84 Å. The cyclic conformation
of the ionophore is stabilized by two head-to-tail intramolecular
hydrogen bonds between O(10)H-O(1) and O(11)H-O(2). In this complex
the Ag^+ cation is completely shielded from bulk solvent by alkyl
substituents which render the complex lipophilic.

The Ba^{2+} complex of lasalocid is an example of a 2:1 carboxylic
ionophore-cation complex (Fig. 3) [22,25]. Two ionophore molecules,
A and B, orient around the Ba^{2+} ion, each with different coordination

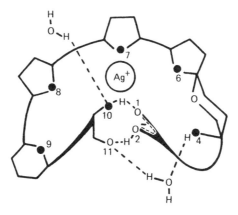

FIG. 2. Schematic representation of the 1:1 Ag^+/monensin complex.
The *intra*molecular head-to-tail hydrogen bonds and the *inter*molecular
hydrogen bonds to an incorporated water molecule are indicated by
dashed lines. The oxygen atoms comprising the liganding cavity are
filled in. (Adapted from Ref. 22.)

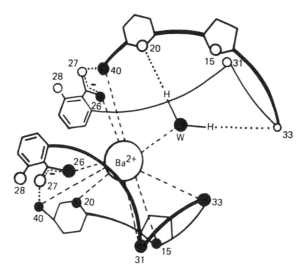

FIG. 3. Schematic view of the 1:2 Ba^{2+}/lasalocid complex. The
*intra*molecular head-to-tail hydrogen bonds and the *inter*molecular
hydrogen bonds to an incorporated water molecule are represented by
dotted lines. Oxygen atoms forming the liganding cavity are filled
in. Ion-dipole bonds are indicated by dashed lines. The carbon
backbone of each of the ionophore molecules is represented by curved
lines, heavier nearer to the viewer and thinner farther away.
(Adapted from Ref. 22.)

geometries with respect to the cation. The Ba^{2+} ion is ninefold
coordinated, lying somewhat closer to one of the lasalocid anions
than the other. The closer of the two anions, molecule A, provides
six coordinating oxygen atoms while the more distant anion, molecule
B, provides only two coordinating oxygens. The ninth ligand is a
water molecule which bridges O(20) and O(33) of molecule B by means
of hydrogen bonds. The oxygen atoms of molecule A involved in
liganding are O(15), O(20), O(26), O(31), O(33), and O(40) and the
oxygen atoms of molecule B involved in liganding are O(26) and O(40).
No intermolecular hydrogen bonds lock the two ionophores together in
the dimeric complex; it is held together primarily by the enclosed
cation. If one refers to the aromatic methylsalicylate ring as the
head of the molecule and the tetrahydropyranyl ring as the tail, the
dimer can be described as head-to-tail. Both molecules in the dimeric

complex adopt similar cyclic conformations stabilized by intramolecular hydrogen bonds: the head-to-tail connection from the C(25) carboxyl to the tertiary O(40)H hydroxyl, the hydrogen bond between the O(25) carboxylate and the secondary O(31)H hydroxyl and the salicylate-type hydrogen bond between the phenolic O(28)H hydroxyl and the C(25) carboxyl group [26].

Not all carboxylic ionophore complexes are neutral zwitterions. We reported that lasalocid forms strong complexes with La^{3+} and Th^{4+} [27] and it is not likely that enough ionophore molecules could crowd about the cation to form a neutral complex. Chen and Springer [28] and Grandjeau and Laszlo [29] have reported that lasalocid forms complexes of varying stoichiometry with lanthanides.

Even though lasalocid tends to form electroneutral complexes with both mono- and divalent cations, comparison of the conduction of current and the rates at which $^{86}Rb^+$ or $^{45}Ca^{2+}$ are transported across lipid bilayers indicates that approximately one out of each thousand transport events carries current [30]. This establishes the existence of rarer, nonelectrically neutral species of lasalocid complex which can transport Rb^+ and Ca^{2+} as electrically charged complexes.

Not all ionophore complexes are true inclusion complexes. Large cations, such as tetraalkylammonium, can form ion pair complexes with the terminal carboxyl of lasalocid without the molecule ever folding about the cation [31]. Lanthanides have a tendency to form analogous outer complexes [28]. To the extent that the cations of such noninclusion complexes are not well shielded from the environment they would not be good candidates for transporting cations across lipid barriers. Conversely, demonstration of complex formation per se does not mean that a given ionophore will transport a given cation efficiently.

2.3. Molecular Attributes of the Carboxylic Ionophores

In order to form stable, lipophilic ion inclusion complexes and also function effectively as mobile carriers in biological membranes, the

ionophores must possess certain structural characteristics: (1) The
molecule must contain both polar and nonpolar groups. The polar
groups function as ligands which replace the solvent molecules in
the primary solvation sphere of the complexed ion. The most stable
ionophore-alkali ion inclusion complexes have a coordination number
of 6, while the large alkaline earth cations prefer a coordination
number of 8 [32]. (2) The molecule must be able to assume a stable
conformation which directs the polar liganding moieties into a central cavity suitable for surrounding a cation [33]. High ion complexation affinities and selectivities are achieved by locking the
coordination sites into a rigid arrangement around the cavity. The
cation that fits most snugly into this cavity is the preferred complexation partner. Rigidity is imparted to the complex by intramolecular hydrogen bonding, particularly the head-to-tail hydrogen
bond characteristic of the carboxylic ionophores, by extended spirane
systems and by substituents which limit rotation about key bonds.
The complexing conformation must also result in the alkyl substituents on the ionophore backbone being directed outward to insulate
the polar internal liganding cavity while crossing an apolar membrane
interior. (3) Ion complexation-decomplexation reactions must proceed
at a sufficiently rapid rate. This is only possible when the ionophore is flexible enough to allow a stepwise rather than a concerted
one-step substitution of liganding moieties for solvent molecules
[34]. Thus in order to be an effective ionophore, a compromise
between ion affinity and ease of displacement of the ion must be
achieved [5,35]. (4) The overall dimensions of the ionophore-cation
inclusion complex must be small enough to ensure adequate mobility
through a lipid barrier.

Owing to natural selection, the carboxylic ionophores produced
by microorganisms meet all the above criteria and are superbly adapted
to perform their carrier function. Although they do not display the
extreme selectivity characteristic of some of the natural and synthetic neutral ionophores, their built-in conformational options
allow the ion complexation and decomplexation steps of the membrane

transport process to be quite rapid thereby facilitating exception-
ally high membrane turnover numbers.

3. TECHNIQUES FOR MEASURING EQUILIBRIUM IONOPHORE-ION AFFINITIES

3.1. Single-Phase Test Systems

The formation of ion inclusion complexes by carboxylic ionophores
in homogenous solvent systems has been studied by a number of techniques including fluorescence, circular dichroism (CD), nuclear
magnetic resonance (NMR), conductimetry, and potentiometry. The
application of the spectral methods in studying complexation is
based primarily on the existence of differences between the conformational and stereoelectronic characteristics of the free molecules
and their cation complexes. Cation binding is measured potentiometrically by means of ion-specific or pH glass electrodes [36,37].
In the case of carboxylic ionophores, the pK_a of the terminal carboxyl group is lowered when complexable cations are present, the
magnitude of the shift reflecting the extent of cation binding.
Conductimetric determinations of stability constants involve the
measurement of bulk electroconductivity as a function of ionophore
concentration [38]. As ionophore is added to a salt solution in
polar media where the cation is not ion paired, the formation of
cation inclusion complexes lowers the measured electroconductivity.
In low-polarity media the salt is generally not dissociated and ionophore complexation may cause an *increase* rather than a *decrease* in
electroconductivity. Table 1 presents a summary of single-phase
complexation data [39].

A commonly used technique for determining complex K_D values
in single phases is optical spectroscopy. The determination of the
K_D values for the monovalent ion complexes of salinomycin by CD is
a good example of the utilization of spectroscopy to determine complex stabilities in homogenous solvent systems [42]. In this case,
the CD spectrum consists of a single positive peak centered at 290 nm

TABLE 1

Single-Phase Stability Constants

Ionophore	Ion selectivity	Method	Ref.
Monensin	$Na^+>K^+>Tl^+>Rb^+>Cs^+$	Fluorometric titration, Tl^+ displacement in MeOH	40
	$Na^+>K^+$	Potentiometric titration in MeOH	41
	$Ag^+,Na^+>K^+>Rb^+>Cs^+>Li^+>NH_4^+$	Potentiometric titration in MeOH, competitive H^+ and M^+ binding	39
Nigericin	$K^+>Na^+$	Potentiometric titration in MeOH	41
Salinomycin	$K^+>Na^+$ $Z>79.6$ $Na^+>K^+$ $Z<79.6$	Spectroscopic titration	42
Ionomycin	Ca^{2+}, Mg^{2+}, K^+, Na^+	Spectroscopic titration (UV) in $CHCl_3$	21
Grisorixin	$Ag^+>Tl^+>K^+>Rb^+>Na^+>Cs^+>Li^+$	Conductometric and potentiometric titrations in MeOH	43
A23187	$Mn^{2+}>Ca^{2+}>Mg^{2+}>Sr^{2+}>Ba^{2+}$	Spectroscopic titration (UV) in water	44
	$La^{3+}>0$	Spectroscopic titration (UV) in ethanol	44
	$NH_4^+>0$	Fluorimetric titration in MeOH	45
X-537A	$Pr^{3+}>Ba^{2+}>Sr^{2+}>Ca^{2+}>Mn^{2+}>$ $Ni^{2+}>Mg^{2+}>K^+>Rb^+>Cs^+>Na^+>Li^+$	Fluorescent and CD titrations in MeOH	46
	$Ba^{2+}>Mg^{2+}>Ca^{2+}>Sr^{2+}>Tl^+>$ $Na^+>K^+>Rb^+>Cs^+$	Fluorometric titration Tl^+ displacement in MeOH	40
	$K^+>Na^+$	Spectroscopic titration (CD) solvent series	31

arising from the $n \to \pi^*$ transition of the C-11 [see Fig. 1(E)] ketone
group. A second $n \to \pi^*$ transition arising from the terminal carboxyl
group could not be used because it falls outside of the transparency
window of several of the solvents used in the study. The magnitude
of the 290 nm CD peak is an extremely sensitive function of the ionization and complexation state of the ionophore (Fig. 4). Titration
of the anionic form of the ionophore with a monovalent cation caused
a graded increase in the peak intensity $[\theta]$. The change in the CD
spectrum with added cation is due principally to a change in the
ionophore conformation as it engulfs the cation (see Sec. 6). However, it is important to note that the cation itself is a significant
vicinal moiety, which by virtue of its charge, polarizability, and
location with respect to the chromophore can modify the CD spectrum
[47].

Analysis of the data obtained from the titration requires a
mathematical analysis which begins with the equation for the dissociation constant of the binding reaction [48]. In the case of salinomycin, the anionic form of the ionophore binds with a single mono-

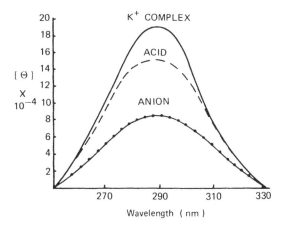

FIG. 4. CD spectrum of salinomycin in methylene chloride as a function of ionization and complexation state. The anionic form was
generated by the addition of excess tri-n-butylamine and the K$^+$
complex by the addition of excess KSCN.

valent cation according to Eq. (1). The equilibrium dissociation
constant K_D for this reaction is defined by Eq. (3):

$$K_D = \frac{[M^+][I^-]}{[M^+I^-]} \tag{3}$$

This equilibrium is such that the fraction of ligands bound is determined by $[M^+]$ and $[I^-]$. Thus, defining x as the fraction of ionophore
complexed, it is clear that

$$x = \frac{[I^-]_{bound}}{[I^-]_{total}} = \frac{[M^+I^-]}{[I^-] + [M^+I^-]} \tag{4}$$

The value of x is determined directly from the CD spectrum. Alliquots
of M^+ are added to a known concentration of I^-. The increase in the
absolute value of $[\theta]$ is indicative of the change in the total fraction of ion inclusion; thus x for a given concentration of cation is
equal to

$$x = \frac{[\theta]_{obs} - [\theta]_{anion}}{[\theta]_{sat} - [\theta]_{total}} = \frac{\Delta[\theta]_{obs}}{\Delta[\theta]_{total}} \tag{5}$$

where $[\theta]_{obs}$ is the observed molecular ellipticity for a given $[M^+]$,
$[\theta]_{anion}$ is the molecular ellipticity of the free anion, and $[\theta]_{sat}$
is the molecular ellipticity of the solution saturated with M^+.
Combining Eqs. (3), (4), and (5) yields:

$$x = \frac{[M^+]}{[M^+] + K_D} = \frac{\Delta[\theta]_{obs}}{\Delta[\theta]_{total}} \tag{6}$$

This equation for determining K_D is the familiar Langmuir saturation
isotherm. It can also be expressed as

$$\frac{1}{x} = \frac{K_D}{[M^+]_{free}} + 1 \tag{7}$$

which indicates that a plot of $1/x$ vs. $1/[M^+]_{free}$ yields a straight
line with a slope of K_D. Figure 5 is a double-reciprocal plot of
$1/x$ vs. $1/[K^+]$ for salinomycin.

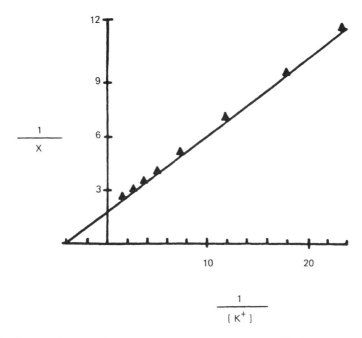

FIG. 5. Double-reciprocal plot of $1/x$ fs. $1/[K^+]$ [see Eq. (7)] for salinomycin anion in methanol. K^+ was added to the methanol solution of salinomycin anion (10^{-4} M) as the thiocyanate salt. The anion of salinomycin was generated by the addition of 3.5 eq tri-n-butylamine.

The terms can be rearranged to give the Scatchard equation:

$$\frac{x}{[M^+]_{free}} = \frac{n}{K_D} - \frac{x}{K_D} \tag{8}$$

where $n = 1$ for a single binding site. Eq. (8) indicates a plot of $x/[M^+]_{free}$ vs. x yields a straight line with a slope of $-1/K_D$ and an x intercept of n/K_D. This equation provides a means of checking the stoichiometry of complexation.

3.2. Two-Phase Partition System

High lipophilicity is a physical characteristic of the ion inclusion complexes of all of the naturally occurring carboxylic ionophores.

In a two-phase system consisting of an aqueous and an organic phase, the ability of the ionophore to increase the distribution of various cations into the organic phase has been utilized to determine the stability of ionophore-cation complexes [39,40]. In this method, a buffered aqueous solution of the test cation is equilibrated with a solution of the ionophore in an immiscible organic solvent by shaking. The amount of cation transferred to the organic phase is monitored by utilizing radiolabeled cation or by atomic absorption. By following the amount of cation transferred to the organic phase as a function of ionophore concentration in the organic phase, cation concentration and pH in the aqueous phase, the stoichiometry and stability of the overall phase transfer reaction can be determined.

Several important factors must be considered when evaluating extraction equilibrium constants [3,50,51]. First, the total cation extracted must be corrected by subtracting the amount of cation extracted into the organic phase in the absence of ionophore. When an aqueous salt solution is shaken against an organic solvent, a small fraction of the salt is extracted into the organic phase. The extent of extraction depends on the nature of the cation and its counterion, their concentration in the aqueous phase, and the composition of the organic phase. The degree of cation-anion pair extraction also increases with increasing aqueous concentration of the cation, increasing solvent polarity and increasing water content in the organic phase. Second, equilibrium extraction constants are highly dependent on the solvent composition and polarity of the organic phase. Haynes and Pressman suggested that this is due to differences in the degree of ligand solvation and to changes induced in ionophore conformation [13].

Extraction by ionophores is described for the monovalent case according to the equation:

$$[HI]_{org} + [M^+]_{aq} \xrightarrow{K_{ext}} [M^+I^-]_{org} + [H^+]_{aq} \qquad (9)$$

and at a pH held constant by buffering:

$$K_{ext} = \frac{[M^+]_{aq}[I^-]_{org}}{[M^+I^-]_{org}} \tag{10}$$

where $[HI]_{org}$ is the concentration of the carboxylic ionophore added to the organic phase, $[M^+]_{aq}$ is the concentration of the test cation added to the aqueous phase, and K_{ext} is the equilibrium extraction constant. For divalent cation binding Eq. (9) becomes:

$$2[HI]_{org} + [M^{2+}]_{aq} \xrightarrow{\quad K_{ext} \quad} [M^{2+}I_2^-]_{org} + 2[H^+]_{org} \tag{11}$$

and

$$K_{ext} = \frac{[M^{2+}]_{aq}[I^-]_{org}^2}{[M^{2+}I_2^-]_{org}} \tag{12}$$

The pH of the aqueous phase was routinely kept around 10 and it has been assumed that because this pH is substantially above reported pK_a values the ionophore exists as the free anion. Taylor et al. [7] point out that this assumption is probably not valid because A23187 remains in the organic phase as the free acid even at aqueous phase pH values above 10.

The equilibrium extraction constant and the stoichiometry of the ionophore-ion complex can be calculated from a double-reciprocal plot of the saturation of the ionophore with the test cation (Fig. 5). The linearity of plots of the reciprocal of the aqueous concentration of cation $[M^+]_{aq}$ against the reciprocal of the organic concentration of cation $[M^+I^-]_{org}$ indicates that saturation of the carboxylic ionophore follows an ideal Langmuir isotherm [49]. The intercept on the ordinate axis establishes the ionophore-ion complex stoichiometry whereas K_{ext} can be determined from the slope of the line. Representative extraction constant data are presented in Table 2. The ionic strengths in the aqueous phases used to calculate the extraction constants often vary from investigator to investigator. The

TABLE 2

Equilibrium Extraction Constants

Ionophore	Ion selectivity	Method	Ref.
Nigericin	$K^+>Rb^+>Na^+>Ca^+>>Li^+$	Toluene/n-butanol (7:3) pH = 10.00, ^{86}Rb displacement	49
	Norepinephrine>ethanolamine	Toluene/n-butanol (7:3) pH = 9.00, ^{3}H, ^{14}C labels	30
Monensin	$Na^+>K^+>Rb^+>Li^+>Cs$	Toluene/n-butanol (7:3) pH = 10.00, ^{86}Rb displacement	49
	Norepinephrine>epinephrine	Toluene/n-butanol (7:3) pH = 9.00, ^{3}H, ^{14}C labels	30
Salinomycin	$K^+>Na^+>Cs^+>Sr^{2+}>Ca^{2+}>Mg^+$	Toluene/n-butanol	52
	$Tl^+>K^+>Rb^+,\ Na^+>Cs^+>Li^+$	Toluene pH = 9.75, radioisotopes	53
Lonomycin	$K^+>NH_4^+\ Na^+>Rb^+>Li^+,\ Cs^+$ $Ba^{2+}>Ca^{2+}\ Mg^{2+}$	Toluene/n-butanol (7:3) pH = 8.30, ^{86}Rb displacement	54
Carriomycin	$K^+>NH_4^+>Rb^+>Na^+>Li^+,\ Cs^+$ $Ba^{2+}>Ca^{2+}>Mg^{2+}$	Toluene/n-butanol (7:3) pH = 8.30, ^{86}Rb displacement	54
Etheromycin	$K^+,\ NH_4^+\ Na^+>Li^+>Rb^+>Cs^+$ $Ba^{2+}>Ca^{2+}>Mg^{2+}$	Toluene/n-butanol (7:3) pH = 8.30, ^{86}Rb displacement	54

X-206	$K^+>Rb^+>Na^+>Cs^+>Li^+$	Toluene/n-butanol (7:3) pH = 10.00, ^{86}Rb displacement	49
Dianemycin	$Na^+>K^+>Cs^+$, $Rb^+>Li^+$	Toluene/n-butanol (7:3) pH = 10.00, ^{86}Rb displacement	49
	$Sr^{2+}>Ca^{2+}>Mg^{2+}$	Toluene/n-butanol (7:3) pH = 8.3, radioisotope	30
	Ethanolamine>norepi>epi	Toluene/n-butanol (7:3) pH = 9.00, ^3H, ^{14}C labels	49
Lasalocid	$Ba^{2+}>Sr^{2+}>Ca^{2+}>Mn^{2+}>Mg^{2+}$ $K^+>Na^+>Rb^+>Cs^+>Li^+$	Hexane spectroscopic	8
	$Ba^{2+}>Sr^{2+}>Ca^{2+}>Mg^{2+}$ $Cs^{2+}>K^+>Rb^+>Na^+>Li^+$	Toluene/n-butanol (7:3) pH = 10.00, ^{86}Rb displacement	49
Lysocellin	$Ba^{2+}>Mg^{2+}$, Mn^{2+}, $Sr^{2+}>Ca^{2+}$ $Rb^+>K^+>Na^+>Li^+$	Toluene/n-butanol (7:3) pH = 8.3, ^{45}Ca, ^{86}Rb displacement	55
6016	$Mg^{2+}>Mn^{2+}>Ca^{2+}>Sr^{2+}>Ba^{2+}$ NH_4^+, $K^+>Na^+$, $Rb^+>Li^+>Cs^+$	Toluene/n-butanol (7:3) pH = 8.3, ^{45}Ca, ^{86}Rb displacement	56
Ionomycin	$Ca^{2+}>Mg^{2+}>Sr^{2+}$, $Ba^{2+}>La^{3+}$	Toluene/n-butanol (7:3) pH = 10.00, ^{45}Ca displacement	57
A23187	$Mn^{2+}>Ca^{2+}$, $Mg^{2+}>Sr^{2+}$ $Li^+>Na^+>K^+=0$	Toluene/n-butanol (7:3) pH = 7.4, atomic absorption	44
	$Mg^{2+}>Ca^{2+}$	Toluene/n-butanol (7:3) pH = 9.8, atomic absorption	58
	$Fe^{2+}>Ca^{2+}$	Toluene/n-butanol (3:1)	59

effects of ionic strength on cation activities are expressed maxi-
mally in the 1-100 mM concentration range normally used in this
technique [7]. Variations in the ionic strength of the aqueous
phase could be responsible for the large variations in equilibrium
constants for the same ionophore obtained by different investigators.

4. CARBOXYLIC IONOPHORE-MEDIATED
ION TRANSPORT

4.1. Electrically Neutral, Ion-Exchange Diffusion Processes

A detailed model for carboxylic ionophore-mediated ion transport has
been proposed which is based on the classical mobile carrier mechanism
of Willbrandt and Rosenberg [60]. The transport cycle normally results in the electrically neutral, potential-insensitive exchange of
cation for cation and/or H^+ depending on the experimental conditions
[2]. Although electrogenic modes of transport have been reported for
lasalocid [30,61], monensin [62], nigericin [62], grisorixin [62],
and alborixin [62], they appear to be of minor importance in comparison with electroneutral transport.

The sequence of events taking place during carboxylic ionophore-
mediated electroneutral transport is depicted in Fig. 6 [2]. (a) At
physiological pH an appreciable fraction of the ionophore exists as
an amphipathic anion trapped at the membrane interface by the charged
carboxylate. (b) When the anionic ionophore encounters a complexable
cation M^+ at the membrane interface, complexation is presumably initiated by the formation of a solvent-separated ion pair between the
fully hydrated cation and the anionic ionophore. The ionophore now
wraps itself around the cation replacing the water molecules in the
primary solvation sphere with the liganding heteroatoms on its carbon
backbone. (c) The zwitterionic complex, its charge internally compen-
sated, is now capable of breaking away from the polar interface toward
the membrane interior. (d) The outward orientation of the lipophilic
alkyl groups on the ionophore backbone enables the complex to diffuse
passively across the apolar membrane interior to the opposite membrane

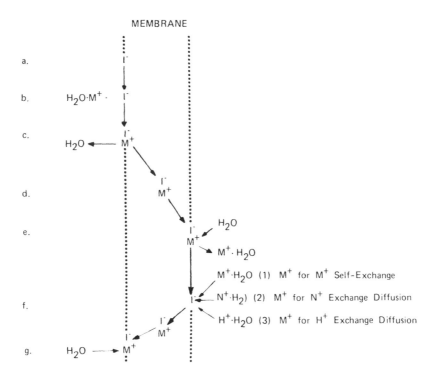

FIG. 6. Steps in the transport cycle catalyzed by carboxylic
ionophores.

interface. (e) The complexed cation is now accessible for resolvation following which it becomes energetically feasible for the cation
to leave the ionophore. (f) The ionophore again trapped at the interface can continue its transport cycle if (1) it complexes another M^+;
this would be termed self-exchange and can be monitored by using a
radiolabeled form of M^+ on one side of the membrane; (2) it complexes
another ion N^+ for an M^+ *for* N^+ exchange termed exchange-diffusion;
(3) its charged carboxylate abstracts an H^+ from the water phase resulting in an M^+ *for* H^+ exchange, an alternative form of exchange-
diffusion. (g) The neutral ionophore complex returns to its starting
interface where its cation is solvated and released to complete the
cycle.

The cation complexation process is more complicated for carboxylic ionophores transporting divalent cations. Initially, a 1:1 complex is formed at the membrane reaction plane by a reaction mechanism analogous to that for complexation of a monovalent cation. This charged 1:1 complex takes on a second ionophore to form a neutral 2:1 complex which is the vehicle for divalent cation transport across the membrane. The 2:1 Mg^{2+}- and Ca^{2+}-binding reactions for A23187 and lasalocid in phospholipid membranes have been described in great detail [56,63]. The ion selectivity of divalent ionophores arises primarily from the intrinsic affinities governing the formation of the initial 1:1 complex.

Predominance of the electrically neutral transport mode appears to be an essential requirement for tolerance of pharmacologically active concentrations of ionophores. Since membrane potentials play an important role in regulating excitable cells, e.g., nerve, muscle, and secretory cells, the neutral ionophores such as valinomycin and nonactin, which alter transmembrane electrical potentials, are capable of creating metabolic chaos in organelles, cells, and animals and are exceedingly toxic. Carboxylic ionophores, however, because of their electroneutral transport mode are well tolerated by most biological systems; hence secondary manifestations of their perturbations of ion gradients can be expressed as nonlethal changes in metabolism.

4.2. Three-Phase Transport Systems

Because the sequence of kinetic events making up the transport process is considerably more complicated than the single on/off rate relationships that determine equilibrium K_D values, transport data derived from dynamic test systems predict more precisely the behavior of the ionophore toward biological membranes. Owing to the low polarity of the organic layer, bulk multiphase liquid systems have permeability properties analogous to those of biomembranes and at the same time are amenable to quantitative study. Such model systems are not complicated by endogenous membrane carrier systems which may mask the

basic ion-transporting activity of the ionophore. Three dynamic
systems which have been used extensively are the vertically stacked
bulk phase system (Fig. 7) [27,64], the U tube (Fig. 8A) [65], and
the Pressman cell (Fig. 9) [66]. Each of these systems consists of
two aqueous phases separated by a water-immiscible organic phase.
The aqueous phases represent the extracellular and intracellular
phases and the organic phase barrier represents the apolar interior
of the membrane bilayer. Ionophore-mediated transport between aqueous
compartments is detected by utilizing radioisotopes or by measuring
ion concentration with atomic absorption or flame photometry (see
Fig. 10). A summary of ion transport selectivities determined in
three-phase systems is given in Table 3.

The vertically stacked bulk transport system as utilized by
Pressman and de Guzman consists of a standard flat-bottomed Klett
tube (12 x 90 mm) containing a Teflon-clad magnetic flea driven by
a synchronous motor (600 rpm) with a small magnet attached at the
shaft [27,64]. The tubes, siliconized to ensure a flat miniscus
between phases, are immersed in a water bath and filled with 1 ml
of an aqueous solution containing 10 mM Tricine buffer (pH 8.0),
10 mM isotopically labeled test cation, and 30% sucrose for density.

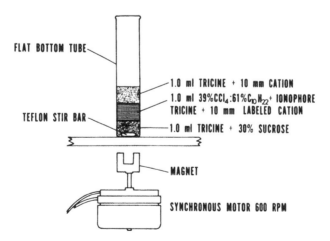

FIG. 7. Vertically stacked bulk phase transport system. (From
Ref. 27.)

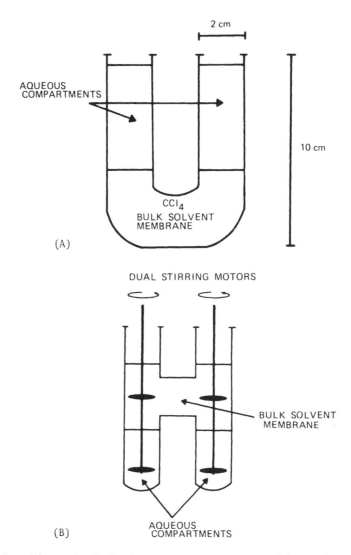

FIG. 8.   (A) U-tube bulk phase transport system; (B) H-tube variation.

Above this is layered 1 ml of an organic phase consisting of 39% CCl_4:
61% n-decane and 10^{-4} M test ionophore. The upper phase consists of
10 mM Tricine buffer (pH 8.0). The omission of sucrose in the third
phase allows it to float above the organic phase barrier. The transport reaction is started by turning on the stirring motor and monitored by counting small aliquots from the upper phase with a

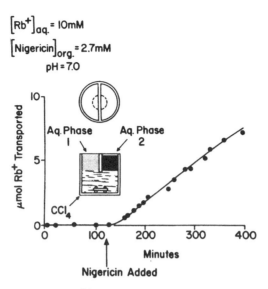

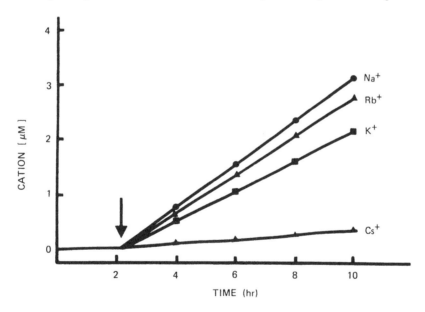

FIG. 9. Translocation of $^{86}Rb^+$ across bulk phase CCl_4 barrier of a Pressman cell by nigericin. The $^{86}Rb^+$ was added to aqueous phase 1 and its appearance in aqueous phase 2 monitored by measuring the radioactivity of aliquots. The Pressman cell [66] was the first bulk phase transport system used to monitor carboxylic ionophore transport.

FIG. 10. Time course for a typical bulk phase transport experiment. The data are for salinomycin- $(10^{-4}$ M) mediated alkali ion transport in a Pressman cell [52]. After the system was allowed to equilibrate for 2 hr, the ionophore was added to the organic phase (indicated by arrow). The transport of cation, in moles, was monitored by the appearance of radioactivity in the alkali ion-free aqueous compartment. (Adapted from Ref. 52.)

TABLE 3

Ion Transport Selectivities Measured in Three-Phase Systems

Ionophore	Transport selectivity	System	Ref.
Monensin	$Na^+>K^+>Rb^+>Li^+$	U tube; $H_2O/CHCl_3/H_2O$	65
Carriomycin	$Rb^+>K^+>Na^+>>Ca^{2+}$	Pressman cell; $H_2O/CCl_4/H_2O$	54
Dianemycin	$Na^+>K^+>Rb^+$	U tube; $H_2O/CCl_4/H_2O$	65
Ionomycin	Ca^{2+}	U tube; $H_2O/CCl_4/H_2O$	67
Lasalocid	Ca^{2+}, $Cs^+>Ba^{2+}>Sr^{2+}>Rb^+>Na^+$	Stacked three-phase; $H_2O/$ (39% CCl_4/61% decane)$/H_2O$	27 64
Br-lasalocid	Ca^{2+}, $Cs^+>Ba^{2+}>Sr^{2+}>Rb^+>Na^+$	Stacked three-phase; $H_2O/$ (39% CCl_4/61% decane)$/H_2O$	64
Dihydrolasalocid	$Ca^{2+}>Ba^{2+}>Sr^{2+}>Cs^+>Rb^+>Na^+$	Stacked three-phase; $H_2O/$ (39% CCl_4/61% decane)$/H_2O$	64
Etheromycin	K^+, $Rb^+>Na^+$, $Cs^+>Ba^{2+}$	Pressman cell; $H_2O/CCl_4/H_2O$	54
Lonomycin	K^+, $Rb^+>Na^+>Ca^{2+}$	Pressman cell; $H_2O/CCl_4/H_2O$	54
Lysocellin	Na^+, $Rb^+>K^+$, $Ca^{2+}>Ba^{2+}$	Pressman cell; $H_2O/CCl_4/H_2O$	55
Nigericin	K^+, $Rb^+>Na^+>>Ca^{2+}$	U tube; $H_2O/CHCl_3/H_2O$	65
Salinomycin	Rb^+, $Na^+>K^+>>Cs^+$, Mg^{2+}, Ca^{2+}, Sr^{2+}	Pressman cell; $H_2O/CCl_4/H_2O$	52
X-14547A	$Ca^{2+}>Rb^+$	U tube; $H_2O/CHCl_3/H_2O$	68

scintillation counter. The system yields reproducible data between replicates of the same or different runs. It lends itself to the simultaneous running of several reaction vessels and indeed we routinely use an apparatus consisting of 20 vessels.

The U tube apparatus is filled with a chloroform solution of the test ionophore to just above the bend. Each side is then filled with an equal volume of aqueous solution [65]. One aqueous compartment contains the test cation. Transport is monitored by the appearance of the test cation in the opposite aqueous compartment. Some care is necessary when filling the tube in order to avoid mixing of the phases. The transport rate in this system is influenced by the tube geometry and mixing efficiency, which precludes comparison of data from replicate vessels.

An interesting variation on the conventional U tube is the H tube [Fig. 8(B)] [69]. In this system, the selectively permeable membrane is modeled by a lighter-than-water 1-hexanol solution of the test ionophore. The two aqueous compartments are contained in the legs of the H. One of the compartments contains the test cation M^+ while the other compartment contains a complexable ion to counter the transport of M^+. In each of the legs there is a stirring shaft with a blade in both the aqueous and organic layers. This system is also influenced by the unique geometry of each H tube employed.

Based on the assumption of electroneutrality in the U-tube system, Ashton and Steinrauf demonstrated that for carboxylic ionophore-mediated ion exchange the equilibrium concentration of ions in the aqueous compartments can be calculated knowing only the initial ion concentrations [65]. At equilibrium the chemical potentials of all species in the system can be described by the Gibbs-Duhem equation:

$$\sum_i n_i d\mu_i^+ = 0 \qquad\qquad\qquad (13)$$

where n_i is the population of ion i and μ_i^+ is the partial molar free energy of ion i. When the barrier between the two aqueous compartments is sharply defined and of negligible mass, only the ion concentrations in the two aqueous compartments need be considered. The differential in Eq. (13), $d\mu_i^+$ simplifies the difference in partial molar free energy between the two compartments:

$$d\mu_i^+ = (\mu_i^+)_L - (\mu_i^+)_R \tag{14}$$

In a system composed of two transportable cations 1 and 2, under the condition of electroneutrality Eq. (14) becomes

$$n_1[(\mu_1^+)_L - (\mu_1^+)_R] + n_2[(\mu_2^+)_L - (\mu_2^+)_R] = 0 \tag{15}$$

When the two ions are transported countercurrently, the ion populations in the two compartments remain constant:

$$(n_1)_L = -(n_1)_R \qquad \text{and} \qquad (n_2)_L = -(n_2)_R \tag{16}$$

Combining Eqs. (15) and (16) gives the expression

$$[(\mu_1^+)_L - (\mu_1^+)_R] - [(\mu_2^+)_L - (\mu_2^+)_R] = 0 \tag{17}$$

Converting the partial molar free energies in Eq. (17) to activities, and assuming $(\mu_1^+)_L = (\mu_1^+)_R$ and $(\mu_2^+)_R = (\mu_2^+)_L$, the equilibrium distribution of complexable cations is given by the expression

$$\frac{(a_2)_L}{(a_1)_L} = \frac{(a_2)_R}{(a_2)_R} \tag{18}$$

The Pressman cell consists of a glass vessel with a septum across the top [66]. The cell is filled with a heavy, low-polarity solvent, e.g., carbon tetrachloride, so as to isolate the two upper compartments which are then filled with the appropriate aqueous buffer. A complexable radioisotope is then added to one upper compartment and the opposite compartment is monitored for the self-exchange of the nuclide across the organic barrier. The vessel is stirred from below by a magnetically driven disc. Early experiments in which a 45-V field was applied across the two aqueous compartments established the fact that ion transport via carboxylic ionophores is for the most part electrically silent, i.e., the rate of ion exchange across the liquid membrane is independent of the magnitude of the electrical field [30].

4.3. Membrane Bilayers and Vesicles

Membrane bilayers and membrane vesicles are still better models of
the biological membrane. Such a system provides a means of assaying
the transport properties of a given carboxylic ionophore in a system
in which the energetic and chemical interconversions associated with
carboxylic ionophore-mediated ion transport can be examined in detail
(c.f. Table 4). Investigating the transport mechanism in a cell-free
system offers considerable advantage over studying the transport
mechanism in intact cells. The carboxylic ionophores can be studied
in a functional state dissociated from complications arising from
unknown internal compartmentalization and intracellular metabolism.
With intact cells the ion-translocating activity of the ionophore
can frequently be attenuated or even completely masked by the activity
of endogenous carriers such as Na^+/K^+-ATPase. In addition, by altering factors such as the polarity profile across the membrane and the
composition of the membrane, an estimation of the effects of these
variables on the transport process can be made.

   Planar membrane bilayers are conventionally formed from films
of a solution of a lipid in a nonpolar solvent which is pipetted so
as to cover a small orifice in a wall separating two aqueous compartments (for reviews on this membrane system, see Refs. 70 and 71).
As the lipid film thins, the interference colors on the membrane dis-
appear when observed with a low-power microscope. This phenomenon
gives rise to the term "black" lipid membrane. The thinning continues
until a uniform bilayer of 50-60 Å is obtained. Although the first
basic studies of the ability of ionophores to act as free carriers in
bimolecular lipid membranes were carried out in this system [72,73],
it is primarily limited to studying neutral ionophores whose effects
on the membrane may be monitored electrometrically. It is seldom
useful with the electrically silent carboxylic ionophores.

   A number of transport selectivities have been determined in
liposomes (Table 4). Although multilamellar vesicles have been used
for determining transport sequences, the system of choice has been
small, unilamellar vesicles (SUVs). SUVs are water-filled micro-

TABLE 4

Carboxylic Ionophore Transport Selectivities Determined
in Black Lipid Membranes and Liposomes

Ionophore	Ion selectivity (transport mode)	System; composition	Ref.
A204	$Fe^{2+}>0$	MLV; lecithin, cholesterol, sterylamine	54
A23187	$Mn^{2+}>Ca^{2+}>Sr^{2+}>Mg^{2+}$, Ba^{2+}	MLV LUV; phosphatidyl-choline, diacetyl phosphate and cholesterol	74
	$Fe^{2+}>0$	MLV; lecithin, cholesterol, sterylamine	70
	$Pr^{3+}>0$	SUV; dipalmitoyl phosphatidylcholine	75
	Ca^{2+}, Mg^{2+} (neutral)	BLM; phosphatidylcholine, cholesterol	76
Alborixin	K^+, Na^+ (electrogenic)	BLM; glyceryl monooleate	62
Dianemycin	$Fe^{2+}>0$	MLV; lecithin, cholesterol, sterylamine	59
	Na^+, K^+, Rb^+, Li^+, Cs^+	MLV; lecithin, diacetyl hydrogen phosphate	77
Lysocellin	$Fe^{2+}>0$	MLV; lecithin, cholesterol, sterylamine	59
Monensin	K^+, Na^+ (electrogenic)	BLM; glyceryl monooleate	62
	$Fe^{2+}>0$	MLV; lecithin, cholesterol, sterylamine	59
Nigericin	K^+, Rb^+, Na^+, Cs^+, Li^+	MLV; lecithin, diacetyl hydrogen phosphate	77
	K^+ (electrogenic)	MLV: lecithin, diacetyl phosphate	78
	K^+, Na^+	BLM; lecithin, cholesterol	79
X-206	$Fe^{2+}>0$	MLV; lecithin, cholesterol sterylamine	59
Lasalocid (X-537A)	$Mn^{2+}>0$	SUV; egg lecithin	80
	$Fe^{2+}>0$	MLV; lecithin, cholesterol, sterylamine	
	$Pr^{3+}>0$	SUV; egg lecithin	81
	$H^+>Cs^+>Rb^+>K^+>Na^+>Li^+$ $Ba^{2+}>Ca^{2+}>Mn^{2+}>Sr^{2+}>Mg^{2+}$	SUV; phospatidylinositide, cholesterol	82
	$Mg^{2+}>Ca^{2+}>Sr^{2+}>Ba^{2+}$ (electrogenic)	BLM; phosphatidylinositide, cholesterol	30

Abbreviations: BLM, black lipid membrane; LUV, large unilamellar vesicle;
MLV, multilamellar vesicle; SUV, small unilamellar vesicle.

vesicles of approximately 250 Å in diameter surrounded by a single lipid bilayer. Transfer of ions and/or amines from the internal aqueous solution to the external phase and vice versa has been monitored using isotope flux [77], CD [80], fluorescence [76], and NMR [80].

Liposomes are particularly amenable to detailed spectroscopic studies and provide a versatile system for studying in detail the mechanism of ionophore-mediated transport across a biological membrane. A detailed analysis of the equilibria and kinetics for the partial reactions of the catalytic cycles of the Ca^{2+} ionophores lasalocid and A23187 has been worked out in liposomes. The intrinsic fluorescence of both ionophores was utilized in fluorescence lifetime, fluorescence stopped-flow, and conventional steady-state experiments.

The transport cycle is complicated in liposomes by the observation that both lasalocid and A23187 bind to the vesicle membrane. Direct interaction of both compounds with vesicle membranes has been detected as an enhancement of intrinsic fluorescence [12,63]. The hyperbolic nature noted for the ionophore-membrane titration curves suggests that a simple 1:1 stoichiometry obtains between the ionophore and a membrane component represented by the reaction:

$$[I] + [PL] \xrightleftharpoons{K_B} [I]_{PL}$$

where $[I]$ is the free ionophore concentration in the bulk phase, $[PL]$ is the concentration of the membrane phospholipid, K_B is the apparent membrane association constant, and $[I]_{PL}$ is the concentration of the membrane-bound ionophore. The binding constant K_B can be calculated from the equation:

$$K_B = \frac{[I]_{PL}}{[I]_{free}[PL]} \tag{19}$$

The total ionophore in the system $[I]_t$ is given by the expression

$$[I]_t = [I]_{free} + [I]_{PL} \tag{20}$$

By rearranging Eq. (20) and substituting for $[I]_{PL}$ in Eq. (19), K_B can be expressed as

$$K_B = \frac{[I]_t - [I]_{free}}{[I]_{free} - [PL]} \tag{21}$$

Defining α as $[I]_{free}/[I]_t$, Eq. (21) becomes:

$$K_B = \frac{1 - \alpha}{\alpha [PL]} \tag{22}$$

Solving for α in terms of K_B and $[PL]$ yields:

$$\alpha = [PL]\left(\frac{1}{1 + K_B}\right) = \frac{1}{1 + K_B[PL]} \tag{23}$$

Recalling the conservation equation for $[I]_t$, it is apparent that

$$\frac{[I]_{PL}}{[I]_t} = 1 - \alpha = 1 - \frac{1}{1 + K_B[PL]} \tag{24}$$

Combining Eqs. (23) and (24) gives the expression:

$$\frac{[I]_{PL}}{[I]_t} = 1 - \frac{1}{(1 - K_B)[PL]} \tag{25}$$

which simplifies to

$$\frac{[I]_{PL}}{[I]_t} = \frac{K_B[PL]}{1 + K_B[PL]} \tag{26}$$

where $[I]_{PL}/[I]_t$ is equal to the ratio of the change in ionophore fluorescence ΔFl induced by a particular concentration of membrane phospholipid to the maximum change in fluorescence ΔFl_{sat}. Thus K_B can be calculated directly from a plot of $\Delta Fl/\Delta Fl_{sat}$ vs. $[PL]$.

Utilizing the fluorescence enhancement which accompanies movement of the salicylate chromophore into the less polar membrane environment, Haynes and Pressman used Eq. (26) to calculate the binding constant of lasalocid to monolayer dimyristoylphosphatidylcholine (DMPC) vesicles to be 1×10^{-4} M^{-1} at a pH of 7.3 [12]. The total membrane-binding reaction is a sensitive function of vesicle membrane composition. Incorporation of dimyristoylphosphatidic acid (DMPA) or dimyristoylphosphatidylethanolamine (DMPE) into DMPC vesicles results in a decrease in ionophore binding.

Fluorescent lifetime experiments have shown the fluorophore of lasalocid to exist in two different environments when bound to the vesicle membrane, possibly reflecting two different modes of ionophore-phospholipid interaction [83]. The fluorescent lifetimes within the two different environments differ by a factor of approximately 2. The longer lived of the two species is calculated to compose between 11 and 20% of the total bound ionophore. Several explanations for the detection of two forms of membrane bound lasalocid are possible: (1) an inward vs. outward orientation of the fluorescent salicylate group; (2) perpendicular vs. parallel orientation of the ionophore with respect to the plane of the membrane; (3) deep vs. shallow insertion of the salicylate group with respect to the glycerol region of the membrane; (4) modes of binding resulting in lesser or greater environmental or ionophore mobility.

The pH of the extravesicular medium was found to have a profound effect on the observed fluorescence of the membrane-bound form of lasalocid [83]. The fluorescence of lasalocid bound to DMPC and DMPA vesicles increased with increasing pH. A 20-fold increase was seen in the fluorescence of lasalocid bound to DMPC vesicles by increasing the pH from 5.3 to 7.3. Under identical observation conditions the fluorescence of lasalocid bound to DMPA vesicles increased by a factor of 2. The pH dependence of the fluorescence signal of the membrane-bound ionophore is interpreted in terms of a protonation equilibrium:

$$I^-_{mem} + H^+ \longleftrightarrow IH_{mem} \qquad (27)$$

in which the fluorescence of the deprotonated form is much greater than that of the free acid. The observation that the apparent pK_a of lasalocid in the membrane environment is several units higher than the pK_a in aqueous solution was interpreted as indicating that the ionophore is bound in an environment of intermediate polarity within the membrane. Two explanations are possible for the decrease in maximum fluorescence with decreasing pH, one chemical and one environmental [83]. Protonation could either weaken or destroy the intramolecular hydrogen bonds of the molecule and thereby increase its ability to undergo conformational and vibrational transitions resulting in a decrease in the excited state lifetime. The environmental effect would involve differences in hydrocarbon chain packing, ionophore polar head group interaction, and water accessibility. Haynes et al. [83] concluded that the influence of the environmental factors would be insufficient to make an appreciable difference on the fluorescence of protonated lasalocid and that the chemical factors are paramount in reducing the fluoresence.

A similar membrane-binding reaction is observed for A23187 [63, 84,85]. Binding constants were calculated by following the enhancement of fluorescence of the benzoxazole ring of A23187 at increasing concentrations of vesicles. To determine the stoichiometry of binding to DMPC membranes, Kauffman et al. [85] plotted the titration data according to the equation:

$$\frac{[DMPC]_t}{[A]_b} = \frac{1}{K_b[A^-] + n} \tag{28}$$

where $[DMPC]_t$ is the total concentration of the phospholipid, $[A]_b$ is the bound A23187, $[A^-]$ is the concentration of the free anionic A23187, K_b is the apparent membrane-binding constant, and n is the number of phospholipid molecules per bound ionophore at saturation. The amount of ionophore bound was determined from Eq. (28). The plot of $[DMPC]_t / [A]_b$ vs. $1/[A^-]$ gives a straight line, and the y intercept n indicated that saturation of the membrane occurred at one molecule of A^- per seven to eight molecules of DMPC [85].

The fluorescence of membrane-bound A23187, like that of lasalocid, shows a pH dependence. Kolber and Haynes [63] reported the ratio of the maximum fluorescence ΔFl_{sat} at pH 7.1-8.7 to be approximately 0.2. The apparent K_B values were also found to be sensitive to pH with a K_B at pH 7.06 of 1.9×10^4 M^{-1} and at pH 8.7 of 0.7×10^4 M^{-1}. The dependence of K_B on $[H^+]$ shows that the neutral form of the ionophore binds better than the charged form. Kauffman et al. [85] found a similar pH dependence for A23187-binding affinities to DMPC vesicles with K_B for the protonated ionophore 10-fold higher than the K_B for the anionic form [85]. The overall difference in membrane affinity is consistent with the predicted higher solvation energy of the anionic A23187 compared with the protonated form in the aqueous phase. However, the binding data have not been corrected for the effects of surface charge on K_B. Presumably this factor could be significant in the case of anionic A23187 binding and may also be in part responsible for the higher affinity of the protonated form. The ability of the ionic strength of the media to mask the surface charge developed upon the binding of anionic A23187 will have to be fully evaluated before the effect of protonation state on K_B can be fully explained.

Fluorescence has also been used to monitor the complexation of alkali and alkaline earth cations by membrane-bound A23187 and lasalocid [26,63]. Cation complexation by the membrane bound form of ionophore proceeds according to the equation:

$$[M^{n+}] + [I^-]_{mem} \longleftrightarrow [MI]^{n-1}_{mem} \tag{29}$$

where n is the charge on the cation and $[I^-]_{mem}$ is the concentration of membrane-bound ionophore. Ion complexation by the membrane-bound form of the ionophore can be observed as a quenching of fluorescence. Thus increasing concentrations of cation cause a decrease in the fluorescence of membrane-bound ionophore according to the equation:

$$\frac{\Delta F1}{\Delta F1_{max}} = \frac{K_{app}[M^{n+}]}{1 + K_{app}[M^{n+}]} \tag{30}$$

where $[M^{n+}]$ is the aqueous divalent cation concentration and K_{app} is the apparent association constant for cation complexation. For divalent ions, the fluorescence quenching due to the formation of 1:2 (M^{2+}/ionophore) complexes are small compared with that of the 1:1 complexes. Table 5 gives values for the intrinsic 1:1 association complexes of A23187 with several divalent cations on DMPC membranes [63]. The cation selectivity sequence for membrane-bound A23187 is identical to the equilibrium selectivity calculated by two-phase

TABLE 5

K_{app} Values for Formation of 1:1 Divalent Ion Complexes by Membrane Bound A23187

Cation	Apparent complexation constant (K_{app})
Ca^{2+}	$1.56 \pm 0.8 \times 10^5$
Mg^{2+}	$6.58 \pm 0.4 \times 10^4$
Sr^{2+}	$8.02 \pm 1.2 \times 10^3$
Ba^{2+}	$4.83 \pm 1.7 \times 10^3$

Source: Ref. 63.

partition studies and the transport selectivities determined in bulk membrane systems. The agreement of the selectivities calculated in the multiphase systems with the rank order of selectivity calculated for 1:1 complex formation on DMPC vesicle membranes suggests that selectivity is determined during this first association.

The composition of the vesicle membrane has profound effects on the cation complexation activity of lasalocid [12,13]. When bound to pure DMPC vesicles, lasalocid binds Ca^{2+} with an apparent complexation constant $K_{app}(Ca^{2+})$ of 1.432×10^3 M^{-1} and K^+ with an apparent complexation constant $K_{app}(K^+)$ of 52.5 M^{-1}. Changing the composition of the vesicle membrane to 50% DMPA/50% DMPC causes a decrease in $K_{app}(Ca^{2+})$ to 40.1 M^{-1} and a decrease in $K_{app}(K^+)$ to 33.0 M^{-1}. Lasalocid bound to pure DMPA vesicles displays a $K_{app}(Ca^{2+})$ of 3×10^3 M^{-1} and a $K_{app}(K^+)$ of 15.7 M^{-1}.

The kinetics of K^+ for M^{2+} exchanges across the vesicular membranes by both lasalocid and A23187 have been studied using stopped-flow fluorimetry [63,83]. Analysis indicates that lasalocid is capable of transporting K^+ across the membrane in a 1:1 complex with a rate constant of 69.5 sec^{-1}. In the presence of divalent cations a slower process involving transport of M^{2+}-lasalocid complexes across the membrane is also observed. The dependence of the divalent ion transport rate on the total ionophore concentration indicates that the transported species is a neutral 1:2 complex. The lower limit for the rate constant for transport of the complex is 35 sec^{-1}. The divalent cation specificity of the overall reaction was shown to be $Mg^{2+} > Ca^{2+} > Sr^{2+} > Ba^{2+}$. The rates of the overall transport at low ionophore concentration are limited by the equilibrium constant for the formation of the neutral 1:2 complex from the $(1:1)^{+}$ complex. Ca^{2+} transport by A23187 compared with that of lasalocid is 67-fold faster. This difference in transport rates reflects the greater ease with which A23187 forms a 1:2 complex from a 1:1 complex at the membrane interface.

The apparent rate k_{app} of A23187-mediated Ca^{2+} transport is a sensitive function of membrane composition. Maximal values of k_{app} were observed for vesicles prepared from pure DMPC (k_{app} = 0.98 sec^{-1}), inclusion of 33% DMPE (k_{app} = 0.217 sec^{-1}), 31% DMPA (k_{app} = 0.057 sec^{-1}), or 32% dipalmitoyl PC (DPPC) (k_{app} = 0.019 sec^{-1}) resulted in lower values for k_{app}. The changes induced in k_{app} by changes in membrane lipid composition were attributed to changes in membrane fluidity [63]. Temperature dependence studies in DMPC membranes indicate, however, that k_{app} is virtually independent of membrane fluidity. A fit of k_{app} for the 1:2 Ca^{2+}-A23187 complex in DMPC vesicles to temperature is linear over a temperature range that includes the lipid phase transition temperature, indicating that dramatic changes in membrane fluidity do not affect the transport rate. Kolber and Haynes [63] offer two possible explanations for the lack of dependence on fluidity: (1) the ionophore fits into both lipid phases, crystalline and gel, equally well; (2) the ionophore disrupts the environment to such an extent that it is insensitive to gross organizational changes in the hydrophobic region of the membrane.

The lack of sensitivity of k_{app} to DMPC membrane fluidity suggests that differences in transport rate encountered upon inclusion of DMPC, DMPA, and DPPC in the vesicle membrane are due to alterations in ionophore-polar head group interactions. Incorporation of DMPA into the DMPC vesicle membrane results in a more tightly packed bilayer membrane due to the smaller area per PE head group. Thighter packing may exclude A23187 from the lipid polar head region thereby decreasing the amount of ionophore available for Ca^{2+} inclusion and transport. The drop in available A23187 is manifested in the decreased k_{app}. It has been speculated that formation of a ternary $A23187-Ca^{2+}-PA^{-}$ complex in DMPC/DMPA vesicles could serve to reduce the amount of A23187 available within the membrane and thereby account for the reduced k_{app} observed upon incorporation of DMPA [63]. Incorporation of DPPC (32%) into DMPC vesicles probably alters the architecture of the membrane interface. The polar head groups of DMPC and DPPC do not lie in the same plane due to the difference in acyl chain length. The decrease in k_{app} upon addition of pure DPPC to pure DMPC vesicles may arise from more extensive membrane binding of A23187 due to the greater surface area created at the interface by the biplanarity. The elucidation of the molecular details of these interactions between the ionophores and the membrane lipid molecules in its immediate vicinity remains an important problem in the understanding of the membrane transport mechanism of ionophores.

5. BIOLOGICAL TEST SYSTEMS

5.1. Introduction

Although a number of biological test systems have been used to investigate carboxylic ionophore-catalyzed transport across biomembranes, isolated erythrocytes and either rat liver or bovine heart mitochondria have been used most extensively owing to their ease of preparation and versatility. Table 6 summarizes in vitro transport data from various biological systems.

5.2. Erythrocytes

Erythrocytes provide an excellent system for evaluating the transport selectivity of the carboxylic ionophores. They are readily

TABLE 6

Cation Transport Selectivities in Biological Test Systems

Ionophore	Selectivity sequence	Method	Ref.
Monensin	$Na^+ > K^+ > Li^+ > Rb^+ > Cs^+$	Mitochondria and erythrocytes	77
	$Na^+, Li^+ > K^+ > Rb^+ > Cs^+$	Erythrocyte ghosts	86
Nigericin	$K^+ > Rb^+ > Na^+ > Cs^+ > Li^+$	Mitochondria and erythrocytes	77
	$K^+ > Na^+ > Rb^+ > Li^+ > Cs^+$	Erythrocyte ghosts	86
Salinomycin	$K^+, Rb^+ > Na^+ > Li^+, Cs^+$	Mitochondria	87
	$K^+ > Na^+$	Erythrocytes	18
Narasin	$Na^+ > K^+, > Li^+ > NH_4^+$ $Rb^+ > Cs^+ > Li^+ > NH_4^+$	Mitochondria	88
	$K^+ > Na^+$	Erythrocytes	18
X-206	$K^+ > Rb^+, Na^+ > Cs^+ > Li^+$	Erythrocyte ghosts	86
	$K^+ > Na^+$	Erythrocytes	18
A-204	$K^+ > Na$	Erythrocytes	18
A23187	$Ca^{2+} >> Mg^{2+}$	Mitochondria	89
	$Fe^{2+} > 0$	Erythrocytes	59
	$Li^+ > Na^+ > K^+$	Chloroplasts	90
	$K^+ > 0$	Mitochondria	91
	$Na^+ > 0$	Erythrocytes	92
Lasalocid	$Cs^+ > Rb^+ > K^+ > Na^+ > Li^+$	Mitochondria	93
	$Cs^+, Rb^+ > Na^+, K^+ > Li^+$	Erythrocyte ghosts	86
	$K^+ > Na^+$	Erythrocytes	18
	$Fe^{2+} > 0$	Erythrocytes	59
Lysocellin	Ca^{2+}, Mg^{2+}	Mitochondria	57
	$Na^+ > K^+$	Erythrocytes	18
Ionomycin	$Ca^{2+} >> Mg^{2+} >> K^+$	Mitochondria	21
Dianemycin	$Na^+, K^+, Rb^+, Li^+, Cs^+$	Mitochondria and erythrocytes	77
	$Li^+ > Na^+ > Rb^+, K^+ > Cs^+$	Erythrocyte ghosts	86
	$Na^+ \geq K^+$	Erythrocytes	18

available and relatively free of complex biological transducers which
might attenuate or mask the activity of the ionophore. In isotonic
media of low alkali ion content, the major effect of carboxylic ionophores on erythrocyte preparations is the exchange of internal alkali
or alkaline earth cations for environmental protons [94]. In media
enriched by metal salts, these compounds induce an exchange of internal for environmental cations. The cation exchange is accompanied
by transient changes in the environmental pH, the direction and magnitude of which depend on the relative positions of the internal and
external cation in the ion selectivity sequence of the ionophore
[18,77].

Figure 11 shows the effect of addition of the K^+ selective
ionophore nigericin on a suspension of human erythrocytes [18].
Because of the transmembrane gradients which are actively maintained

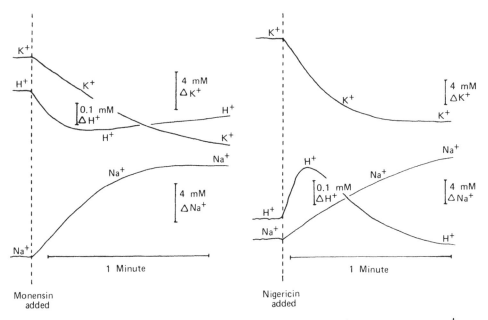

FIG. 11. Ionophore-mediated exchange of internal K^+ for external Na^+
and H^+ in a suspension of human erythrocytes (10% packed cell volume)
[18]. The changes in the activities of K^+, Na^+, and H^+ were monitored
by ion-specific electrodes in the extracellular medium. A downward
deflection indicates movement of ions from the cells into solution.
Monensin and nigericin (10 μM) were added at times indicated.

under physiological conditions (i.e., high intracellular K^+, low Na^+; low extracellular K^+, high Na^+;), the monovalent carboxylic iono- phores principally promote an exchange of internal K^+ for external Na^+. Following addition of the ionophore, there is an immediate egress of K^+ countered by a stoichiometric influx of H^+ and Na^+. As a consequence, the transmembrane K^+ gradient is quickly dissipated while the Na^+ activity gradient lingers. In order to facilitate the complete dissipation of the Na^+ gradient, the direction of nigericin- mediated H^+ flux reverses to balance the continued ingress of Na^+. Thus there is a bidirectional proton flux associated with nigericin- mediated K^+ for Na^+ exchange which is a function of the difference between net K^+ egress and net Na^+ ingress. With monensin, the protons move out of the cell with K^+ to counter the ingress of the more highly selected for Na^+. The movement of H^+ out of the cell is a reflection of the propensity of monensin to select for Na^+ over K^+ [18].

The superposition of the equilibrium affinity ratios of monen- sin and nigericin over the membrane permselectivity which they induce is in accord with their operating within the *equilibrium domain* [49, 86]. Since the complexation-decomplexation reactions between car- boxylic ionophores and cations occur rapidly enough to be considered at equilibrium relative to the rate of movement of the ion-carrier complexes across the membrane interior, the ionophore-induced permea- bilities simply reflect the product of the equilibrium ion selectivity intrinsic to the ionophore and the mobility ratios of the cation in- clusion complexes. If the size and electronic nature of the cation complexes of a given ionophore are similar, the mobility ratios of the complexes approach unity and the permeability ratios reflect equilibrium affinities.

5.3. Mitochondria

The effect of carboxylic ionophores on passive ion transport in resting mitochondria has been elaborately studied by monitoring their ability to accelerate swelling in 100-150 mM alkali salt solutions. The carboxylic ionophores alone cause minor mitochondrial swelling in solutions of the K^+ salts of impermeant strong acids ($NCS^- > NO_3^- > Cl^-$) [95]. However, their effect is considerably increased in the presence of various protonophores [77]. When the incubation medium is a solution of the K^+ salt of a permeant weak acid, the carboxylic ionophores induce substantial swelling even in the absence of a protonophore [96]. Thus in the presence of an agent which makes protons available in the intramitochondrial compartment, the activity of the ionophores can be rationalized in terms of their ability to stimulate nonelectrogenic M^+-H^+ exchange [77,96]. In accord with their operating in the *equilibrium domain*, no qualitative differences between the ion selectivity patterns for the induced swelling of mitochondria and equilibrium ion selectivity patterns measured in one- or two-phase test systems have been noted.

The nature of the effects of ionophores on metabolically active mitochondria is more complex. The chemiosmatic theroy of oxidative phosphorylation suggests that electron transport results in the build-up of an electrical potential difference (negative inside) across the inner membrane of respiring mitochondria [97]. According to the chemiosmotic mechanism, the hydrogen and electron carriers of the respiratory chain are looped in such a way that oxidation-reduction processes are accompanied by the translocation of protons from one side of the membrane to the other. The resulting asymmetric distribution of protons generates an electrochemical potential difference. In response to this membrane potential, the charge transferring neutral ionophores such as valinomycin can promote an influx of K^+ down the electrical potential gradient and up the K^+ concentration gradient. The dissipation of the membrane potential due to positive charge translocation by valinomycin has been proposed to accelerate the proton pump thereby increasing the transmembrane pH gradient [77]. Since the inwardly directed transport of substrate anions depends on the magnitude of Δ pH across the membrane, the valinomycin-induced proton expulsion has been found to facilitate substrate anion accumulation and thus to increase the rate of mitochondrial respiration [98].

The carboxylic ionophores reverse the effects caused by the neutral ionophores on respiring mitochondria. They induce an outflow of accumulated cations from the mitochondria, proton uptake, and shrinking [1]. In low concentrations, nigericin and dianemycin have been observed to accelerate the respiration induced by valinomycin or the nactins, but as their concentration is increased they become inhibitors [99]. The inhibition can be counteracted by increasing the K^+ or substrate anion concentration in the extramitochondrial media. The basic process underlying the activity of the carboxylic ionophore is again a K^+ for H^+ exchange. This exchange can lower or even reverse the transmembrane pH gradient established by the respiratory chain. If the fall in intramitochondrial pH is not very significant, dissipation of Δ pH leads to acceleration of respiration; if, on the other hand, it is large, then respiration can be inhibited due to arrest of the substrate anion inflow. The cation selectivity of this effect is in general consistent with the equilibrium complexing selectivity of the ionophore.

However, caution should be used in the quantitative and in some cases the qualitative interpretation of these data (cf. Refs. 18 and 88). If the pH gradient across the inner mitochondrial membrane changes during the course of the response to ionophores, the effect of this change on the respiratory and phosphorylation rate can be mediated through shifts in the steady-state intramitochondrial concentration of substrate anions and/or phosphate. Since these steady-state concentrations are determined by numerous transport equilibria, many of the ionophore-induced effects are very intricately dependent on the nature of the substrate anions and their accompanying cations. An excellent example of substrate effects on the cation dependence of respiratory acceleration has been noted by Estrado-O et al. for the neutral ionophore beauvericin [100]. If the substrate is glutamate, respiration in the presence of beauvericin is promoted most strongly by K^+, but if the substrate is succinate or α-hydroxybutyrate, respiration is best stimulated by Na^+. A relation of this sort may be assumed to be the result of the superposition of the cation selectivities for two processes, beauvericin-mediated cation transport and cation-dependent substrate uptake.

6.   CONFORMATIONAL ASPECTS OF ION CAPTURE
AND MEMBRANE TRANSPORT

The ion capture and release steps of the membrane transport mechanism involve a rearrangement of the liganding heteroatoms on the ionophore backbone. The molecular details of this rearrangement have been the focus of a great deal of research, in solution by various spectro- scopic techniques and in the solid state by x-ray diffraction. Two mechanisms have been proposed: (1) In the highly polar environment in which an ionophore exists prior to ion capture, it is in an open, acyclic conformation which is radically different from the cyclic conformation of the ion inclusion complex. Conversion to the open conformation, which is energetically favored in polar environments, facilitates the quick release of cations at the membrane interface [19]. (2) The conformations of a number of carboxylic ionophores, whether free acids or ion inclusion complexes, are isomorphic in the solid state. Consequently, it is felt that the liganding cavity is basically performed and ion capture and release involves minimal conformational change. This hypothesis is supported by the conforma- tions of a number of carboxylic ionophores being isomorphic in the solid state whether or not they are complexed to cations [26].

Conformational studies conducted in solution support the first hypothesis. It appears that there are specific regions in the back- bone of carboxylic ionophores termed *hinges* which allow a significant amount of backbone flexibility [101]. Hinges are sterically unhin- dered regions between rigid ring systems and highly substituted carbon-carbon bonds. Extensive CD, ^{1}H NMR, ^{13}C NMR, and molecular modeling studies conducted on the carboxylic ionophores A23187, lasalocid [18,19,102], salinomycin [42], and narasin [103] indicate that there is a conformational equilibrium involving rotation about the hinge bonds. These hinge bonds not only play a role in the mechanism of ion capture but also modulate the effects that membrane microenvironments have on the precomplexation conformation of the ionophore.

Hinge bonds were first proposed for the carboxylic ionophore A23187 [Fig. 1(D)] in a detailed structural analysis of the A23187-Ca^{2+} inclusion complex [101]. ^1H NMR provided vicinal coupling constants which were analyzed using a Karplus relationship [104] to provide approximate values of backbone rotational angles. Three regions of the molecule, the spiroketal (C10-C18), the benzoxazole ring (C2-C8), and the pyrole ring (C21-C24) are rigid. Consequently, there is no conformational change in these moieties associated with Ca^{2+} inclusion or release. However, a rotation of up to 40° does occur around the C9-C10 bond upon capture of Ca^{2+}. This enables the carboxyl group on the benzoxazole ring to form a head-to-tail hydrogen bond with the NH group of the pyrrole ring. Prior to the rotation, the carboxyl group and the NH group are not close enough to hydrogen bond. A 60° rotation is also possible around the C19-C20 bond. Although it appears that free rotation is possible around the C8-C9 bond, rotation is hindered by the C11 methyl group in the free acid form. It is only after rotation around the C9-C10 bond that adjustments about the C8-C9 bond become sterically possible. Thus the conversion of A23187 from the uncomplexed, open form to the cyclic, Ca^{2+}-complexed form involves significant rotations about three bonds, C8-C9, C9-C10, and C19-C20.

The conformational equilibrium of the carboxylic ionophores can be shifted toward the noncomplexing, acyclic form by increasing solvent polarity. CD studies using the n → π^* transition of the C11 ketone groups of salinomycin [42] and narasin [103] [see Fig. 1(E)] indicate that the conformational state of both of these ionophores is shifted by altering the polar and protic properties of the solvent. Figure 12 illustrates the effect of solvent polarity, as indicated by Kosower's Z value, on the CD spectrum of salinomycin free acid and salinomycin anion. The rotational strength $-R_0^T$ of the anion

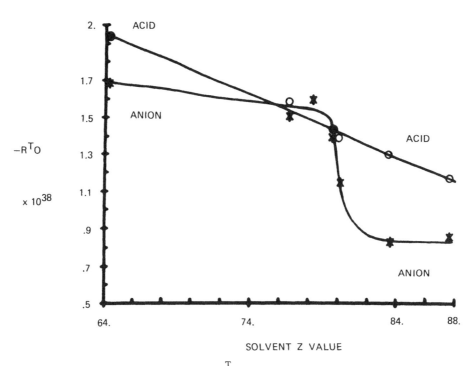

FIG. 12. Rotational strength $-R_O^T$ of the free acid of salinomycin (o) and of the free anion of salinomycin (✱) as a function of solvent Z value. The anion was generated by the addition of 2 eq of tri-n-butylamine. (From Ref. 42.)

drops sharply between Z values of 80 and 83, remaining stable above and below these values. Preliminary molecular modeling studies indicate that the increase in $-R_O^T$, upon lowering the solvent polarity, is consistent with the conformation shifting from an open form in which the liganding heteroatoms are highly solvated to a narrow pitch helix in which the liganding cavity is well defined [42].

There is a correlation between precomplexation conformation and the ability of the ionophore to recognize and bind cations. CD was utilized to calculate K_D values of salinomycin for Na^+ and K^+ [see Eqs. (6) and (7)] [42]. A plot of $K_D(K^+)/K_D(Na^+)$ vs. Z indicates a sharp shift in selectivity between Z values of 80 and 83 (Fig. 13). This shift is coincident with the shift in the conformational equilibrium of salinomycin anion seen in Fig. 12. Apparently,

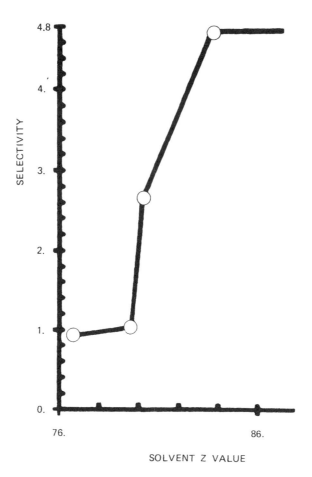

FIG. 13. Cation selectivity, $K_D(K^+)/K_D(Na^+)$, as a function of solvent
Z value. (From Ref. 42.)

the ability of the salinomycin anion to discriminate between cations
is influenced by the position of the precomplexation conformational
equilibrium.

Detailed ^{1}H NMR studies have been conducted on the effects of
solvent polarity and ion inclusion on the conformational equilibrium
of narasin [Fig. 1(E), R = CH_3] [103,105]. The chemical shifts, δ, of
most of the protons showed little change ($|\delta| < 0.1$ ppm) as a function
of the empirical solvent polarity parameter E_T. However, protons 10,

12, 13, 14, and 25 showed significantly larger chemical shift changes ($0.17 \leq |\delta| \leq 0.48$ ppm) on going from the solvent of lowest (cyclohexane-d_6) to highest (CD$_3$OD) E_T values. Furthermore, these changes appeared concerted with the largest change occurring between E_T values of 39.1 (CDCl$_3$) and 42.2 (acetone-d_6). The shifts observed for protons 10, 12, and 13 are likely due to an alteration in the relative diamagnetic shielding by the C11 ketone group. The change occurs due to rotations around the C10-C11 and C11-C12 hinge bonds. The changes in chemical shifts reported by Caughey et al. [103] for narasin as a function of E_T are comparable to differences in the shift values reported between protons of salinomycin and its 17-epi analog [106]. Due to the epimerization at C17, the epianalog is forced into an elongated noncyclic conformation.

The H12-H13 vicinal coupling constant is also consistent with the C12-C13 bond acting as a hinge [103]. The ^3J value increases from <0.5 Hz in CDCl$_3$ to 2.5 Hz in CD$_{3OD}$. This change can be due to an alteration of 20-40° in the dihedral angle and/or an alteration in the exchange rate between rotamers.

CD studies of lasalocid [Fig. 1(F)] revealed a similar polarity-dependent conformational equilibrium. Lasalocid not only has an $n \rightarrow \pi^*$ transition arising from the ketone group (C12) but in addition has three $\pi \rightarrow \pi^*$ bands arising from the aromatic ring. Each of these electronic transitions gives rise to CD bonds at corresponding wavelengths. The lowest CD band at 210 nm arising from the $A_{1g} \rightarrow E_{1u}$ $\pi \rightarrow \pi^*$ transition was not used because it fell outside of the transparency range of several of the solvent systems in our polarity continuum. The CD band at 245 nm (peak I in Fig. 14) arises from the $A_{1g} \rightarrow B_{1u}$ $\pi \rightarrow \pi^*$ transition; the CD band centered at 300 nm (peak II in Fig. 14) is a composite peak arising from the C12 ketone $n \rightarrow \pi^*$ transition at 294 nm and the $A_{1g} \rightarrow B_{2u}$ $\pi \rightarrow \pi^*$ transition at 317 nm. The molar ellipticity [θ] of each of the bands is extremely solvent-dependent. The [θ] values for both peaks decrease linearly between E_T values of 40 and 54. Above and below these values, [θ] remains unchanged for both peaks. Thus lasalocid, like salinomycin and narasin, exists in a conformational equilibrium which can be shifted between limiting states by altering solvent polarity.

The structural factors which mediate the interaction of lasalocid free acid with solvent have been studied by molecular modeling [19]. Conformer structures were generated under two limiting sets of modeling conditions. In a highly polar environment it is probable that the oxygen atoms on the ionophore backbone are highly solvated and the likelihood of forming any intramolecular hydrogen bonds is low. In a low-polarity environment, intramolecular hydrogen bonding is much more likely. The two limiting conditions primarily affect the stability of two hydrogen bonds, O(26)-HO(40) and O(26)-HO(31). A third intramolecular hydrogen bond, the O(28)H-O(27) salicylate

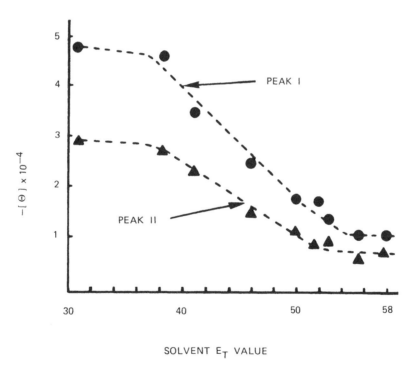

FIG. 14. $[\theta]_I$ and $[\theta]_{II}$ of lasalocid anion as a function of solvent E_T value. (From Ref. 19.)

hydrogen bond, is extremely stable and polarity-independent. The
occurrence and stability of these hydrogen bonds in solution has
been confirmed by NMR studies and infrared studies in which the
exchangeability of the hydroxyl protons was monitored as a function
of solvent polarity [10,107].

Under high-polarity modeling conditions, energy minimization
of lasalocid using consistent force field calculations yielded the
acyclic quasilinear conformation shown in Fig. 15 [19]. The tetra-
hydropyran ring is in a chair conformation with the C22 ethyl group
and the C16-C19 bond equitorial, and the C22 hydroxyl and the C21
methyl groups transdiaxial. The tetrahydrofuran ring is in a twist

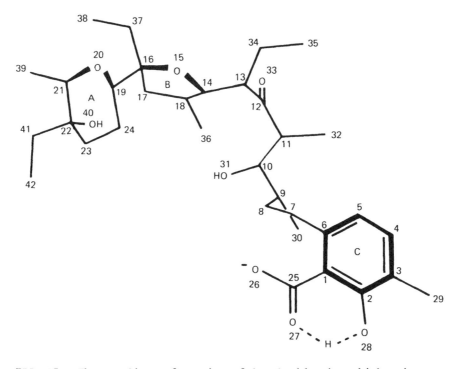

FIG. 15. The acyclic conformation of lasalocid anion which exists
in high-polarity media. QCFF energy minimization calculations indicate that this conformer has the lowest intrinsic energy [18,19].
Solvent effects were modeled using an empirical specific site
approach [108].

conformation with C18 above and C14 below the plane determined by
the other three atoms. The aromatic ring is planar. The C25 carboxylate is twisted out of the plane of the aromatic ring very
slightly (3°) to minimize unfavorable steric interactions with the
ionophore backbone at C6. The carbon-carbon backbone bonds between
C6 and C14, are all in their lowest energy, staggered conformation.

Formation of the cyclic conformer (Fig. 16) under apolar modeling conditions preceeds without significant conformational changes
in any of the rings [19]. The C25 carboxylate, however, twists 26°
out of the plane of the aromatic ring during formation of the head-
to-tail O(40)H-O(26) hydrogen bond due to the steric bulk of the

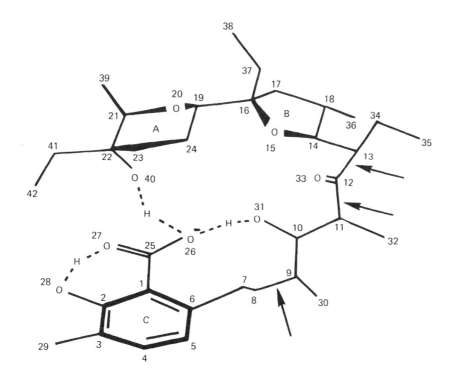

FIG. 16. The cyclic conformer of lasalocid anion which predominates
under low-polarity conditions and in ion inclusion complexes. Dashed
lines indicate intramolecular hydrogen bonds. Rearrangement from the
cyclic to the acyclic forms occurs by rotation around the bonds indicated by arrows. (After Refs. 18 and 19.)

carbon backbone at the C6 position. The major conformational changes during cyclization occur by rotation around the C8-C9, C11-C12, and C12-C13 carbon-carbon bonds. The C7-C8-C9-C10 torsion angle rotates from anti to gauche, thereby moving the C25 carboxylate into the vicinity of the O(31)H hydroxyl and O(33) carboxyl liganding moieties. The carboxylate is now in a position to form the O(26)-HO(31) intra-molecular hydrogen bond. Cyclization is completed by reduction of the O(33)-C12-C11-C10 torsional angle from 74 to 32°, which focuses the THP and THF rings into the central liganding cavity and places the O(40)H hydroxyl in a position to form the second structure-stabilizing hydrogen bond, O(40)H-O(26). Van der Waals strain aris-ing from 1,3-steric interactions between the C32 methyl group and the C34, C35 ethyl group is minimized by a decrease in the O(33)-C12-C13-C14 torsional angle from 60 to 29°.

NMR studies of lasalocid conducted in a solvent polarity con-tinuum analogous to that used in the CD study confirm the presence of a solvent-mediated conformational equilibrium (Fig. 17). The E_T dependency of the chemical shifts is consistent with two conforma-tional states existing in solution, one predominating in high-polarity media and the other in low-polarity media. All solvent-dependent chemical shifts occur between the same narrow range of E_T values, 42.1 (acetone) to 41.2 (dichloromethane), and remain constant above and below these values. Since ^{13}C NMR spin lattice relaxation mea-surements have shown lasalocid free acid to exist as a monomer both in methanol and in chloroform [107,109], this concerted shift must be indicative of a change between limiting conformational states of the monomeric ionophore rather than a monomer-dimer equilibrium. Figure 17 also reveals that the solvent polarity-induced chemical shifts are localized in three regions of the ionophore backbone, other regions being relatively unaffected. Protons 21, 41u, and 41d on the THP ring are shifted downfield on going from polar to apolar media and protons adjacent to the C12 ketone, H11. H13, H34d, and $CH_3(32)$ are all shifted upfield. On the aromatic side of the mole-cule, C1-C10, one H7, and one H8 proton shift upfield while the other

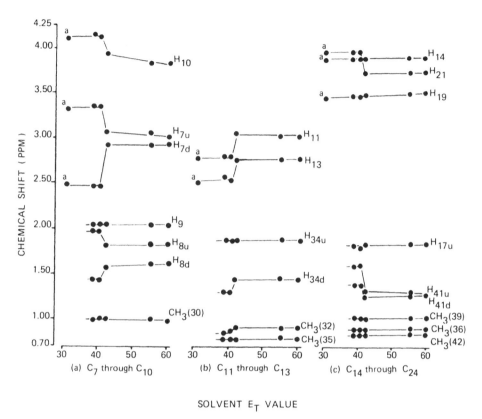

FIG. 17. Proton chemical shifts (ppm) of lasalocid free acid as a
function of E_T. The shifts are grouped according to the segment of
the ionophore backbone which the proton or alkyl group is on. The
solvents employed in this study were (1) 5% D_2O in CD_3OD, E_T = 60.2;
(2) CD_3OD, E_T = 55.5; (3) CD_3COCD_3, E_T = 42.2; (4) CD_2Cl_2, E_T = 41.1;
(5) $CDCl_3$, E_T = 39.1; (6) Cyclo-C_6D_{12}, E_T = 30.9.

H7 and H8 protons and H10 shift downfield. The chemical shifts of

protons on the THF ring and on C9 are solvent-independent and sepa-

rate the regions of solvent dependency from each other.

Coupling constant and nuclear Overhauser enhancement (nOe) data

recently obtained by us in low-polarity media are in accord with the

cyclic structure found in x-ray crystal studies [22]. However, as

the solvent polarity increases, the NMR data are consistent with the

shift of the conformational equilibrium to the acyclic form predicted by CD studies and computer modeling [19].

In the cyclic conformation, the carbon backbone of lasalocid
is locked into a single rigid orientation (see Fig. 16). The plane
of the aromatic ring is parallel to the C7-H7u and C7-C8 bonds. The
carbon backbone between C6 and C10 folds back on itself via an anti-
periplanar relationship between C6 and C9 and a gauche relationship
between C7 and C10 [102]. The methylene protons on C7 and C8 are
oriented differently with respect to the magnetically anisotropic
methyl salicylate group and consequently experience very different
magnetic environments. This is manifested by the large chemical
shift differences of 0.76 ppm between H7u and H7d and of 0.52 ppm
between H8u and H8d (see Fig. 17). However, above an E_T value of
42.1 the collapse of the chemical shift differences between the
members of the geminal pair suggests that an increased freedom of
rotation about C6-C7 and C7-C8 results in both protons in each of
the geminal pairs experiencing a more similar environment.

The conformations we observe for the THP and THF rings in
methanol are virtually identical to those reported in chloroform
[102]. However, protons 21, 41u, and 41d are shifted upfield relative to their resonance positions in chloroform. Since in chloroform lasalocid is locked into a macrocycle by intramolecular hydrogen
bonds between the C25 carboxyl and the HO(31) and HO(40) hydroxyls,
the aromatic ring is proximal to these protons. Thus, while the
chemical shifts of H21, H41u, and H41d in polar media can be calculated by the additive effects of the shielding constants of directly
bonded and vicinal groups, the chemical shifts observed in chloroform
are consistent with an appreciable ring anisotropy. The fact that
these protons do not sense the aromatic ring in methanol is consistent with a backbone rearrangement requiring the breaking of the
intramolecular hydrogen bonds [19].

The J values we observe for protons on C9-C13 of lasalocid are
virtually identical in methanol and chloroform. This indicates that
the torsional angles about C9-C10, C10-C11, and C13-C14 have not
changed. However, previous CD and computer-modeling studies have

suggested that a significant backbone rearrangement may still occur in this region by using the sp^2-hybridized C12 as a hinge [19]. Such a rearrangement would be expected to alter the relative diamagnetic shielding of adjacent protons. Indeed, as solvent polarity is raised above that of chloroform, concerted downfield shifts are observed for H13, H34u, and CH_3(32). Proton 34d and CH_3(35) are insensitive to solvent polarity changes, possibly because they are directed away from the carbonyl shielding cone.

Rotations around C11-C12 and C12-C13 alter the spatial arrangement of the two halves of the molecule and are detected by changes in the magnitudes of the nOe across the C12 hinge. In chloroform positive nOe's are observed between H11 and H13, and H11 and H14, consistent with the O(33)-C12-C11-C10 dihedral angle of 32° and the O(33)-C12-C13-C14 dihedral angle of 29°. The three protons are in a triangular spatial arrangement. In methanol an nOe is observed across the hinge from H11 to H13, but is no longer detected between H11 and H14. This is consistent with the changes predicted from computer modeling [19].

By integrating the spectroscopic data into the transport model depicted in Fig. 6, a more detailed picture of ion transport emerges. The transport cycle begins with the ionophore anion confined to the reaction plane near the membrane interface. The conformational equilibrium of the ionophore is shifted toward the acyclic form. Under these conditions the liganding heteroatoms on the ionophore backbone are strung out and highly solvated. The ionophore does not present a focused inducible dipole system capable of strong interaction with solution cations. The most likely mechanism for initial interactions between the acyclic ionophore and solution cations appears to be ion pairing to the ionophore carboxylate. Interaction between the carboxylate and a solution cation is inversely proportional to the *square* of the distance between them while the strength of interaction between a given liganding heteroatom on the ionophore backbone and a solution cation is inversely proportional to the *cube* of their separation distance.

Once a complexable cation has been ion paired to the terminal
carboxylate, ion-induced dipole interactions with adjacent liganding
heteroatoms become more favorable and the resulting forces reorient
·the ionophore backbone about the hinge bonds into the cyclic conformer. As molecular reorientation proceeds, the heteroatoms move
closer to the optimal ligand-cation bond distance permitted by structural constraints and form a liganding field capable of stabilizing
the cation relative to the bulk solvent. The cyclic conformer, with
the polar liganding groups focused into the cation binding cavity
and the lipophilic alkyl groups shielding the exterior, is the form
most compatible with the apolar membrane interior. Since this conformation has a reduced capability to interact with the polar environment at the membrane interface, it readily enters the lipophilic
membrane interior where it is further stabilized. Upon diffusion
to the opposite membrane interface, the complex is again subjected
to interaction with a polar environment. Because electrostatic
factors no longer prevail over unfavorable steric factors, the conformational equilibrium shifts back toward the acyclic form releasing
the cation. The acyclic anionic ionophore is now confined to the
opposite membrane interface ready to initiate the return leg of the
catalytic transport cycle.

7. BIOLOGICAL APPLICATIONS OF IONOPHORES

As intriguing as these exquisitely designed, highly selective complexation and transport machines are to physical chemists, the ionophores have had an even greater impact on cell biologists. This
derives from the important role of ion gradients, especially that
of Ca^{2+}, in regulating biological activity. The biological effects
of ionophores have been reviewed elsewhere in depth [2,110,111] and
will be briefly highlighted here.

As discussed in the opening chapter of this book, the transport property of ionophores was first recognized from their ability
to perturb the energy-dependent uptake of cations by isolated mito-

chondria, stimulated by neutral ionophores, e.g., valinomycin, and reversed by carboxylic ionophores, e.g., nigericin. Combining ionophores of each class induces a short circuit of ion movement which can totally dissipate the metabolically generated energy of these subcellular organelles [1].

The ionophores affect bacterial transport [112] as well as the light-driven cation transport of chloroplasts [113] and bacterial chromatophores [114]. Erythrocytes have been already discussed as representative biological membranes for studying the selectivity of ionophore-induced transport.

The utilization of ionophores as probes for perturbing the metabolism of intact eukaryotic cells stems from the recognized role of Ca^{2+} as a *second messenger* in activating cells [115]. Many cells are quiescent until stimulated to express their full metabolic potentiality, either electrically or by the triggering of specialized receptors on their surface with appropriate agonists. For most activatable cells the stimulus is transmitted within the cell by a rise in cytosolic Ca^{2+} activity from resting levels of approximately 10^{-7} M to approximately 10^{-6}-10^{-5} M. Examples of cellular activation believed mediated through elevated Ca^{2+} activity include: (1) contraction of skeletal, heart, and smooth muscle; (2) secretion by promoting exocytosis, i.e., discharge of intracellular secretory vesicles; (3) synaptic transmission of neural impulses; (4) initiation of the prostaglandin cascade by stimulation of phospholipase A which hydrolyses arachidonate from phospholipids; (5) response of lymphocytes to antigens, e.g., "capping"; (6) initial events following the penetration of egg cells by sperm.

In 1970 Paul characterized the structure of lasalocid by x-ray crystallography of its *barium* salt [25]. This suggested that lasalocid would by analogy complex Ca^{2+} and transport it across membranes which we confirmed experimentally [30]. When lasalocid was added to isolated vesicles of sarcoplasmic reticulum, the intracellular Ca^{2+} storage organelles of muscle, Ca^{2+} was released affirming its competency to raise intracellular Ca^{2+} activity [116]. Lasalocid also affected a more integrated cellular system, stimulating the beating

of the isolated perfused rabbit heart [30]. At an even higher level
of cellular integration, the intact dog, lasalocid strongly stimulated cardiac contractivity and blood output [27,64,117]. Paradoxi-
cally, lasalocid also induces *relaxation* of vascular smooth muscle,
particularly of the coronary vessels, i.e., coronary dilation, by a
mechanism which is not completely understood.

Despite the apparent verification of our prediction that the
Ca^{2+}-transporting ionophore lasalocid would perturb Ca^{2+}-controlled
processes at the cellular and tissue level, the prompting rationale
proved inaccurate. Other carboxylic ionophores, selective for monovalent cations and devoid of the capacity to complex or transport
Ca^{2+}, proved even more potent in evoking the same cardiovascular
effects. Among these are monensin, salinomycin, narasin, X-206,
A-204, dianemycin, grisorixin, and alborixin; one can presume the
list would include all monovalent carboxylic ionophores [64].

A biological synergism between Ca^{2+} and Na^+ has long been
recognized although the underlying mechanisms have only recently
been understood. Na^+ can make Ca^{2+} available to intracellular Ca^{2+}
receptors either by bringing extracellular Ca^{2+} into the cell by an
exchange-diffusion carrier, or by displacing intracellularly seques-
tered Ca^{2+} [118]. Na^+-selective ionophores are surprisingly potent
in stimulating cellular processes presumed to be Ca^{2+}-controlled,
suggesting that there are biological control mechanisms by which Na^+
modulates the availability of Ca^{2+} to intracellular receptors [67,
119]. This is consistent with the long recognized increase in cardiac contractility by *digitalis*, which inhibits the outwardly directly
Na^+ pump thereby increasing the net inward leak of Na^+ into heart
cells [120]. We find that digitalis and monovalent ionophores exert
synergistic effects on cellular processes, e.g., capping of cultured
lymphocytes [121], and at the same time elevate intracellular Ca^{2+}
activity [122] as detected by the intracellular fluorescent Ca^{2+}
probe Quin-2 [123].

These results indicate that the dual ability of lasalocid to
transport both Ca^{2+} and Na^+ into cells makes it ambiguous as a probe

of biological mechanisms. We prefer monensin, a readily available monovalent ionophore, for probing the role of alkali cations in cellular control.

In addition to its aforementioned effects in increasing cardiac contractility, dialating coronary arteries, and promoting lymphocyte capping, monensin also stimulates the release of epinephrine from adrenals in the intact animal [117] as well as isolated chromaffin cells [124], stimulates the release of prostaglandins from the renal medulla [125], the release of serotonin from platelets [125], but inhibits the secretion of procollagen from cultured fibroblasts [126] and enzymes from pancreatic acinar cells [127]. These latter inhibitions are attributed to a disturbance of the protein packaging function of the Golgi system.

Of particular economic importance is the use of carboxylic ionophores (monensin, lasalocid, salinomycin, and narasin) as agricultural feed additives. They control coccidiosis, a disease caused by a pathogenic protozoan in poultry, promote a more efficient feed-to-food conversion in cattle by inhibiting growth of less favorable flora in the rumin, and improve feed-to-food conversion in swine by a more obscure mechanism [128]. In view of the ready absorption of ionophores from the gut [129], it is difficult to eliminate participation of systemic effects such as induced secretion of hormones, e.g., growth hormone.

Despite the diversity of effects of monovalent carboxylic ionophores, we would ascribe their ultimate mechanism exclusively to their ability to transport the biologically significant ions Na^+, K^+, and H^+ across membranes. This is based on the fact that all monovalent carboxylic ionophores have equivalent biological properties and for the most part are chemically inert molecules were it not for their unique ability to transport cations across membranes. Subtle differences in biological properties may accrue from the relative rates at which a given monovalent cation-selective ionophore transports Na^+, K^+, and H^+ [119].

The use of carboxylic ionophores for probing biological systems has been extended by the discovery that A23187 is much more selective

for polyvalent than for monovalent ions, thus providing an analagous
agent for perturbing intracellular Ca^{2+} activity directly [58]. At
the moment this is probably the most extensively employed ionophore
in biological research. Ionomycin may be slightly superior to A23187
in potency, divalent selectivity, and nonspecific detrimental effects
on membranes [130], but it is currently not as easy to obtain. In
principle, the polyvalent ionophores X-14547A [68] should also be
usable for raising intracellular Ca^{2+} activity but it has not been
available long enough for extensive biological applications to appear
in the literature.

8. OVERVIEW

The ionophores are truly wonderful compounds. They utilize molecular
strategies for complexing and discriminating between cations that
challenge our most sophisticated technology, including CD, NMR, x-ray
crystallography, and computer modeling. An altogether different
approach is required to elucidate the dynamic diffusional mechanisms
by which they transport ions across membranes. We have here been
concentrating on the *carboxylic* subgroup of ionophores since they are
well tolerated by living systems, but in order to use them rationally
in biology their precise molecular properties must be catalogued. In
this chapter we have attempted to assemble data from many sources
which will provide the reader with some feel for their diverse properties.

   For those who are disinclined to succumb to the lure of the intri-
cate mechanisms by which the ionophores conduct their business, there
is still their value as tools for biological research. The biologist
is challenged to construct strategies for utilizing ionophores to control metabolism and to explore what inherent ionic control processes
reside within the cell. Even industry shares in the ionophore bonanza
with respect to their use as livestock feed additives. As we look
into the future we may anticipate suitably selected ionophores as
drugs to alleviate pathological conditions such as shock and heart

failure. But caution is also advised lest the widespread dissemination of these powerful agents provide a livelihood for environmental toxicologists.

ACKNOWLEDGMENT

The preparation of this manuscript and unpublished work referred to was supported in part by NIH Grant HL-23932.

REFERENCES

1. B. C. Pressman, E. J. Harris, W. S. Jagger, and J. H. Johnson, *Proc. Natl. Acad. Sci. USA, 58,* 1949 (1967).

2. B. C. Pressman, *Ann. Rev. Biochem., 45,* 501 (1976).

3. C. Liu, in *Polyether Antibiotics,* Vol. 1 (J. W. Westley, ed.), Marcel Dekker, New York, 1983, p. 43.

4. J. W. Westley, in *Antibiotics,* Vol. 4 (J. W. Corcoran, ed.), Springer Verlag, Berlin, 1981, p. 41.

5. H. Diebler, M. Eigen, G. Ilgenfritz, G. Maass, and R. Winkler, *Pure Appl. Chem., 20,* 93 (1969).

6. J. Burgess, *Metal Ions in Solution,* John Wiley and Sons, New York, 1978.

7. R. W. Taylor, R. F. Kauffman, and D. R. Pfeiffer, in *Polyether Antibiotics,* Vol. 1 (J. W. Westley, ed.), Marcel Dekker, New York, 1983, p. 103.

8. H. Degani and H. L. Friedman, *Biochemistry, 13,* 5022 (1974).

9. H. Degani, R. M. D. Hamilton, and H. L. Friedman, *Biophys. Chem., 4,* 363 (1976).

10. C. Shen and D. J. Patel, *Proc. Natl. Acad. Sci. USA, 73,* 4277 (1976).

11. D. R. Pfeiffer, R. W. Taylor, and H. L. Lardy, *Ann. NY Acad. Sci., 307,* 402 (1978).

12. D. H. Haynes and B. C. Pressman, *J. Membr. Biol., 16,* 195 (1974).

13. D. H. Haynes and B. C. Pressman, *J. Membr. Biol., 18,* 1 (1974).

14. G. Eisenman, S. Ciani, and G. Szabo, *Fed. Proc., 27,* 1289 (1968).

15. Yu. A. Ovchinnikov, V. T. Ivanov, and A. M. Shkrob, *Membrane Active Complexones*, B.B.A. Library, Vol. 12, Elsevier, New York, 1974.

16. D. H. Bush, K. Farmery, V. Goedken, V. Katovic, A. C. Melnyk, C. R. Sperati, and N. Tokel, *Adv. Chem. Ser., No. 100*, 14 (1971).

17. L. Fabbrizzi, C. Paoletti, and R. M. Clay, *Inorg. Chem., 17*, 1042 (1978).

18. G. R. Painter and B. C. Pressman, in *Top. Curr. Chem., Vol. 101*, Springer Verlag, Berlin, 1982, p. 83.

19. G. R. Painter, R. Pollack, and B. C. Pressman, *Biochemistry, 21*, 5613 (1982).

20. B. K. Toeplitz, A. I. Cohen, P. T. Funke, W. L. Parker, and J. Z. Gougoutas, *J. Am. Chem. Soc., 101*, 3344 (1979).

21. R. F. Kauffman, R. W. Taylor, and D. R. Pfeiffer, *J. Biol. Chem., 255*, 2735 (1980).

22. E. Duesler and I. C. Paul, in *Polyether Antibiotics* (J. W. Westley, ed.), Marcel Dekker, New York, 1983, p. 87.

23. A. Agtarap, J. W. Chamberlin, M. Pinkerton, and L. Steinrauf, *J. Am. Chem. Soc., 89*, 5737 (1967).

24. D. L. Ward, K. T. Wei, T. G. Hoogerheide, and A. I. Popov, *Acta Crystallogr. 334*, 110 (1978).

25. S. M. Johnson, J. Herrin, S. J. Liu, and I. C. Paul, *J. Am. Chem. Soc., 92*, 4428 (1970).

26. M. Dobler, *Ionophores and Their Structures*, John Wiley, New York, 1981).

27. B. C. Pressman and N. T. de Guzman, *Ann. NY Acad. Sci., 227*, 380 (1974).

28. S. T. Chen and C. S. Springer, *Bioinorg. Chem., 9*, 101 (1978).

29. J. Grandjeau and P. Laszlo, *J. Am. Chem. Soc., 106*, 1472 (1984).

30. B. C. Pressman, *Fed. Proc., 32*, 1698 (1973).

31. G. R. Painter and B. C. Pressman, *Biochem. Biophys. Res. Commun., 97*, 1268 (1980).

32. L. Pauling, *The Nature of the Chemical Bond*, Cornell University Press, New York, 1960.

33. W. E. Morf and W. Simon, *Helv. Chim. Acta, 54*, 2683 (1971).

34. W. E. Morf and W. Simon, in *Progr. Macrocyclic Chem.*, Vol. 1 (R. M. Izatt and J. J. Christensen, eds.), John Wiley and Sons, New York, 1979, p. 1.

35. W. Burgermeister and R. Winkler-Oswatitsch, *Topics in Curr. Chem.*, Vol. 69, Springer Verlag, Berlin, 1977, p. 204.

36. C. Moore and B. C. Pressman, *Biochem. Biophys. Res. Commun.*, *15*, 562 (1964).

37. B. C. Pressman, *Methods Enzymol.*, *10*, 714 (1967).

38. H. K. Friensdorff, *J. Am. Chem. Soc.*, *93*, 1 (1971).

39. P. O. G. Gertenbach and A. I. Popov, *J. Am. Chem. Soc.*, *97*, 4738 (1975).

40. G. Cornelius, W. Gartner, and D. H. Haynes, *Biochemistry*, *13*, 3052 (1974).

41. W. K. Lutz, H. K. Wipf, and W. Simon, *Helv. Chim. Acta*, *53*, 1741 (1970).

42. G. R. Painter and B. C. Pressman, *Biochem. Biophys. Res. Commun.*, *91*, 1117 (1979).

43. P. Gachon and A. Kergomard, *J. Antibiot.*, *28*, 351 (1975).

44. D. R. Pfeiffer, P. W. Reed, and H. A. Lardy, *Biochemistry*, *13*, 4007 (1974).

45. D. T. Wong, *FEBS Lett.*, *71*, 175 (1976).

46. H. Degani, H. L. Friedman, G. Navon, and E. M. Kossower, *Chem. Commun.*, 432 (1973).

47. D. W. Urry, *J. Am. Chem. Soc.*, *94*, 77 (1974).

48. R. A. Alberty and F. Daniels, *Physical Chemistry*, 5th ed., Wiley and Sons, New York, 1979.

49. B. C. Pressman, *Fed. Proc.*, *27*, 1283 (1968).

50. G. Eisenman and G. Szabo, *J. Membr. Biol.*, *1*, 294 (1969).

51. B. C. Pressman and D. H. Haynes, in *Molecular Basis of Membrane Function* (D. C. Tosteson, ed.), Prentice-Hall, Englewood Cliffs, N.J., 1969, p. 221.

52. M. Mitani, T. Yamanishi, and Y. Miyazaki, *Biochem. Biophys. Res. Commun.*, *66*, 1231 (1975).

53. B. C. Pressman and G. A. Vallega, unpublished results.

54. M. Mitani and N. Otake, *J. Antibiot.*, *31*, 750 (1978).

55. M. Mitani, T. Yamanishi, E. Ebata, N. Otake, and M. Koenuma, *J. Antibiot.*, *30*, 186 (1977).

56. N. Otake and M. Mitani, *Agric. Biol. Chem.*, *43*, 1543 (1979).

57. M. Mitani and N. Otake, *J. Antibiot.*, *31*, 888 (1978).

58. P. W. Reed and H. A. Lardy, in *The Role of Membranes in Metabolic Regulation* (M. A. Mehlman and R. W. Hanson, eds.), Academic Press, New York, 1972, p. 111.

59. S. P. Young and B. D. Gomperts, *Biochim. Biophys. Acta*, *469*, 281 (1977).

60. W. Willbrandt and T. Rosenberg, *Pharmacol. Rev.*, *13*, 109 (1961).

61. H. Celis, S. Estrado-O, and M. Montal, *J. Membr. Biol.*, *18*, 187 (1974).

62. R. Sandeaux, P. Seta, G. Jeminet, M. Alleaume, and C. Gavach, *Biochim. Biophys. Acta*, *511*, 499 (1978).

63. M. A. Kolber and D. H. Haynes, *Biophys. J.*, *36*, 369 (1981).

64. B. C. Pressman and N. T. de Guzman, *Ann. NY Acad. Sci.*, *264*, 373 (1975).

65. R. Ashton and L. K. Steinrauf, *J. Mol. Biol.*, *49*, 547 (1970).

66. B. C. Pressman, *Anal. NY Acad. Sci.*, *147*, 753 (1969).

67. C. H. Liu and T. E. Hermann, *Biol. Chem.*, *253*, 5892 (1978).

68. C. Liu, T. E. Hermann, D. N. Bull, N. J. Palleroni, B. T. Prosser, J. W. Westley, and P. A. Miller, *J. Antibiot.*, *32*, 95 (1979).

69. K. Hiratani, *Chem. Lett.*, 21 (1981).

70. F. A. Henn and T. E. Thompson, *Ann. Rev. Biochem.*, *38*, 241 (1969).

71. A. D. Bangham, *Ann. Rev. Biochem.*, *41*, 753 (1972).

72. P. Mueller and D. O. Rudin, *Biochem. Biophys. Res. Commun.*, *26*, 398 (1967).

73. A. A. Lev and E. P. Buzhinsky, *Tsitologiya*, *9*, 102 (1967).

74. G. Weisman, P. Anderson, C. Serhan, E. Samuelson, and E. Goodman, *Proc. Natl. Acad. Sci. USA*, *77*, 1506 (1980).

75. G. R. A. Hunt, L. R. H. Tipping, and M. R. Belmont, *Biophys. Chem.*, *8*, 341 (1978).

76. G. D. Case, J. M. Vanderkooi, and A. Scarpa, *Arch. Biochem. Biophys.*, *162*, 174 (1974).

77. P. J. F. Henderson, J. D. McGiven, and J. B. Chappell, *Biochem. J.*, *111*, 521 (1969).

78. E. P. Bakker and K. Van Dam, *Biochim. Biophys. Acta*, *339*, 285 (1974).

79. M. Toro, C. Gomez-Lojero, M. Montal, and S. Estrado-O, *Bioenergetics*, *8*, 19 (1976).

80. H. Degani, *Biochim. Biophys. Acta*, *509*, 364 (1978).

81. M. S. Fernandez, H. Celis, and M. Montal, *Biochim. Biophys. Acta*, *323*, 600 (1973).

82. H. Celis, S. Estrada-O, and M. Montal, *J. Membr. Biol.*, *18*, 187 (1974).

83. D. H. Haynes, V. C. K. Chiu, and B. Watson, *Arch. Biochim. Biophys.*, *203*, 73 (1980).

84. J. S. Pushkin, A. I. Vistnes, and M. T. Coene, *Arch. Biochem. Biophys.*, *206*, 164 (1981).

85. R. F. Kauffman, J. C. Clifford, and D. R. Pfeiffer, *Biochemistry, 22,* 3985 (1983).

86. M. J. Heeb and B. C. Pressman, in *Molecular Mechanisms of Antibiotic Action on Protein Biosynthesis and Membranes* (E. Munez, P. Garcia-Fernandez, and D. Vasquez, eds.), Elsevier, Amsterdam, 1972, p. 603.

87. M. Mitani, T. Yamanishi, Y. Miyazaki, and N. Otake, *Antimicrob. Agents Chemother., 9,* 655 (1976).

88. D. T. Wong, D. H. Berg, R. H. Hamill, and J. R. Wilkinson, *Biochem. Pharmacol., 26,* 1373 (1977).

89. P. W. Reed and H. A. Lardy, *J. Biol. Chem., 247,* 6970 (1972).

90. P. M. Sokolove, *Biochim. Biophys. Acta, 545,* 155 (1979).

91. D. R. Pfeiffer and H. A. Lardy, *Biochemistry, 15,* 935 (1976).

92. P. Flatman and V. L. Lew, *Nature, 270* (1977).

93. S. Estrado-O, H. Celis, E. Calderon, G. Gallo, and M. Montal, *J. Membr. Biol., 18,* 201 (1974).

94. E. J. Harris and B. C. Pressman, *Nature, 216,* 918 (1967).

95. A. Azzi and T. F. Azzone, *Biochim. Biophys. Acta, 131,* 468 (1967).

96. E. J. Harris, M. Hofer, and B. C. Pressman, *Biochemistry, 6,* 1348 (1967).

97. P. Mitchell, *Adv. Enzymol., 29,* 33 (1967).

98. M. Hofer and B. C. Pressman, *Biochemistry, 5,* 3919 (1966).

99. S. Estrada-O, S. N. Graven, and H. A. Lardy, *Fed. Proc., 26,* 610 (1967).

100. S. Estrada-O, C. Gomez-Lojero, and M. Montal, *J. Bioenergetics, 3,* 417 (1972).

101. C. M. Deber and D. R. Pfeiffer, *Biochemistry, 15,* 132 (1976).

102. M. J. O. Anteunis, *Bioorg. Chem., 5,* 327 (1976).

103. B. Caughey, G. Painter, B. C. Pressman, and W. A. Gibbons, *Biochem. Biophys. Res. Commun., 113,* 832 (1983).

104. M. Karplus, *J. Am. Chem. Soc., 85,* 2870 (1963).

105. M. J. O. Anteunis and N. A. Rodios, *Bull. Soc. Chim. Belg., 90,* 715 (1981).

106. M. J. O. Anteunis and N. A. Rodios, *Bull. Soc. Chim. Belg., 90,* 471 (1981).

107. D. J. Patel and C. Shen, *Proc. Natl. Acad. Sci. USA, 73,* 1786 (1976).

108. V. Madison and K. D. Kopple, *J. Am. Chem. Soc., 102,* 4855 (1980).

109. J. Y. Lalemand and V. Michon, *J. Chem. Res., S*, 2081 (1978).

110. B. C. Pressman and M. Fahim, *Ann. Rev. Pharmacol. Toxicol., 22*, 465 (1982).

111. P. W. Reed, in *Polyether Antibiotics*, Vol. 1 (J. W. Westley, ed.), Marcel Dekker, New York, 1983, p. 185.

112. B. C. Pressman, in *Wirkungsmechanism von Fungiziden und Antibiotics*, Academie-Verlag, Berlin, 1967, p. 3.

113. N. Shavit, H. Degani, and A. San Pietro, *Biochim. Biophys. Acta, 216*, 208 (1970).

114. M. Nishimura and B. C. Pressman, *Biochemistry, 9*, 1360 (1969).

115. H. Rasmussen and D. B. P. Goodman, *Physiol. Rev., 57*, 421 (1977).

116. A. H. Caswell and B. C. Pressman, *Biophys. Biochem. Res. Commun., 49*, 292 (1972).

117. N. T. de Guzman and B. C. Pressman, *Circulation, 69*, 1072 (1974).

118. C. S. van Breemen, P. Aaronson, and R. Loutzenhiser, *Pharmacol. Rev., 30*, 167 (1979).

119. B. C. Pressman and G. R. Painter, in *Biochemistry of Metabolic Processes* (D. L. F. Lennon, F. W. Stratman, R. N. Zahlten, eds.), Elsevier Biomedical, New York, 1983, p. 41.

120. G. A. Langer, *Ann. Rev. Physiol., 35*, 55 (1973).

121. L. Y. W. Bourguignon and B. C. Pressman, *J. Membr. Biol., 43*, 91 (1983).

122. K. Trevorrow and B. C. Pressman, unpublished results.

123. R. Y. Tsien, *Biochemistry, 19*, 2396 (1980).

124. S. Suchard, R. W. Rubin, and B. C. Pressman, *J. Cell. Biol., 94*, 531 (1982).

125. H. R. Knapp, O. Oelz, L. J. Roberts, B. J. Sweetman, J. A. Oates, and P. W. Reed, *Proc. Natl. Acad. Sci. USA, 74*, 4251 (1977).

126. N. Uchida, H. Smilowitz, and M. L. Tanzer, *Proc. Natl. Acad. Sci. USA, 76*, 1868 (1979).

127. A. Tartakoff and P. Vassalli, *J. Cell. Biol., 79*, 694 (1978).

128. B. C. Pressman and M. Fahim, in *Myocardial Injury* (J. J. Spitzer, ed.), Plenum, New York, 1983, p. 543.

129. B. C. Pressman, G. R. Painter, and M. Fahim, *A.C.S. Symposium Series, 140*, 3 (1980).

130. W. Liu, D. S. Slusarchyk, G. Astle, W. H. Trejo, W. E. Brown, and E. Meyers, *J. Antibiot., 31*, 815 (1978).

Chapter 10

COMPLEXES OF D-CYCLOSERINE AND RELATED AMINO ACIDS
WITH ANTIBIOTIC PROPERTIES

Paul O'Brien
Department of Chemistry
Queen Mary College
London, United Kingdom

1. INTRODUCTION

The antibiotic most commonly known as D-cycloserine (I) (D-4-amino-oxazolidin-3-one) was first discovered in 1955 [1]. It was isolated from cultures of *Streptomyces orchidaceus* and given the trade name seromycin. The same substance was found to be produced by both

Streptomyces garyphalus [2] and *Streptomyces lavendulae* [3]; and termed oxamycin and PA-94, respectively.

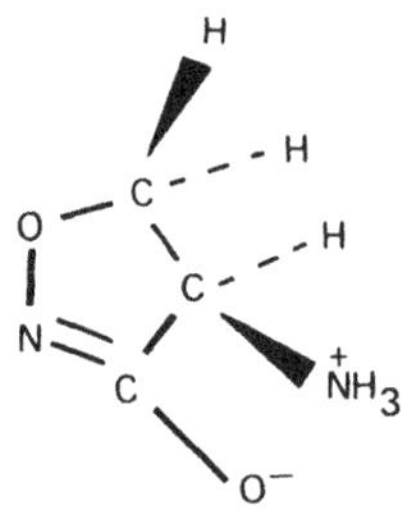

(I)

The structure of D-cycloserine (I) was correctly deduced from derivatization and degradation studies [4]; the total synthesis of D-cycloserine from DL-serine was achieved [5]. Some physical properties of D-cycloserine are summarized in Table 1 [5,6].

   A number of derivatives of D-cycloserine have also been used as antibiotics. These include the product of the condensation of D-cycloserine and terepthalaldehyde "terizidone" (II) (1,4-bis-D-oxo-4-isoxazolidinyliminomethyl)benzene [7] and (R)-4-[(1-methyl-3-oxo-1-butenyl)amino]-3-isoxazolidinone (III) produced by condensing 2,4-pentanedione and D-cycloserine [8].

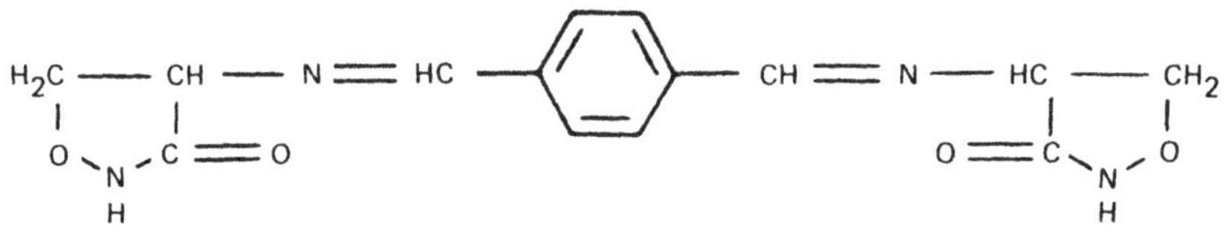

(II)

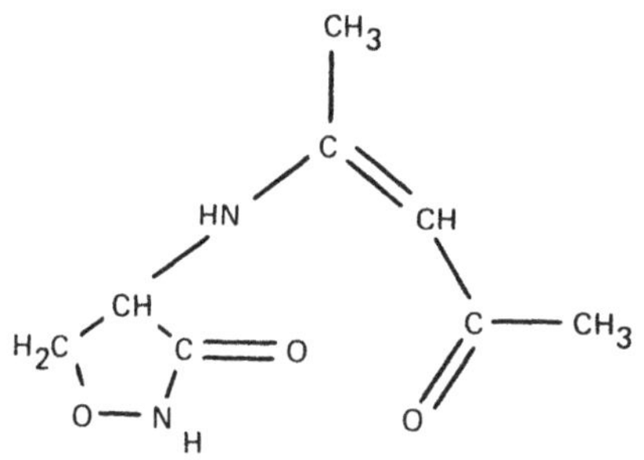

(III)

TABLE 1

Properties of D-Cycloserine

pK_1	4.4
pK_2	7.4
M.P.	154-156
$[\alpha]_D^{25}$	73.7° (c,1 in water)
Mol wt.	102

These derivatives function in vivo by releasing D-cycloserine and are used to avoid the adverse side effects of high in vivo D-cycloserine concentrations vide infra.

2. MODE OF ACTION

D-cycloserine belongs to a group of antibiotics which interfere with the incorporation of D-alanine into bacterial cell walls [9]. Other related amino acids include O-carbamyl-D-serine (IV) [10], β-fluoro-D-alanine (V) [11], and β-chloro-D-alanine (VI) [12]. Alafosfalin (VII) [13] is not an amino acid but functions in a similar way to the D-alanine analogs.

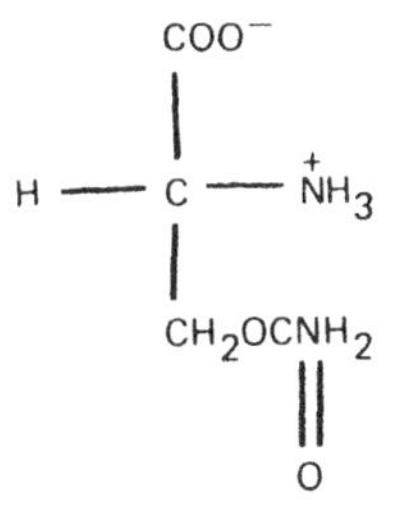

(IV) 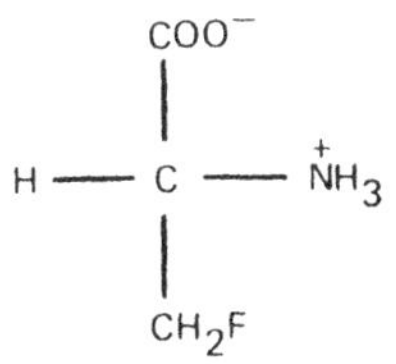   (V) 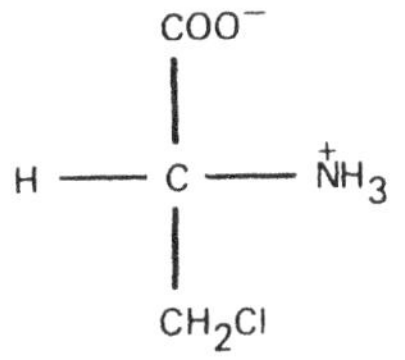   (VI)

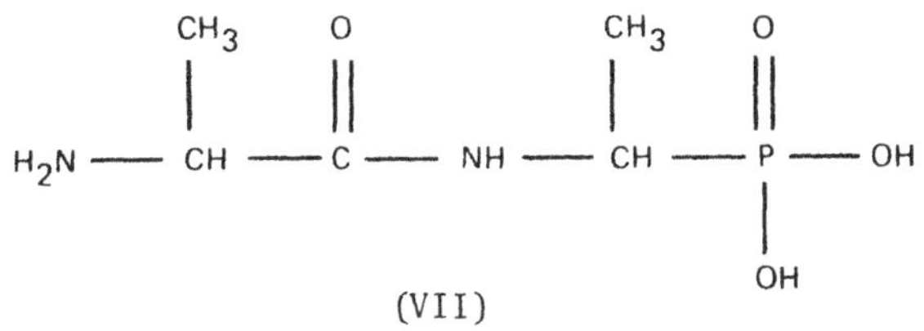

(VII)

A comprehensive review of this kind of antibiotic has recently been published [9].

The incorporation of D-alanine into the bacterial cell wall depends critically on two enzymes, alanine racemase (EC 5.1.1.1) and D-alanine:D-alanine ligase (ADP) (EC 6.3.2.4). Strominger and co-workers [14] first demonstrated that D-cycloserine was an effective inhibitor of these enzymes.

The following discussion will concentrate on alanine racemase inhibition. Most racemases are pyridoxal-dependent enzymes. These antibiotics initially function as competitive inhibitors of the enzyme and form Schiff's bases with enzyme-bound pyridoxal. There is also a subsequent time-dependent irreversible inactivation of the enzymes.

Among the above amino acids two different mechanisms for the irreversible inhibition can be identified. D-cycloserine causes the irreversible formation of substituted oximes and/or the acylation of enzyme amino acid residues. O-carbamyl-D-serine and the halogen-substituted amino acids function by alkylating the enzyme nucleo-phile.

The proposed mechanism for the irreversible inhibition of pyridoxal-linked enzymes by D-cycloserine is illustrated in Fig. 1 [15]. Behavior of this kind is termed suicide substrate inhibition [16]. A detailed discussion of all the above reactions is provided by Neuhaus and Hammes [9].

3. USES OF D-CYCLOSERINE

D-cycloserine is inhibitory for *M. tuberculosis* at concentrations of 5-20 μg/ml in vitro [17]. The drug has been used in the treatment of tuberculosis and of severe urinary tract infections. The typical dose is 250 mg twice daily, this may be doubled in extreme cases. A major advantage of D-cycloserine is that there is no cross-resistance of bacteria between it and other tuberculostatic drugs. The combina-tion of D-cycloserine and fluoro-D-alanine has been termed a "universal

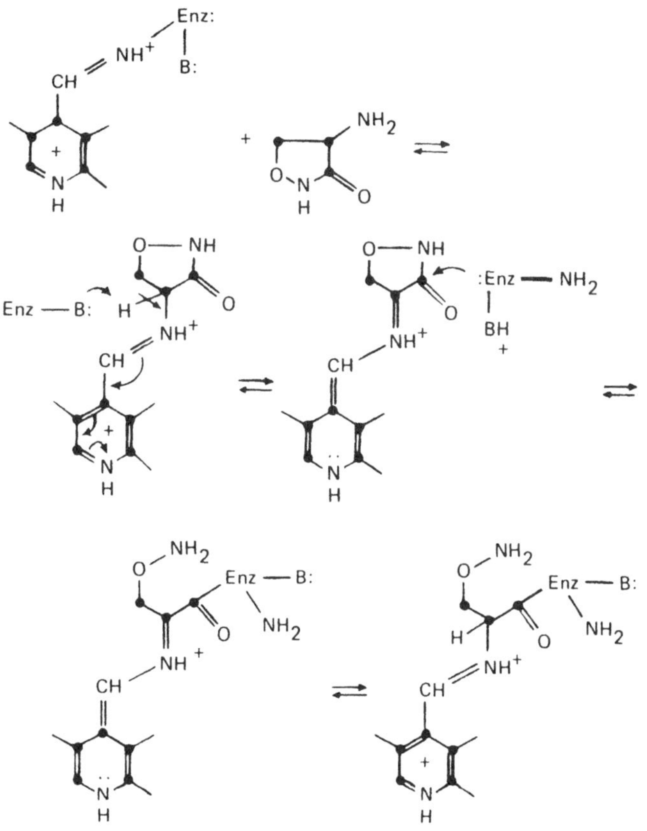

FIG. 1. Irreversible inhibition of pyridoxal-linked enzymes by cycloserine. The proposed scheme is based in part on experiments which show that D-cycloserine is an irreversible inhibitor of alanine racemase from *B. subtilis*. (Reprinted with permission from Ref. 15.)

antibiotic" [18] in that no bacterial strain has been found to be resistant to this combination.

Such advantages are tempered by the adverse side effects of D-cycloserine. The central nervous system (CNS) [9,17] is often disrupted, various minor symptoms can be recognized, but in cases of a serious reaction grand mal seizures may result. The adverse side effects have limited the use of the drug, e.g., at present there appears to be relatively little or no use of D-cycloserine

in the United Kingdom. One way of overcoming these problems is the
use of a derivative such as terizidone (II). The slow release of
D-cycloserine from this drug avoids high in vivo concentrations of
D-cycloserine and consequent tubular reabsorption; adverse effects
on the CNS are thus minimized. The various other D-alanine analogs
mentioned in the introduction [10-13], and a number of derivatives
of cycloserine [7,8] have been developed for similar reasons.

The use of complexation to detect D-cycloserine in pharmaceutical preparations has been reported [19,20]. The assay of this
drug is particularly important because it is degraded fairly rapidly
(e.g., 50% in one year [19]).

The veterinary use of D-cycloserine has been reported to combat
shipping fever [21]. A preparation containing zinc bacitracin, D-
cycloserine, and O-carbamyl-D-serine has been patented as a promoter
of animal growth [22].

4. ORGANIC CHEMISTRY OF D-CYCLOSERINE

Cycloserine is easily converted into a dimer (VIII) (+)-3,6-bis(amino-
methyl)-2,5-piperazinedione [23], this reaction occurs even in the
solid state [24].

$$2, R=CH_2ONH_2$$

(VIII)

The equilibrium between monomer and dimer is reached rapidly in the
pH range 1-2 at ambient temperatures [23]; the important steps in
this reaction are illustrated in Fig. 2. A detailed study of the
hydrolysis of the D-cycloserine dimer has also been reported [25].

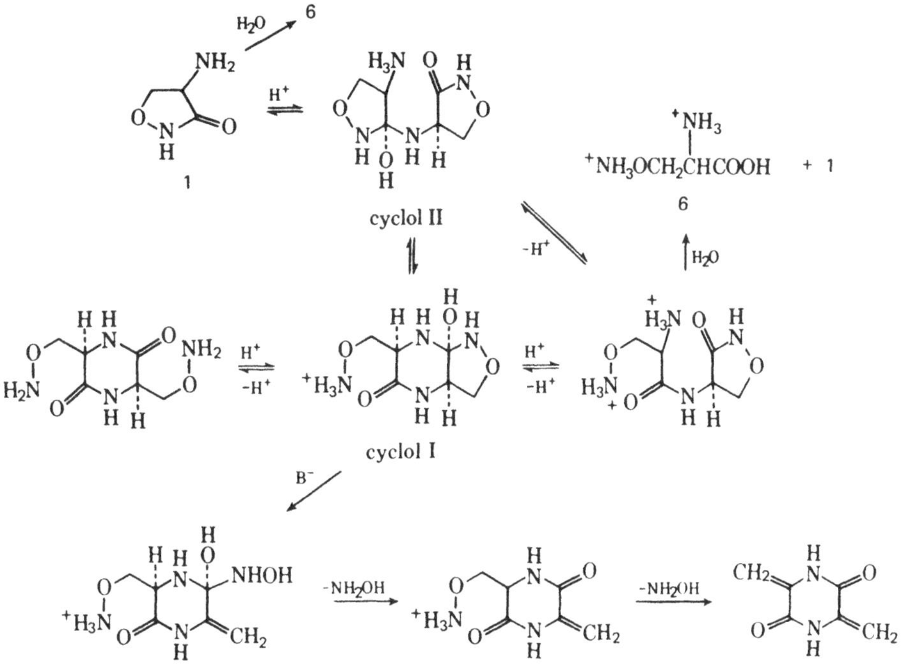

FIG. 2. Reaction scheme for the dimerization of D-cycloserine, also showing degradation products. (Reprinted with permission from Ref. 23.)

These reactions need to be considered in studies of metal ion complexation and the dimer also shows biological activity.

5. METAL ION COORDINATION BY D-CYCLOSERINE

5.1. Solution Equilibria

The protic equilibria important in understanding the behavior of D-cycloserine in aqueous solution are summarized in Fig. 3. Although the microscopic equilibrium constants are not known for this system the main route for ionization has been suggested to be [28] as shown over leaf.

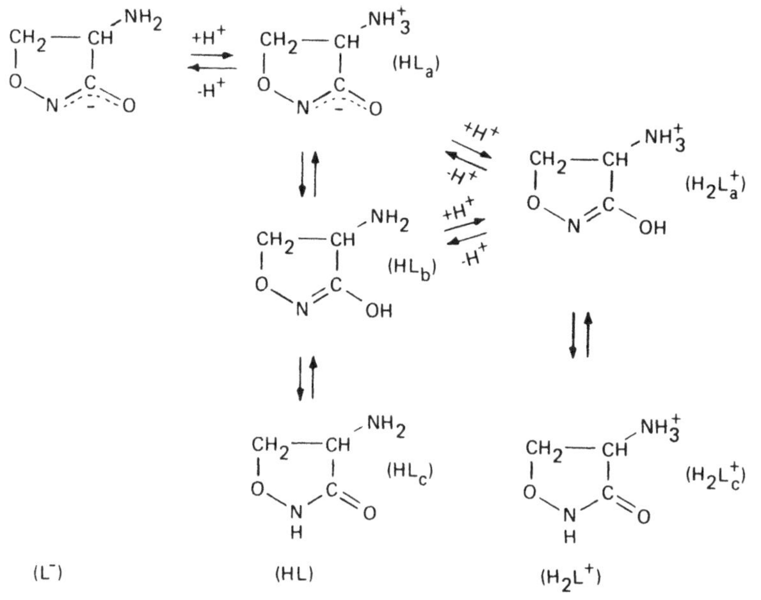

FIG. 3. Ionization scheme for D-cycloserine. (Reproduced with permission from Ref. 28.)

$$L^- + H^+ = HL_a$$

$$HL_a = HL_c$$

$$HL_a + H^+ = H_2L_a^+$$

There have been a number of studies of the complexation of metals by D-cycloserine in aqueous solution [26-28]. The metals which have been investigated include Co(II), Ni(II), Cu(II), Zn(II), and Hg(II); some results are summarized in Table 2. The structures considered for metal complexes in solution have included those listed below:

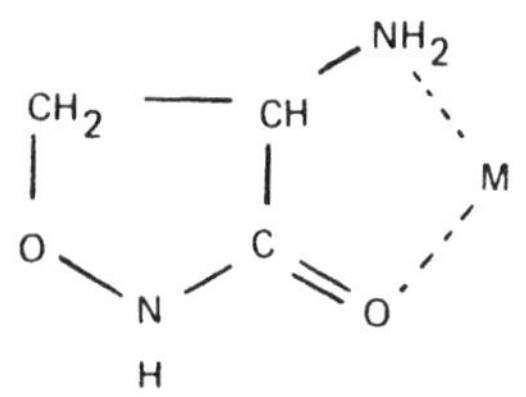

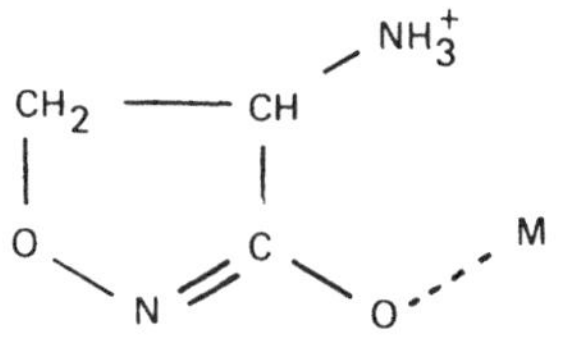

TABLE 2

Stability Constants of D-Cycloserine Complexes[a]

Metal	$\log K_{ML}^{M}$	$\log B_{ML_2}^{M}$	Ref.
Co(II)	3.38	5.91	28
	-	5.7	26
Ni(II)	4.02	7.23	28
Cu(II)	6.20	10.53	28
	5.55	10.25	27
	-	9.70	26
Zn(II)	3.48	6.09	28
	-	6.00	26

[a]All at 25°C and I = 0.1, except [27] 30°C, dilute solution.

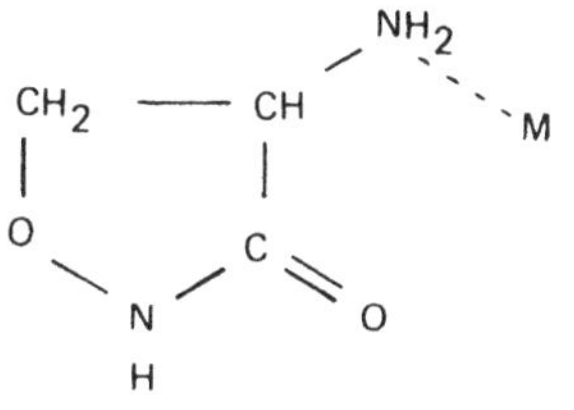

(XI) 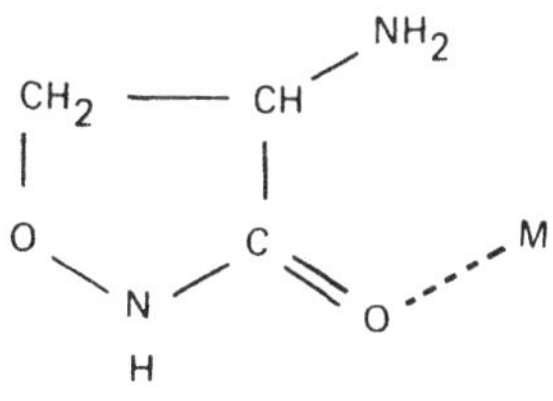   (XII)

In the most recent study of complexation in aqueous [28] solution the chelated complexes of the kind (IX) were held to be dominant. Equilibrium constants were similar to those measured for other metal complexes containing a hydrazide, and Raman results [29] for nickel(II) and zinc(II) complexes also support the dominance of chelated (N-O) structures in neutral solutions. However, when protonated species were considered, complexation by the deprotonated imine group of structure (X) was not suggested. This is surprising as the protonated form of the first copper complex, CuHL$^+$, has considerable stability. Copper(II) is well known to form particularly stable complexes with nitrogen containing ligands and would undoubtedly

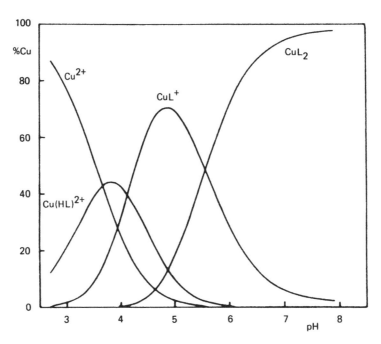

FIG. 4. Distribution diagram for the copper(II)/D-cycloserine
system. (Reproduced with permission from Ref. 28.)

bind more strongly to the imine group than to the adjacent carboxylate. A typical distribution diagram, that of the copper(II) system,
is illustrated in Fig. 4.

5.2. Main Group Complexes

Perhaps the first report of what might justifiably be called a complex of D-cycloserine was the preparation of the silver salt carried
out during the original characterization of the amino acid [4]. The
salt was used in assigning infrared bands, notably N-H stretches.
Subsequently the preparation of calcium, magnesium, barium, and
strontium salts of D-cycloserine was described and patented [30].
In a later infrared study of potassium and silver salts, the metals
were concluded to be O-bonded [31].

There are a number of reports concerning zinc complexes; the results of potentiometric studies are summarized in Table 2. Complexes of the form $[Zn(D-CS)_3X_2] \cdot nH_2O$ have been synthesized [32] by reacting the amino acid with a zinc halide in water at 40°C. On the basis of infrared spectra a pentacoordinate structure was suggested for these complexes. A Raman [29] study of the complexation of zinc in neutral aqueous solution produced evidence for a tetrahedral chelated (N_2O_2) complex. Titration of zinc-containing solutions with D-cycloserine complexation was followed by monitoring the Raman spectra; $\bar{n}$ values greater than 2.0 could not be obtained. This suggests that for zinc the highest complex formed in solution is the bis species.

The synthesis of various cadmium and mercury complexes of D-cycloserine has been reported [32]. Solid cadmium and mercury chloride and mercury bromide complexes of 1:1 stoichiometry were synthesized by the same method as that described above for the zinc complexes. On the basis of far-infrared spectra, polymeric structures were suggested, whether or not the bridging ligand, which might be D-cycloserine or a halogen, could be determined. In contrast for cadmium bromide 1:2 and 1:3 complexes were isolated. These were suggested to have C_{2V} and pentacoordinated geometries, respectively. A solution study [33] of mercury(II) complexation led to a value of 17.5 of $\log\beta_{ML_2}^M$ (25°C, I = 0.4). The complexes involved were suggested to be O-bonded and not chelated.

5.3. Transition Metal Complexes

Octahedral complexes of the type $[M(D-CS)_3X_3]$ containing monodentate O-bonded cycloserine have been synthesized [34] for chromium(III), rhodium(III), and iridium(III) and the related manganese(II) complex. The preparation involved refluxing the metal chloride with a suspension of D-cycloserine in chlorobenzene. Infrared and electronic spectra were reported, together with magnetic moments. On the basis

of the splitting of metal-ligand vibrations, the complexes are held
to be meridional isomers.

Similar complexes $[M(D-CS)_2X_2]$ have been synthesized [32] with
divalent cobalt and nickel, on the basis of spectroscopic methods a
variety of tetrahedral and pentacoordinate geometries involving O-
bonded D-cycloserine are suggested for these and related halogen
species. Complexes with platinum(II), palladium(II), and copper(II)
$[M(D-CS)_2X_2]$ have been reported [35] (M = Pd(II), Pt(II) and X = Cl
with n = 2; M = Pd(II), Pt(II) and X = Br, I with n = 2; M = Cu(II),
and X = Cl, Br with n = 2). Infrared evidence is again used to
suggest that all these complexes contain O-bonded D-cycloserine.
The 1:2 species are held to be square planar and monomeric; the 1:3
complexes to have a dimeric octahedral structure. The infrared
results were interpreted with the help of an earlier study of metal
isoxazole complexes [36]. All the above complexes are of a remarkably different structure to the (N-O) chelated form of complex [26-
29] believed to dominate in aqueous solution.

The results of spectroscopic studies of solutions also provide
evidence for chelated complexes. An early qualitative study [37] of
the complexes with cobalt, copper, nickel, and zinc using electronic
spectra concluded that chelate complexes were dominant. A purple
copper(II) complex $[Cu(D-CS)_2] \cdot 2H_2O$ precipitates from neutral solutions of copper(II) and D-cycloserine (pH = 6.99) over several days
[38,39]. Such solutions are initially green and may contain either
monomeric "ring-open" complexes (XIII) [38] or polymeric complexes
(XIV) [39]. These complexes were studied by EPR and electronic

(XIII)   (XIV)

spectroscopy and circular dichroism. The room temperature solution
EPR spectra of the purple complex showed N-superhyperfine splitting
attributable to two equivalent nitrogens [38]; this is compelling
evidence for a chelate complex (IX). The other spectroscopic properties of the purple complex were all in accord with such a structure.

The situation is much less clear for the green complex initially
present in these solutions. A solid 1:1 complex $[Cu(D-CS)] \cdot ClO_4 \cdot 4H_2O$
has been isolated from such green solutions and this may be a relevant
observation. However, the precise chemistry of this system is not
understood and is currently under investigation [40]. It is most
unusual for the formation reaction of a copper(II) complex to proceed slowly and there must be some unusual chemistry involved in
these reactions. The widespread occurrence of ring-open complexes
with kinetically inert metals reflects some similar chemistry [34].
For the nickel(II) complex in aqueous solution Raman evidence [29]
also supports the formation of bis-chelated (N,O) D-cycloserine
complexes; trans-water molecules complete an octahedron.

There are a limited number of observations concerned with the
biological activity of metal complexes. Copper(II) has been reported
[41] to suppress the inhibition of *Pseudomonas fluorescens* by D-
cycloserine; iron(II) and iron(III) have been found to inactivate
D-cycloserine [42]. In contrast, the lytic action of D-cycloserine
on *S. aureus* was unaffected by manganese(II), iron(II), cobalt(II),
or copper(II) [43].

6. METAL ION COORDINATION BY
D-CYCLOSERINE DERIVATIVES

There have been a number of studies of the complexation of transition metals by "terizidone" (II), 4,4'[1,4-phenylenebis(methylidyne-
nitrilo)]bis(isoxazolidin-3-one). Complexes with divalent cobalt,
nickel, copper, zinc, and cadmium of empirical formula $M(trz)_2X_2 \cdot nH_2O$
have been synthesized [44]. The complexes were prepared by reacting
the hydrated metal salt with a suspension of the ligand in acetone
or ethanol and refluxing for 1 hr. Spectroscopic studies were under-

taken and it was concluded that the complexes were monodentate
0-bonded or bidentate 0,0'-bonded through the oxygen of the keto
group.

In a later report [45] a large number of complexes of terizidone with trivalent chromium, ruthenium, and rhodium were investigated. The preparation procedure was similar to that described
above; however, by varying the ratio of ligand-to-metal complexes
of 1:1, 1:2, 1:3, and 1:6, ligand/metal stoichiometries could be
isolated. All the complexes were octahedral; a wide range of spectroscopic and thermal measurements were reported. Complexes were
held to be monodentate and either 0-bonded e.g., $[Cr(trz)(H_2O)_2Cl_3]$
(CrO_3Cl_3) or N-bonded e.g., $[Ru(trz)_6]Br_3(RuN_6)$.

In contrast, in a series of complexes [46] of iron(II),
iron(III), manganese(II), or palladium(II) terizidone was found
to be N-bonded by the imino nitrogen of the isoxazole ring. The
difference in coordination suggested between these complexes and
those described in the above study is surprising as the complexes
were all prepared in a similar manner [44].

7. CONCLUSIONS

There have been a number of studies of the metal ion complexes of
D-cycloserine. It is frequently pointed out in such work that coordination of an antibiotic to a metal ion may markedly alter the
effectiveness of the drug both in vitro and in vivo [32,34,35], and
there is much work to support this as a general suggestion [41,47-
49]. In addition, the mode of action of D-cycloserine has been
described as being better understood than that of any other antibiotic [49]. It is hence quite surprising that there has been little
or no serious study of D-cycloserine complexes as antibiotics. There
are only a number of early and disjointed observations and speculations [37,41-43,50]. Complexation may alter the toxicity, bioavail-
ability, pharmacokinetics, and even mode of action of a drug. There

is hence scope for studies of the microbiology of well-characterized D-cycloserine complexes.

This leads to a second point. There is a remarkable lack of unequivocal structural information about D-cycloserine complexes, e.g., there do not appear to have been any crystal structure determinations. The majority of papers dealing with solid state complexes report O-bonded monodentate complexes [32,34,35], and although these reports seem reliable, a far wider range of chelate complexes (IX) might yet be synthesized. There appear to be no cobalt(III) complexes, a quite startling omission from the coordination chemistry of any ligand. The coordination sites of the molecule include the imino nitrogen; coordination by this site has been reported [46,39] but often appears to be ignored when structures are considered [28]. This site is probably particularly important in the copper(II) system, which shows unusual properties.

The newer D-amino acid analogs, such as β-D-fluoroalanine, would obviously readily form complexes with metals; to date it appears that complexation by these antibiotics has not been studied.

The antibiotic D-cycloserine was first isolated nearly 30 years ago, and it is an excellent ligand for many metal ions. The history of its clinical use has been dogged by adverse side effects. In an attempt to produce a drug which minimizes such adverse interactions, many organic derivatives have been synthesized. However, the therapeutic potential and toxicity of metal ion D-cycloserine complexes has not been investigated. There is hence still considerable scope for further studies of the metal ion complexes of D-cycloserine, particularly for more biological work on well-characterized compounds. There are apparently no studies of the coordination chemistry of the newer D-amino acid analogs.

ABBREVIATIONS

D-CS	D-cycloserine	trz	"terizidone"
M	a metal ion	X	a halogen

NOTE ADDED IN PROOF

During the time this article has been in press a number of complexes
of deprotonated D-cycloserine have been reported [51]. Complexes of
Cr(III), Mn(II), Fe(II), Fe(III), Co(II), Ni(II), Zn(II), Zr(IV),
Pd(II), Ag(I), Cd(II), Os(III), Pt(II) and Hg(II) have been prepared;
the majority of the complexes were suggested to be N,O chelates [51].

ACKNOWLEDGMENTS

I should like to thank Prof. W. Hunter and Dr. J. Walker (Department
of Pharmacy, Chelsea College) for helpful discussions. Correspondence
with the following has been particularly valuable: Ms. V. Campbell
(Roche), Ms. A. C. Boddy (Eli Lilly), Prof. C. H. Stammer (University
of Georgia), and Prof. F. C. Neuhaus (Northwestern University).

REFERENCES

1. D. A. Harris, M. Ruger, M. A. Reagen, F. J. Wolf, R. L. Beck,
    H. Wallick, and H. B. Woodruff, *Antibiot. Chemother.*, 5, 183
    (1955).

2. R. P. Buhs, I. Putter, R. Ormond, J. E. Lyons, L. Chaiet, F. A.
    Kuehl Jr., F. J. Wolf, N. R. Trenner, R. L. Peck, E. Howe, B. D.
    Hunnewell, G. Downing, E. Newstead, and K. Folkers, *J. Am. Chem.
    Soc.*, 77, 2344 (1955).

3. G. M. Shull and J. L. Sardinas, *Antibiot. Chemother.*, 5, 395
    (1955).

4. P. H. Hidy, E. B. Hodge, V. V. Young, R. L. Harned, G. A. Brewer,
    W. F. Phillips, W. F. Runge, H. E. Stavely, A. Pohland, H. Boaz,
    and H. R. Sullivan, *J. Am. Chem. Soc.*, 77, 2345 (1955).

5. C. H. Stammer, A. H. Wilson, F. W. Holly, and K. Folkers, *J. Am.
    Chem. Soc.*, 77, 2346 (1955).

6. J. W. Lamb, *Analytical Profiles of Drug Substances*, Vol. 1
    K. Flory, ed.), Academic Press, New York, 1972, p. 53.

7. F. Bonati, L. Bertoni, G. Rosati, and V. Zanicheli, *Farmaco.
    Prat.*, 20, 381 (1963).

8. N. P. Jensen, J. J. Friedman, H. Kropp, and F. M. Kahan, *J.
    Med. Chem.*, 23, 6 (1980).

9.   F. C. Neuhaus and W. P. Hammes, *Pharm. Ther.*, *14*, 265 (1981).

10. E. Wang and C. Walsh, *Biochemistry, 17,* 1313 (1978).

11. J. Kollonitsch and L. Barash, *J. Am. Chem. Soc.*, *98,* 5591 (1976).

12. J. M. Manning, N. E. Merrifield, W. M. Jones, and E. C. Got-schlich, *Proc. Natl. Acad. Sci. USA, 71, 417* (1974).

13. J. G. Allen, F. R. Atherton, M. J. Hall, C. H. Hassall, S. W. Holmes, R. W. Lambert, L. J. Nisbet, and P. S. Ringrose, *Nature, 272,* 56 (1978).

14. J. L. Strominger, E. Ito, and R. H. Threnn, *J. Amer. Chem. Soc., 82,* 998 (1960).

15. R. R. Rando, *Acc. Chem. Res., 8,* 281 (1975).

16. R. H. Abeles, *Chem. Eng. News, 61*(38), 48 (1983).

17. Goodman and Gilman, *The Pharmacological Basis of Therapeutics,* 6th ed. (A. Goodman, ed.), Macmillan, New York, 1980, p. 1210.

18. Anonymous, *Chem. Eng. News, 53* (Oct. 6th, 1975), p. 6.

19. I. Korol, *Farmatsevt. Zh. (Kiev), 18,* 44 (1963); *Chem. Abstr., 61,* 5463a.

20. L. R. Jones, *Anal. Chem., 28,* 39 (1955).

21. D. R. Bright and R. D. Williams, U.S. Patent 4,271,174, June 1981.

22. Commercial Solvents, British Patent 1,134,788, 1966; *Chem. Abstr. 70,* 65793.

23. F. Lassen and C. H. Stammer, *J. Org. Chem., 36,* 2631 (1971).

24. M. Ya. Karpeskii, Yu. N. Brensov, R. M. Khomutov, E. S. Severin, and O. L. Polyanovskii, *Biochemistry, 28,* 280 (1963).

25. J. L. Miller, F. C. Neuhaus, F. O. Lassen, and C. H. Stammer, *J. Org. Chem., 33,* 3908 (1968).

26. J. B. Neilands, *Arch. Biochem. Biophys., 62,* 151 (1956).

27. P. J. Niebergall, D. A. Hussar, W. A. Cressman, E. T. Sugita, and J. T. Dolusio, *J. Pharm. Pharmacol., 18,* 729 (1966).

28. A. Braibanti, F. Dallavalle, and G. Mori, *Ann. Chim. (Rome), 71,* 223 (1981).

29. H. Takahashi, H. Yamada, and T. Igarashi, *Spectrochim. Acta, 37A,* 247 (1981).

30. Commercial Solvents, British Patent 787,741, 1957; *Chem. Abstr., 52,* 7623d.

31. V. G. Vinokurov, V. S. Troitskaya, and N. K. Kochetkov, *Zhur. Obshscei Khim., 31,* 205 (1961). *Chem. Abstr., 55,* 22288.

32. C. Preti, G. Tosi, and P. Zannini, *Z. Anorg. Allg. Chem., 453,* 173 (1979).

33. N. G. Lordi, *J. Pharm. Sci., 52,* 397 (1963).

34. C. Preti and G. Tosi, *Austr. J. Chem., 33* (1980).

35. C. Preti and G. Tosi, *J. Coord. Chem., 9,* 125 (1979).

36. C. Preti, G. Tosi, M. Massacesi, and G. Ponticelli, *Spectrochim. Acta, 32A,* 1779 (1976).

37. E. Neuzil and J. C. Breton, *Bull. Med. A.O.F., 3,* 149 (1958).

38. H. Sakuri, C. Shibata, and H. Matsuura, *Inorg. Chim. Acta, 56,* L25 (1981).

39. P. O'Brien, *Inorg. Chim. Acta, 78,* L37 (1983).

40. P. O'Brien, unpublished results.

41. E. D. Weinberg, *Bacteriol. Rev., 47,* 52 (1957).

42. Personal communication Harned to Weinberg; cited by F. C. Neuhaus in *Antibiotics,* Vol. 1 (D. Gottlieb and P. D. Shaw, eds.), Springer-Verlag, Berlin, 1968, p. 40.

43. J. L. Smith and E. D. Weinberg, *J. Gen. Microbiol., 28,* 559 (1962).

44. C. Preti and G. Tosi, *J. Coord. Chem., 10,* 209 (1980).

45. C. Preti, L. Tassi, G. Tosi, P. Zannini, and A. F. Zanoli, *Austr. J. Chem., 35,* 1829 (1982).

46. C. Preti, L. Tassi, G. Tosi, P. Zannini, and A. F. Zanoli, *J. Coord. Chem., 12,* 177 (1983).

47. A. J. Thompson, R. J. P. Williams, and S. Reslova, *Structure Bond., 11,* 1 (1972).

48. S. Kirschner, Y. K. Wei, D. Francis, and D. J. Bergman, *J. Med. Chem., 9,* 368 (1966).

49. A. Albert, *Selective Toxicity,* 6th ed., Chapman and Hall, London, 1979.

50. F. Ito, T. Akoi, M. Yamamoto, M. Yuasa, H. Mizobata, and K. Tone, *Med. J. Osaka Univ., 9,* 23 (1958).

51. F. Forgheri, C. Preti, G. Tosi, and P. Zanni, *Austr. J. Chem., 36,* 1125 (1983).

Chapter 11

IRON-CONTAINING ANTIBIOTICS

J. B. Neilands
Department of Biochemistry
University of California
Berkeley, California

J. R. Valenta
Smith Kline and French Laboratories
Philadelphia, Pennsylvania

1. INTRODUCTION

1.1. Scope

This brief review will be restricted to a family of microbial compounds which, when isolated, contain iron, or, because of a particular chemical constitution, can be expected to exist in vivo at least partially in an iron complex form. This means that our scope will be confined to certain *siderophores* (Gr: "iron bearers"), such as albomycin, and to a few other compounds with antibacterial activity which happen to be particularly effective ligands for Fe(III) or Fe(II). For the former ion, this will mean hydroxamic acid residues R-CO-N(OH)R'; for the latter ion, mainly cyclic 1-4 diimines related to bipyridyl. The focus will be on the chemical nature of the compounds.

The red-brown, iron-containing products isolated from *actinomyces* were originally dubbed *siderochromes* and classed as either *sideromycins* or *sideramines*, depending on whether they displayed antibiotic or growth factor activity [1]. This nomenclature was not entirely satisfactory since the biological activity of a particular compound proved to be species-specific. Subsequently, Lankford [2] proposed the general term siderophore for all microbial products specifically binding and transporting Fe(III).

For an historical account of siderophores with antimicrobial activity, the reader is referred to reviews by Gause [3], Prelog

[1], Keller-Schierlein et al. [4], Nüesch and Knüsel [5], Emery [6], and Keller-Schierlein [7].

1.2. Background

Interest in iron-containing antibiotics derives, in the first instance, from the unique physical and chemical properties of iron and the possible universal requirement of the metal for all living species. Dating from the aerobic phase of evolution, environmental iron has existed, in the main, in the higher oxidation state. Given the solubility product constant of $\sim 10^{-38}$ M for $Fe(OH)_3$, the concentration of free Fe^{3+} present in solution at biological pH will be limited to about 10^{-18} M. This small concentration of free ferric ion is beneficial from the standpoint of its toxicity in an aerobic environment (see below), but it has required the perfection of specialized mechanisms for assimilation and metabolism which are able effectively to compete with hydroxyl ion. The siderophores and their transport systems constitute the high-affinity iron assimilation pathway common to virtually all aerobic and facultative anaerobic microorganisms. Meanwhile, as we know from the classic work of Schade and Caroline [8], animal tissues have become equipped with singularly effective ligands for binding and control of Fe(III), viz. serum transferrin, ovotransferrin, and lactoferrin. These ligands assure that the animal survives in competition with microorganisms for the iron on which they both depend for reactions ranging from DNA synthesis to energy metabolism.

1.3. Mechanisms of Antibiosis Related to Iron

At this time it appears that there are at least three distinct means whereby iron and antibiosis may be correlated.

Type I. In this mechanism there is a simple competition among species for iron. O'Brien and Gibson [9] pointed out that siderophores effective for only one species will endow that life form with

a competitive advantage in the struggle for limiting resources of
soluble iron. There is no doubt that this ecological principle is
an important one in the microbial world and is effected by structural variation, which may range from a specific chirality in the
ferrisiderophore as a coordination compound which affects recognition of the transport component of the system [10], to variations
in the thermodynamic stability of the complexed metal ion. The
prime example of this type of antibiosis is that due to deferriferrichrome A. This compound, although acting as a siderophore in the
producing fungi [11], denies iron to enteric bacteria stripped by
mutation of their own siderophore and hence it very effectively
inhibits the growth of such species [12]. In contrast to ferrichrome
A, ferrichrome serves as a potent source of iron for many bacterial
species, which typically carry specific receptors for its transport.
Several other examples can be cited. A hydroxamic acid from *Asper-
gilli* and *Penicillia*, active against *Proteus* sp., turned out to be
the common fungal siderophore deferritriacetylfusigen [13]. A
variety of siderophores of the linear catechol genre were shown to
exhibit bacteriostatic and fungistatic action [14]. Iron deprivation
is at least in part the basis of the bacteriostatic action of transferrin, lactoferrin, and ovotransferrin [15].

The principle just enunciated can be extended to purely synthetic compounds such as ethylenediaminedi-(o-hydroxyphenylacetic acid)
(Fig. 1), which stresses severely most bacteria lacking efficient
indigenous siderophore systems. Hutner [16] drew attention to this
compound, which may also be described as N,N'-ethylenebis[2-(ortho-
hydroxyphenyl)]glycine, pointing out that its growth-promoting effects
for plants had not been explored in microbes. Rogers [17] then showed
that exposure of *E. coli* to the chelator, generally abbreviated as
EDDA or EDDHA, invokes expression of siderophore synthesis. EDDA is
reported [18] to have a stability constant of $10^{33.9}$ for Fe(III) but
has substantial avidity for other bionutritious metal ions and is
hence less specific than transferrin or deferriferrichrome A for
iron(III). Miles and Khimji [19,20] devised an agar plate test for
resistance to EDDA, a putative index of the capacity of the cell to

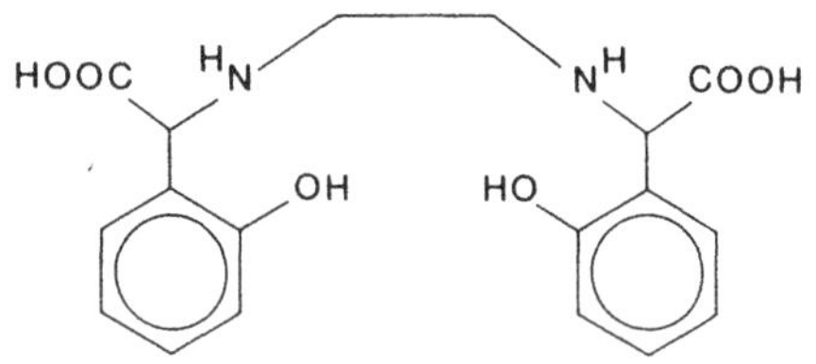

FIG. 1. Structure of ethylenediaminedi-(o-hydroxyphenylacetic acid), EDDA or EDDHA; a synthetic chelating agent widely used to stress microorganisms for iron [19,20].

produce siderophores. The crux of the assay is the use of a light seeding in conjunction with the correct level of EDDA, the latter determined by trial and error, which should be of the order of 1 mg/ml or less. The readout is a constellation of small, discrete colonies rather than a halo of confluent growth.

Type II. In the second mechanism, a toxic substance is smuggled into the cell on a pathway designed for iron. Thus, illicit transport on the ferrichrome receptor accounts for the uptake of albomycin (see below) [21]. This receptor also serves as the common binding site for bacteriophages T1, T5, ϕ80, UC-1, and for colicin M. Additional examples of this type of competition are the ferric enterobactin-colicin B and ferric aerobactin-cloacin combinations, thus leading to the conclusion that the iron pores are favorite channels for penetration of lethal agents. There is no evidence to suggest that such agents use any part of the high-affinity iron uptake system other than the surface receptor and once inside the cell the mechanism of antibiosis may be completely unrelated to iron metabolism [22].

Type III. The third mechanism of antibiosis by iron is based on the propensity of free ferrous or ferric ions to exacerbate the toxicity of partially reduced oxygen species, such as superoxide anion and hydrogen peroxide. Only catalytic levels of the free metal ions are required to generate the inordinately active oxidizing species $OH^\bullet$.

$$O_2^{\bullet -} + Fe^{3+} = Fe^{2+} + O_2$$
$$H_2O_2 + Fe^{2+} = Fe^{3+} + OH^\bullet + OH^-$$

Ligands which bring the metal ion into close contact with sensitive biomolecules, such as DNA, need not be specific for one oxidation state of iron or even for this element in order to achieve this type of lethal action. Antibiotics such as streptonigrin appear to operate on this principle because their potency is directly correlated with the intracellular concentration of iron [23]. Additional examples, such as the bleomycins [24], are discussed in Chapter 4 of this volume.

1.4. Prospects for Chemotherapeutic Intervention

Given the fact that the high-affinity iron assimilation systems are generally switched on at about micromolar concentrations of iron, it can be assumed with confidence that pathogenic organisms are fully derepressed when growing in the tissues of the host. If the high-affinity iron absorption pathway is correlated with virulence, as seems to be the case [25], and if this pathway is unique to the pathogen and is not duplicated in the host, it is obvious that a rational approach to chemotherapy is possible based on exploitation of some aspect of microbial iron metabolism. The possibilities would seem to be:

(1) Control by a more powerful ligand for iron. Antibiosis via iron starvation may be difficult to achieve in practice owing to the widespread presence of indigenous siderophores. Thus starving a pathogen for iron will in all likelihood only cause it to excrete higher levels of its own siderophore. Since the microbial protoplasm affords a thermodynamic sink for iron, relatively feeble chelators, such as aerobactin, can extract iron from more powerful ones, such as deferriferrichrome A [26]. In any event, this strategy would not be expected to be more successful than that already existing in the case of the iron-binding proteins transferrin, lactoferrin, and ovotransferrin. All of these are bacteriostatic and undoubtedly impede growth initiation from small inocula but are not able to prevent entirely proliferation of extraneous organisms.

(2) *Illicit transport* [27] *of a toxic moiety.* Use of siderophore receptor-mediated uptake of an antimicrobial substance is sound
in principle and is the basis of action of albomycin. Siderophore
receptors are known to have a very high affinity [28] for the transported solute and would be able to pick these up at very low concentrations in body fluids. While the chemistry of this approach could
be quite straightforward, viz. synthesis of a ferric trihydroxamate-
β-lactam complex, generation of intracellular toxicity may require
scission of the poison from the vehicle. One could only speculate
as to what type of amide bond, for instance, might be cleaved by the
target cell. The principle is further illustrated by the toxicity
of the scandium derivative of enterobactin [29].

(3) *Inactivation of a siderophore receptor.* Microbial cells
can be expected to have a more limited capacity to synthesize the
surface receptors for siderophores. Accordingly, a monoclonal anti-
body directed at such receptors could be quite effective in providing
protection against specific infections.

In a recent editorial [30], entitled "Siderophores as Antimicrobial Agents," it is concluded that the field does indeed have
potential but is plagued by the cross reactivity of microbial iron
carriers. Thus an agent antagonizing an iron assimilation system
of one microbial species may actually serve to supply iron to a
second species. Thus we need to know more about the individual high-
affinity systems and the specificity parameters of their receptors
before an effective agent can be devised. Direct extrapolation from
laboratory media to the animal organism will have to take into account
certain differences in milieu. For example, enterobactin is more
efficient than aerobactin in removing iron from iron transferrin when
the transfer is performed in hepes buffer at neutral pH. The same
reaction carried out in the presence of serum, however, indicates
the acyclic hydroxamic acid aerobactin to be superior [31]. The
highly aromatic character of the catechol enterobactin causes it
to adhere to serum albumin and other proteins, and this binding
apparently interferes with the ability of the siderophore to acquire

iron from the transferrin. The redox balance of the cell may also
play a role, as evidenced by the potentiation of the antibacterial
activity of deferriferrioxamine B in the presence of ascorbic acid
[32]. Finally, siderophore ligands have been observed to display a
synergistic effect when administered with other antibiotics [33].
Thus, deferriferrioxamine B was found to enhance the antibacterial
activity of gentamycin, chloramphenicol, cephalothin, cefotiam, and
cefsulodin. Deferriferrichrome A should be screened in this way
because few bacteria seem able to extract iron from this fungal
siderophore.

2. IRON(III) COMPLEXES

2.1. Albomycin and Grisein

The most important iron-containing antibiotic is albomycin. It was
first reported in 1947 under the name grisein [34], the product of
a strain of *Streptomyces griseus* studied in the Waksman laboratory.
In 1951 Soviet workers [35] described an antibiotic under the name
albomycin from *Streptomyces subtropicus*. The two substances were
compared carefully by Stapley and Ormond [36], who concluded the
antibiotics to be closely related, possibly identical, in all biological and chemical properties. In a companion paper, entitled
"Penalty of Isolationism," Waksman [37] commented: "The repetitious
and the frequently unjustified creation of 'new species' of anti-
biotic-producing organisms and of 'new antibiotics' can be avoided
only by close collaboration among the scientific workers throughout
the world."

The early structural work was done in Prague by the Šorm group
using material furnished by the Soviet colleagues. The Czech chemists
[38] established the general character of the main component, δ_2 as
a peptide containing Fe(III) complexed to tri-N^δ-acetyl-N^δ-hydroxy-
ornithine. They reported the molecule to possess three residues of
serine contained within a cyclohexapeptide moiety and, in addition,
they noted the presence of a pyrimidine sulfonic acid of unusual
structure. The overall constitution was thus established as an

elaborate form of the ferrichrome compounds, which had just been
characterized as ferric trihydroxamatocyclohexapeptides.

A perfect analogy with the ferrichromes was, however, questioned when the x-ray structure of ferrichrome A [39] revealed the
presence in the cyclohexapeptide ring of a single residue of glycine
critical to the formation of the β turn. Furthermore, a decade later,
Maehr [40] isolated from *Streptomyces griseus* var. X-2455 three antibiotics, Ro 5-2667, Ro 7-7730, and Ro 7-7731, which were identical in
properties with albomycins δ_2, δ_1, and ε and which had properties
inconsistent with the assigned structure of albomycin. The compounds
could contain at most a single serine residue and, in addition, the
structure of the sulfur-substituted pyrimidine was doubted.

Recently, Benz et al. [41] reported a revised structure for
albomycin (Fig. 2). These workers isolated from *Streptomyces* sp.
WS 116 (DSM 1692) three sulfur-containing antibiotics identical in
properties with albomycins δ_1, δ_2, and ε. They showed by spectroscopic and other methods that the sulfur is contained within a thionucleoside moiety appended as a substitution on the β carbon of a

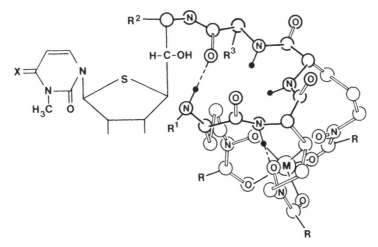

FIG. 2. Revised structure for albomycin. The x-ray structure of
ferrichrome has been modified to accommodate the thionucleoside
residue of albomycin. Retention of the H bonding and conformation
of the peptide, as shown, have not been verified experimentally.
Substituent X = N(C=O)NH$_2$, NH and O for albomycins δ_2, ε and δ_1,
respectively [41]. For all albomycins, R = CH$_3$; R^1 = H; R^2 = COOH;
R^3 = CH$_2$OH; M = Fe^{III}.

seryl residue at the terminal position of a *linear* pentapeptide.
The molecule has been drawn here in a way that shows its formal
analogy with the known x-ray and solution structure of ferrichrome
A, although this point has not been verified experimentally.

Albomycin displays a broad spectrum of antibiotic activity at
a minimum inhibitory concentration rivaling that of the β-lactam
antibiotics. Furthermore, the antibiotic is generally effective
against bacteria resistant to the common chemotherapeutic agents
such as penicillin, streptomycin, tetracycline, and erythromycin.
On the other hand, sensitive populations show a propensity to
acquire resistance. Cells are readily protected from albomycin by
correspondingly low concentrations of ferrichrome; this can be demonstrated conveniently by application of the agar strip diffusion cross-
test (Bonifas) and has enabled identification of the ferrichrome
receptor in the outer membrane of *Escherichia coli* [21]. This
receptor, designed by *E. coli* for an exogenous siderophore common
to a large number of fungi, serves also as a common binding site
for albomycin, bacteriophages T1, T5, φ80, UC-1, and colicin M [42].

Albomycin-resistant mutants lack the 78,000-Da outer membrane
receptor for ferrichrome which maps in an hydroxamate siderophore
operon at minute 2.5 on the *E. coli* chromosome [43]. The 78-kDa
receptor is iron-regulated, but the mechanism of control differs
from that seen in other iron-regulated outer membrane proteins [44].
The speed with which resistant mutants appear has diminished clinical
interest in albomycin. Such loss of a high-affinity iron assimilation
system, provided it were essential to virulence, would prevent growth
in vivo in any case. However, bacteria do not synthesize ferrichrome
and hence cannot depend on this particular uptake system for acquisi-
tion of iron from host tissues. Use of aerobactin as a vehicle might
prove effective against forms of *E. coli* causing disseminating infections since in these strains the aerobactin system, shown to be opera-
tional in serum [26], is thought to be a determinant of virulence.
Albomycin is inactive against fungi and the literature apparently
does not contain citations designed to probe the reason, in spite

of the fact that ferrichrome-type siderophores are produced and used
by many fungal species.

2.2. Ferrimycin A_1'

After albomycin the most important ferric trihydroxamate antibiotic
is the ferrimycin group from species of *Streptomyces* such as *S.
griseoflavus*. The group is characterized by maximum instability in
the pH range 5-7. Bickel et al. [45] showed ferrimycin A_1 to be an
elaborated form of ferrioxamine B in which the amino group of the
siderophore is linked to an aminohydroxybenzoic acid residue which,
in turn, is bonded to an iminoester-substituted lactam. New information obtained by ^{13}C and 300-MHz ^{1}H NMR spectroscopy has confirmed
the structure of the ferrioxamine and aromatic moieties but suggests
the presence of a seven-membered oxygen-bridged cycle (Fig. 3) [46].

   The ferrimycins display a narrower range of action than albomycin and are essentially effective only against Gram-positive bacteria. They are antagonized by the ferrioxamine siderophores.

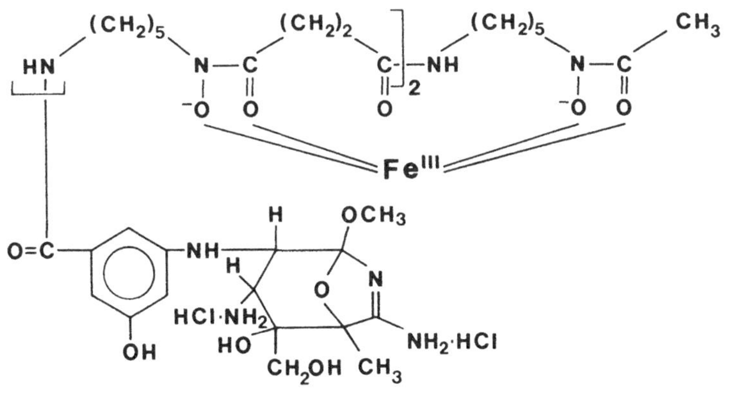

FIG. 3. Ferrimycine A_1 dihydrochloride [46,71].

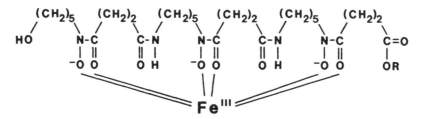

FIG. 4. Danomycin [46]. In danoxamine the toxic residue at R is replaced by H [71].

2.3. Danomycin

A group of antibiotics comprising danomycin from *Streptomyces alba-duncus*, A22765 from *Streptomyces aureofaciens,* and succinimycin from *Streptomyces olivochromogenes* are difficult or impossible to separate from one another by paper chromatography [5] and are related in structure to the ferrimycins. Unlike the latter, however, the former group shows maximum stability in the pH range 5-7. Most structural work has been performed on danomycin [46], which has been shown to contain a ferrioxamine moiety, dubbed danoxamin, composed of a residue of 5-hydroxyamino-1-pentanol, two residues of 1-amino-5-hydroxyaminopentane, and three residues of succinic acid. The nature of the alcohol ester-ified to the terminal succinic remains to be elucidated (Fig. 4).

Like the ferrimycins, danomycin, A22765 and succinimycin affect only gram positive microorganisms.

3. IRON(II) COMPLEXES

3.1. Ferroverdin and Viridomycin

A green, iron-containing pigment, designated ferroverdin, was iso-lated by Chain et al. [47] from a species of *Streptomyces* and the structure eventually established as the p-vinyl phenyl ester of 4-hydroxy-3-nitrosobenzoic acid [48] (Fig. 5). An x-ray structure

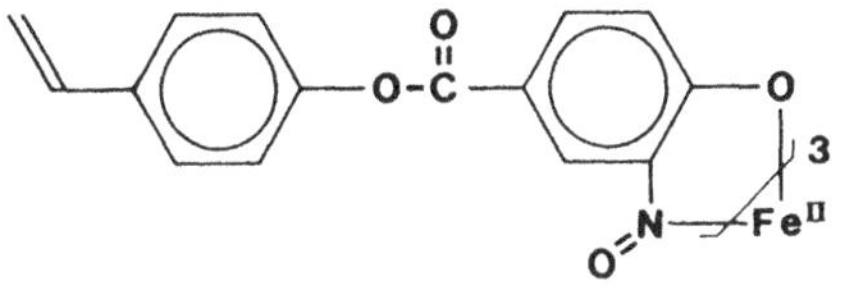

FIG. 5. Ferroverdin: three ligand molecules are coordinated to iron(II).

of the iron complex, which was shown to be in the ferrous state by measurement of the magnetic moment, disclosed the metal ion as coordinated to three ligand molecules through their aromatic hydroxyl and nitroso substituents [49].

Subsequently a group of antibiotics, called viridomycins, from *Actinomyces viridaris* was characterized by synthesis as containing 4-hydroxy-3-nitrosobenzaldehyde [50]. The compounds, which can be viewed as the ligand segment of ferroverdin, were less active against *E. coli* than they were against either *Sarcina lutea* or *Bacillus subtilis*. Viridomycin A, the iron(II) complex (Fig. 6), was shown to be tris rather than bis(4-hydroxy-3-nitrosobenzaldehydato-N^3,O^4)ferrate(II) [51]. In *E. coli* the mechanism of action was postulated to be directed at the structure and integrity of the bacterial envelope.

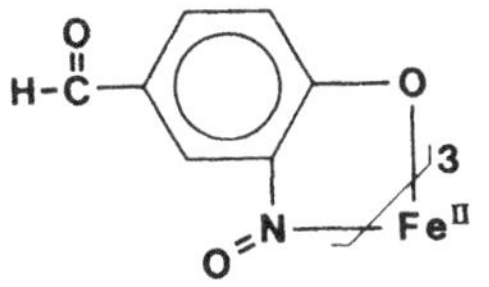

FIG. 6. Viridomycin A, the iron(II) complex.

3.2. Ferropyrimine, Ferrorosamine, and Siderochelin

Bacteria form a family of 1,4-diimines which act as powerful ligands
for iron(II) to afford magenta colored complexes reminiscent of the
hue of ferrous 1,10-phenanthroline or ferrous bipyridyl.

The first of these to be described chemically was pyrimine,
L-2-(2-pyridyl)-Δ^1-pyrroline-5-carboxylic acid, from *Pseudomonas* GH
[52]. Unlike the trihydroxamates, only very slightly higher levels
of the product are obtained by culture of the organism at low iron.
However, pyrimine may be readily prepared in optically pure form by
Claisen condensation of methyl picolinate with N-trityl glutamic
ester. It exists in an open chain form in aqueous mineral acid and
can be crystallized as the dihydrochloride of L-5(2-pyridyl)-2-amino-
5-ketopentanoic acid. Independent work by a French group with a
pigment, ferrorosamine [53], from *Bacillus roseus fluorescens,* later
reclassified as a *Pseudomonas* sp., yielded an amorphous powder from
which the base 2-(2-pyridyl)-1-pyrroline, the decarboxylation product
of pyrimine, was isolated by distillation.

Recently, ferropyrimine has been obtained from the phytopatho-
genic bacterium *Erwinia rhapontici* [54] (Fig. 7). It was speculated
that iron complexation could be the basis of plant disease caused by
the organism, such as crown rot of rhubarb.

An antibiotic designated siderochelin has been obtained from
Nocardia sp. SC 11,340 and shown by x-ray diffraction analysis to be
trans-3,4-dihydro-4-hydroxy-5-(3-hydroxy-2-pyridinyl)-4-methyl-2H-

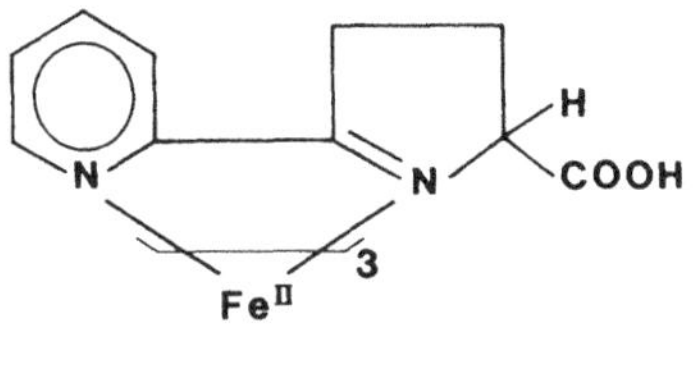

FIG. 7. Ferropyrimine [52].

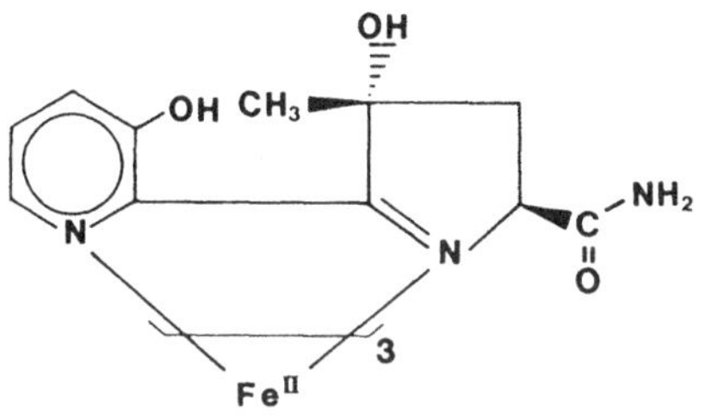

FIG. 8. Trans-3,4-dihydro-4-hydroxy-5-(3-hydroxy-2-pyridinyl)-4-methyl-2H-pyrrole-2-carboxamide (siderochelin A [55]). The Fe(II) complex is shown.

pyrrole-2-carboxamide [55] (Fig. 8). It is thus a derivative of pyrimine bearing additional substituents on both heterocyclic rings. Siderochelin is weakly active against a broad spectrum of bacteria, fungi, and protozoa.

4. MISCELLANEOUS COMPOUNDS

4.1. Aspergillic Acids

Aspergillic acid, a cyclic hydroxamic acid with broad spectrum antibiotic activity, has the distinction of being the first natural product to be shown to contain the RCON(OH)R' link [56]. The aspergillic acids are produced by strains of *Aspergillus flavus*.

4.2. Pulcherrimin

The red pigment of the yeast *Candida pulcherrima* and of special strains of *Bacillus cereus* and *Bacillus subtilis* has been characterized as the iron(III) complex of the cyclic anhydride of N^{α}-hydroxyleucine [57]. Pulcherriminic acid is the oxidation product. Iron does not regulate the biosynthesis of these compounds and their function is apparently unknown [58].

4.3. Mycelianamide

Mycelianamide, from the mycelium of *Penicillium griseofulvum*, is the
only known example among the natural diketopiperazines which is a
cyclic arylidene bishydroxamic acid [59].

4.4. Antibiotic G1549

The culture broth of *Pseudomonas alcaligenes* has been shown to contain a cyclic hydroxamic acid, antibiotic G1549, believed to be
1-hydroxy-5-methoxy-6-methyl-2(1H)-pyridinone [60]. Although toxic
G1549 protected guinea pigs against cutaneous infection with *Micro-
sporum canis*. It is moderately active against Gram-positive bacteria,
fungi, and *Trichomonas vaginalis*.

4.5. ASK-753 and NRCS-15

Iron-containing antibiotics from *Streptomyces* ASK-753 and NRCS-15
have been purified but their structures are apparently unknown [61,
62].

4.6. Hadacidin

N-formyl-N-hydroxyglycine (hadacidin) from *Penicillium aurantio-
violaceum* [63] inhibits purine synthesis and has weak antitumor
activity [64].

4.7. N^{δ}-hydroxyarginine

An antibiotically active amino acid from the fungus *Nannizzia gypsea*
has been characterized by synthesis as N^{δ}-hydroxyarginine [65].

4.8. Lipoxamycin

Streptomyces virginiae has yielded a lipid-soluble C-16 acyclic
ketone terminating in a hydroxyamine residue acylated by the carboxyl
group of serine [66]. The compound, lipoxamycin, has antifungal
activity.

4.9. Fosfomycinens

Strains of *Streptomyces* have been shown to produce a family of antibiotics, the fosfomycinens, which comprise a phosphoric acid function
linked via a three-carbon bridge to a hydroxamic acid [67].

4.10. Actinonin

The broad-spectrum antibiotic known as actinonin, from *Streptomyces*
sp. *Cutter* C12 (NCIB 8845) or *Streptomyces roseopallidus,* is unique
in having a primary, rather than a secondary, hydroxamic acid structure [68]. It is a pseudopeptide containing L-prolinol, L-valine,
D-pentyl succinic acid, and hydroxylamine.

4.11. Thiohydroxamic Acids

N-methyl-N-thioformylhydroxylamine and its Cu(II) and Fe(III) complexes were isolated independently by two groups in Japan from *Pseudomonas* sp. and named fluopsin [69] and thioformin [70].

5. GENERAL CONCLUSIONS

Microorganisms form selective ligands for iron(III) and iron(II),
some of which display antibiotic activity. The mechanism of antibiosis varies from a simple starvation for iron to an exploitation

of specific transport receptors, designed for iron compounds, in the
membrane of the sensitive species. The development of a line of
antimicrobial agents based on iron complexation is an attractive
possibility which remains to be explored. Thus ferrichrome might
be used with profit as a vehicle for drug delivery in fungi and,
provided problems of resistance could be overcome, in bacteria.
Similarly, analogs of the ferrioxamine-type siderophores might prove
effective against Gram-positive bacteria.

REFERENCES

1. V. Prelog, *Pure Appl. Chem.*, *6*, 327 (1963).

2. C. E. Lankford, *Crit. Rev. Microbiol.*, *2*, 273 (1973).

3. G. F. Gause, *Brit. Med. J.*, *2*, 1177 (1955).

4. W. Keller-Schierlein, V. Prelog, and H. Zähner, *Prog. Chem. Org. Nat. Prod.*, *22*, 279 (1964).

5. J. Nüesch and F. Knüsel, in *Antibiotics*, Vol. 1 (D. Gottlieb and P. D. Shaw, eds.), Springer Verlag, New York, 1967, p. 499.

6. T. Emery, *Adv. Enzymol.*, *35*, 135 (1971).

7. W. Keller-Schierlein, in *Development of Iron Chelators for Clinical Use* (W. F. Anderson and M. C. Hiller, eds.), DHEW Publ. No. (NIH) 76-994, Washington, D.C., 1976, p. 80.

8. A. L. Schade and L. Caroline, *Science, 164,* 340 (1946).

9. I. G. O'Brien and F. Gibson, *Biochim. Biophys. Acta, 215,* 393 (1970).

10. J. B. Neilands, T. Erickson, and W. H. Rastetter, *J. Biol. Chem., 256,* 3831 (1981).

11. D. J. Ecker and T. Emery, *J. Bacteriol., 155,* 616 (1983).

12. R. Wayne, K. Frick, and J. B. Neilands, *J. Bacteriol., 126,* 7 (1976).

13. A. J. Middleton, D. S. Cole, and K. D. Macdonald, *J. Antibiotics, 31,* 1110 (1978).

14. R. J. Bergeron, G. T. Elliott, S. J. Kline, R. Ramphal, and L. St. James III, *Antimicr. Agents and Chemoth., 24,* 725 (1983).

15. R. R. Arnold, M. F. Cole, and J. R. McGhee, *Science, 197,* 263 (1977).

16. S. H. Hutner, *Ann. Rev. Microbiol., 26,* 314 (1972).

17. H. J. Rogers, *Infect. Immun.*, *7*, 445 (1973).

18. G. Anderegg and F. L'Eplattenier, *Helv. Chim. Acta*, *47*, 1067 (1964).

19. A. A. Miles and P. L. Khimji, *J. Med. Microbiol.*, *8*, 477 (1975).

20. P. L. Khimji and A. A. Miles, *Br. J. Exp. Pathol.*, *59*, 137 (1978).

21. R. Wayne and J. B. Neilands, *J. Bacteriol.*, *121*, 497 (1975).

22. V. Braun, K. Günther, K. Hantke, and L. Zimmermann, *J. Bacteriol.*, *156*, 308 (1983).

23. H. N. Yeowell and J. R. White, *Antimic. Agents and Chemoth.*, *22*, 961 (1982).

24. L. W. Oberley and G. R. Buettner, *FEBS Lettr.*, *97*, 47 (1979).

25. E. D. Weinberg, *Microbiol. Rev.*, *42*, 45 (1978).

26. K. Konopka, A. Bindereif, and J. B. Neilands, *Biochemistry*, *24*, 6503 (1982).

27. B. N. Ames, G. F.-L. Ames, J. D. Young, D. Tsuchiya, and J. Lecocq, *Proc. Natl. Acad. Sci. USA*, *70*, 450 (1973).

28. E. H. Fiss, P. Stanley-Samuelson, and J. B. Neilands, *Biochemistry*, *21*, 4517 (1982).

29. D. S. Plaha and H. J. Rogers, *Biochim. Biophys. Acta*, *760*, 246 (1983).

30. R. L. Jones and R. W. Grady, *Eur. J. Clin. Microbiol.*, *2*, 411 (1983).

31. K. Konopka and J. B. Neilands, *Biochemistry*, *23*, 2122 (1984). (1984).

32. B. S. van Asbeck, J. H. Marcelis, J. J. M. Marx, A. Struyvenberg, J. H. van Kats, and J. Verhoef, *Eur. J. Clin. Microbiol.*, *2*, 426 (1983).

33. B. S. van Asbeck, J. H. Marcelis, J. H. van Kats, E. Y. Jaarsma, and J. Verhoef, *Eur. J. Clin. Microbiol.*, *2*, 426 (1983).

34. D. M. Reynolds, A. Schatz, and S. A. Waksman, *Proc. Soc. Exp. Biol. Med.*, *64*, 50 (1947).

35. G. F. Gauze and M. G. Brazhnikova, *Novosti Med.*, *23*, 3 (1951).

36. E. D. Stapley and R. E. Ormond, *Science*, *125*, 587 (1957).

37. S. A. Waksman, *Science*, *125*, 585 (1957).

38. J. Turková, O. Mikeš, and F. Šorm, *Coll. Czech. Chem. Commun.*, *30*, 118 (1965).

39. A. Zalkin, J. D. Forrester, and D. H. Templeton, *J. Am. Chem. Soc.*, *88*, 1810 (1966).

40. H. Maehr, *Pure Appl. Chem.*, *28*, 603 (1971).

41. G. Benz, T. Schroder, J. Kurz, C. Wünsche, W. Karl, G. Steffens, J. Pfitzner, and D. Schmidt, *Angew. Chem. Suppl.*, 1322 (1982).

42. J. B. Neilands, *Microbiol. Sci.*, *1*, 9 (1984).

43. L. Fecker and V. Braun, *J. Bacteriol.*, *156*, 1301 (1983).

44. P. E. Klebba, M. A. McIntosh, and J. B. Neilands, *J. Bacteriol.*, *149*, 880 (1982).

45. H. Bickel, P. Mertens, V. Prelog, J. Seibl, and A. Walser, *Tetrahedron*, Suppl. 8, 171 (1966).

46. P. Huber, Doctoral dissertation, *Eidg. Tech. Hochsch.* (1982), Zürich, Switzerland.

47. E. B. Chain, A. Tonolo, and A. Carilli, *Nature*, *176*, 645 (1955).

48. A. Ballio, H. Bertholdt, A. Carilli, E. B. Chain, V. DiVittorio, A. Tonolo, and L. Vero-Barcellona, *Proc. Roy. Soc. B*, *158*, 43 (1963).

49. S. Candeloro, D. Grdenic, N. Taylor, B. Thompson, M. Viswamitra, and D. Crowfoothodgkin, *Nature*, *224*, 589 (1969).

50. A. S. Khokhlov and I. N. Blinova, *Akad. Nauk. SSSR Doklady Ser. Biol.*, *215*, 1493 (1974).

51. C. C. Yang, and J. Leong, *Antimic. Agents and Chemoth.*, *20*, 558 (1981).

52. R. Shiman and J. B. Neilands, *Biochemistry*, *4*, 2233 (1965).

53. M. Pouteau-Thouvenot, A. Gaudemer, and M. Barbier, *Bull. Soc. Chim. Biol.*, *47*, 2085 (1965).

54. G. Feistner, H. Korth, H. Ko, G. Pulverer, and H. Budzikiewicz, *Current Microbiol.*, *8*, 239 (1983).

55. W. C. Liu, S. M. Fisher, J. S. Wells, C. S. Ricca, P. A. Principe, W. H. Trejo, D. P. Bonner, J. Z. Gougoutos, B. K. Toeplitz, and R. B. Sykes, *J. Antibiot.*, *34*, 791 (1981).

56. J. C. MacDonald, in *Antibiotics*, Vol. 2 (D. Gottlieb and P. D. Shaw, eds.), Springer Verlag, New York, 1967, p. 43.

57. A. J. Kluyver, J. P. vander Walt, and A. J. van Triet, *Proc. Natl. Acad. Sci. USA*, *39*, 583 (1953).

58. D. G. Kupfer, R. L. Uffen, and E. Canale-Parola, *Archiv. Mikrobiol.*, *56*, 9 (1967).

59. N. Shinmon and M. P. Cava, *J. Chem. Soc. Chem. Commun.*, 1020 (1980).

60. W. R. Barker, C. Callaghan, L. Hill, D. Nobll, P. Acred, P. B. Harper, M. A. Sowa, and R. A. Fletton, *J. Antibiot.*, *32*, 1096 (1979).

61. E. Z. Khafagy and B. M. Haroun, *J. Antibiot.*, *27*, 874 (1974).

62. B. M. Haroun, *J. Antibiot., 27,* 14 (1974).

63. T. F. Emery, in *Antibiotics,* Vol. 2 (D. Gottlieb and P. D. Shaw, eds.), Springer Verlag, New York, 1967, p. 17.

64. H. T. Shigeura and C. N. Gordon, *J. Biol. Chem., 237,* 1937 (1962).

65. J. Widmer and W. Keller-Schierlein, *Helv. Chim. Acta, 57,* 657 (1974).

66. H. A. Whaley, *J. Am. Chem. Soc., 93,* 3767 (1971).

67. Y. Kuroda, M. Okuhara, T. Goto, M. Okamoto, H. Terrano, M. Kohsaka, H. Aoki, and H. Imanaka, *J. Antibiot., 33,* 29 (1980).

68. J. J. Gordon, J. P. Devlin, D. E. Wright, and L. Ninet, *J. Chem. Soc. Perkin, 1819* (1975).

69. K. Shirahata, T. Deguchi, T. Hayashi, I. Matsubara, and T. Suzuki, *J. Antibiot., 23,* 546 (1970).

70. Y. Egawa, K. Umino, Y. Ito, and T. Okuda, *J. Antibiot., 24,* 124 (1971).

Chapter 12

CATION-IONOPHORE INTERACTIONS: QUANTIFICATION OF THE
FACTORS UNDERLYING SELECTIVE COMPLEXATION BY
MEANS OF THEORETICAL COMPUTATIONS

Nohad Gresh and Alberte Pullman
Institut de Biologie Physico-Chimique
Laboratoire de Biochimie Théorique associé au CNRS
Paris, France

1. INTRODUCTION

The specificity of cation complexation by an ionophore is a dominant
factor in the selectivity of ionophore-enhanced cation translocation
across membranes [1].

Within the overall free energy balance for complexation ΔG,
the enthalpy contribution ΔH often dictates the complexation selectivities (see, e.g., Refs. 2-6). The elements involved in the overall enthalpy balance for cation complexation have been recognized for
some time [7-10]. They consist essentially of the cation-ionophore
interaction energy, the desolvation energy of the cation and of the
ionophore, the variation of the intramolecular energy of the ionophore as it adapts its shape to accommodate the cation, and the
resolvation energy of the cation-ionophore complex formed.

The important problem is, of course, the quantification of the
relative weights of these elements, quantification which, if accomplished, would not only ensure the understanding of the structural
and dynamic mechanisms governing the selectivity of known natural
or synthetic ionophores, but may be expected also to help in the
design of new ionophores with improved complexation selectivities.

Some attempts at such a quantification were made earlier,
particularly by Eisenman et al. [11-13] and by Morf and Simon [8,
14-16]. Useful as they were, the scope of these treatments remains
restricted on account of the rather limited value of the potential
functions utilized and inherent approximations, inevitable at the
time when these first attempts were made.

In the approach presented in this review we have undertaken to
compute, on a firmer basis, the main components of the association
and their interplay for a certain number of representative ionophore-
ligand complexes. An important element ensuring an appreciable improvement with respect to previous treatments is the use of a refined
methodology for the treatment of intermolecular interactions, the
inclusion, when necessary, of a variety of contributions, and the
use of known structural data pertaining to the investigated complexes,
which have recently become available.

2. OUTLINE OF THE METHODOLOGY

The basis of the computations is a refined additive procedure elaborated in our laboratory [17] for the fast computation of intermolecular interaction energies. Two important developments were decisive in this elaboration. The first one, theoretical, concerns the refinement and extension of the methods of quantum chemistry to the point that it has become standard routine to compute nonempirically the binding energy between a cation and a small ligand of the type of the nucleophilic constituents of the ionophores. The second one, experimental, is the establishment of refined methods using high-pressure mass spectrometry, flowing afterglow, and ion cyclotron resonance spectroscopic techniques for measuring in the gas phase the enthalpies of binding of cations to small ligands, not only at the single-ligand level but also for the stepwise addition of ligands around an ion [18].

These simultaneous developments have led to a considerable progress in the current knowledge of ion-ligand interaction. The binding enthalpies measured in the gas phase provided a unique way to test the accuracy of the theoretical computations of binding energies. The interest of disposing of such a stringent test of the theory is that aside from the binding energies, the computations yield also as subproducts complementary results which cannot be attained as yet by direct measurements, e.g., the equilibrium position of the ion with respect to the ligand (distances, angles), the lability of the binding (i.e., the amount of energy lost upon moving the ion away from its preferred position), and also information on the nature of the interaction, since it is possible to define and calculate explicitly, inside the total binding energy, the separate values of its main components: electrostatic, induction, repulsion, charge transfer, and dispersion [19].

Thus in the case of alkali cations, a significant amount of gas phase data have been accumulated on their binding enthalpies to ligands such as water, ammonia, methylamines, aldehydes, ketones,

esters, ethers, amides, etc., and numerous computations of variable
accuracy of the corresponding energies are equally available.

What has been learned upon comparison of the two sets of data
can be summarized briefly as follows:

(1) The obtention of accurate energy values requires the utilization of *ab initio* computations of the self-consistent type. Semiempirical methods may yield artifacts [20].

(2) If the computations are pushed toward the Hartree-Fock
limit (using very extended atomic basis sets), the values of the
measured gas phase enthalpies can be reproduced within less than
1 or 2 kcal/mol. An illustration of this situation for the case of
Li^+, Na^+, and K^+ binding to water and ammonia [21] is given in Table 1.

(3) The main component of the binding energy is the electrostatic interaction E_C (see, for instance, Ref. 21). It is important
to underline, however, that this is not equivalent to saying that the
order of the binding energies to a given ion follows the order of the
dipole moments of the ligands. This is very clearly illustrated by
the fact that ammonia, although having a *smaller* dipole moment (1.4 D)
than water (1.8 D), has always a *larger* value of E_C and of ΔE to a
given ion than water. This reflects the fact that E_C is the value of
the electrostatic interaction calculated from the exact electronic
distribution [22]. It can be approximated by at least an extended
multipole expansion but not by only the dipole term.

TABLE 1

Computed (ΔE) and Measured Equilibrium Values (ΔE_{exp}) of
Binding Enthalpies of Li^+, Na^+, K^+ to H_2O and NH_3
(values in kcal/mol)

Value	Li^+		Na^+		K^+	
	H_2O	NH_3	H_2O	NH_3	H_2O	NH_3
ΔE	-33.8	-38.5	-24.8	-27.8	-17.2	-18.3
ΔE_{exp}	-34.0	-39.1	-24.0	-29.1	-16.9	-17.9

Note: See Ref. 21 for a detailed discussion.

(4) The conditions under which less refined calculations can reproduce the correct results have been defined [23,24]. It was shown, in particular, that satisfactory values of the energies and of the other essential characteristics of the binding could be obtained using a minimal basis of atomic orbitals, provided it be appropriately selected.

This accumulated basic knowledge has been of fundamental importance for the treatment of the interaction of the cations with ionophores. The ideal situation would, of course, be to apply to this problem the methods which have proven satisfactory for the small ligands, but cation-ionophore systems are yet too large to be dealt with by such a rigorous procedure. For such systems the possible treatment is to utilize an expression of the energy of interaction consisting of a sum of terms representing its various parts, namely, the electrostatic ("charge-dipole"), polarization ("charge-induced dipole"), repulsion, charge transfer, and dispersion (London) components, the parameters of which have been calibrated so as to reproduce the results of the *ab initio* calculations on small ligands.

The utilization (based on perturbation theory) of a sum of such components (or of some of them only) has been at the basis of the calculations of intermolecular interactions for a long time, but the long-standing difficulties of this procedure have always been twofold, related in the first place to the choice of reliable analytic expressions of each term and, on the other hand, to the appropriate choice of the constants or parameters inside each term, in a correct and consistent way ensuring the obtention of reliable overall energies. Examples of the first dilemma are, for instance, the representation of the electrostatic term by charge-charge or charge-dipole or charge-multipole interactions, the choice of the charges, their locations or that of the multipoles, etc. Another example is the choice between atom-atom and bond-bond repulsions, the distance dependence of the repulsions (R^{-12} or exponential), and the like.

The developments described above together with recent theoretical developments of perturbation theory [25] have made possible at

the same time *an improved choice of the analytical formulas for the
individual components and the fitting of the parameters so as to
reproduce the results of ab initio calculations on small systems,
themselves tested on the experimental energies.*

In this fashion, if the fitting of a *restricted* number of
parameters is done carefully on a large number of systems, taking
care of reproducing not only the binding energies but also the distances and angles, the lability of the binding , and so on, one may
obtain a satisfactory calibration of the methodology applicable to
larger entities. The essential features of the computational procedure which we utilize are the following:

The interaction energy ΔE between two molecular species is
evaluated as a sum of four components:

$$\Delta E = E_{MTP} + E_{pol} + E_{rep} + E_{dl} \tag{1}$$

E_{MTP} is the electrostatic interaction energy, obtained as a
sum of multipole-multipole interactions between the involved molecules. For each molecule considered, the multipole expansion utilized is a multicenter multipole expansion of its *ab initio* charge
density distribution, in which each overlap distribution is represented by a monopole, a dipole, and a quadrupole. This "overlap
multipole" development (OMTP) originally described in [26] leads to
a multipole center at each atom and in the middle of all pairs of
atoms, whether chemically linked or not. The use of such an elaborate development was shown to ensure a satisfactory representation
of the molecular electrostatic potential generated by a molecule
[27] and of the electrostatic contribution to the binding energy
[28].

E_{pol} is the energy due to the polarization of one molecule by
the field of its partner and vice versa. This term is computed in
a way consistent with the use of multipoles to represent the electron density distribution of each entity, namely, by calculating the
field $\vec{\mathcal{E}}_i$ created at each atom and bond center of molecule A by the
multipoles of molecule B and reciprocally. A polarizability α_i is

affected to each atom and bond center using experimental bond polar-
izabilities, partitioned consistently into pure atomic and pure bond
contributions (see Ref. 17 for details), and E_{pol} is computed as the
summation

$$-\frac{1}{2} \Sigma \ \alpha \ |\vec{\mathcal{E}}_i|^2 \tag{2}$$

E_{rep} is a repulsion contribution, computed as a sum of bond-
bond interactions, a formulation adopted on the basis of the existing
dependence of the repulsion on the square of the overlap between the
involved bond orbitals [17].

E_{dl} is a dispersion-like term, which contains the dispersion
and also the charge transfer contribution when present (see Ref. 17
for a discussion).

The parameters of the procedure were calibrated so as to reproduce the results of *ab initio* SCF computations (energy and equilibrium
distances) in model cases, i.e., the interaction of a water molecule
with H_2O, Na^+, K^+, and NH_4^+. The accuracy of the procedure was subsequently tested and was shown to reproduce satisfactorily the results
of *ab initio* supermolecule computations in several representative
cases, and in particular for the interactions of alkali, alkaline-
earth, and ammonium ions with anionic ligands (phosphate, carboxylate)
as well as neutral ligands (amide or ester carbonyl oxygens) [17,29,
30]. Its ability to reproduce experimental results when available
was also shown [31,32].

The important characteristics of the above-described procedure,
which have proven decisive in ensuring the accuracy [17,29-32] of the
results, is the utilization of multipole expansions of *ab initio* electron density distributions. This raises no difficulty for the computation of intermolecular interactions between molecular species of
small or medium size for which an *ab initio* wave function is readily
computable, but for interactions involving very large or macromolecules which cannot yet be computed *in toto* the multipolar expansions
cannot be derived in this way. The solution adopted, which constitutes the second important element of our methodology, relies on a

general strategy [33,34] devised in our laboratory for the evaluation
of the molecular electrostatic potential of macromolecules, in which
the macromolecule is built from subunits chosen in such a way as to
minimize the electrostatic perturbations caused by the subdivision,
and on which an *ab initio* computation can be performed and thus an
accurate multipole expansion obtained. Examples of the fragmentation
utilized will be indicated in the examples treated.

We shall present results of cation-binding studies performed
on four representative ionophores: valinomycin, nonactin, ionophore
A23187, and gramicidin A.

3. SELECTIVE BINDING OF ALKALI CATIONS
BY VALINOMYCIN

Valinomycin is a cyclic depsipeptidic antibiotic which selectively
enhances potassium permeability across membranes. It was isolated
in 1955 by Brockmann and Schmidt-Kastner [35] and its chemical structure was elucidated a few years later [36,37]. Valinomycin was first
demonstrated to uncouple oxidative phosphorylation in mitochondria
[38]. It was subsequently shown that this uncoupling effect results
from the active transport of potassium to the inside of membranes
coupled with proton extrusion [39,40].

Valinomycin can enhance potassium permeation through artificial
as well as natural membranes [41-43]. The transport of alkali ions
is strongly discriminatory. The sequence $Rb^+ > K^+ > Cs^+ > Na^+$ was
derived for induced permeation through mitochondrial [41], erythrocyte [43], or synthetic [44] membranes. This sequence is the same
as that of the binding constants of the valinomycin complexes with
these ions, measured in polar solvents [45,46]. In addition, the
K^+/Na^+ selectivity ratio, which amounts to 10,000/1 [47], is the
largest known to date. This situation explains why alkali-ion complexation by valinomycin has lent itself to a wealth of physicochemical studies: infrared absorption spectrometry [48,49], ^{1}H NMR
[50], or ^{13}C NMR [49], and circular dichroism [46]. Kinetic studies

of membrane transport processes were carried out with the help of
electric [51,52] or ultrasonic [53] relaxation experiments.

The chemical formula of valinomycin is recalled in Fig. 1a.
Until very recently, only one of the metal complexes of valinomycin,
that of K^+, had been crystallized and analyzed through x-ray crystallography [54,55]. In the complex the molecule takes up a "bracelet-
like" structure in which all the amide NH and CO groups form a tight
array of successive hydrogen bonds [54,55], leaving the six ester
carbonyls free to encage the ion, three from above and three from
below (Fig. 1b). Taking into consideration the rigidity conferred
to the molecule by the sequence of hydrogen bonds and experimental
indications on the permanence of a bracelet-like structure in the
other alkali complexes (vide infra), we have attempted to account
for the observed complexation order by considering the two most conspicuous components of the energy balance involved [56], namely, the
interaction energy between the ions and valinomycin, assumed to keep
the geometry of the K^+ complex, and the desolvation energies of the
ions. For the latter term, we have utilized the negative of the
available experimental values of the hydration enthalpies of the ions
[4]. The use of hydration enthalpies as representative of the solvation enthalpies is certainly justified in the case of small alcoholic
solvents for which the binding enthalpies to the alkali ions should
not be very different from the binding enthalpies in water (see, e.g.,
Refs. 57 and 58).

The subdivision utilized to derive the multipolar expansions on
the ionophore is indicated in Fig. 1a. The *ab initio* SCF wave functions of the fragments were computed with the basis set indicated
above [24] and the interaction energies of the cation with the ionophore reconstituted from the fragments were computed by the methodology described above, optimizing the position of the cation inside the
cavity by means of an energy minimization procedure (see Ref. 56).

The values of the intermolecular interaction energies of Na^+,
K^+, Rb^+, and Cs^+ with valinomycin at the outcome of the minimization
are reported in Table 2 together with the set of the six final carbonyl oxygen-cation distances. The following observations can be
made:

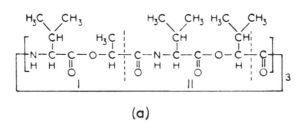

(a)

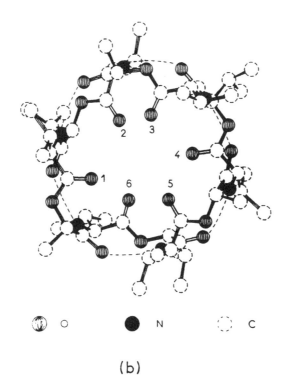

(b)

FIG. 1. Valinomycin. (a) Chemical formula. The vertical dashed
lines represent the positions where the separations between two
consecutive fragments were made. (b) Bracelet conformation of its
complex with K^+ (ion and hydrogen atoms are omitted for clarity),
seen from the side of the three methyl groups. O_1, O_3, and O_5
denote the inward-directed ester carbonyl oxygens from subfragments
of type I and O_2, O_4 and O_6 the corresponding oxygens from subfragments of type II. Hydrogen bonds between an amide carbonyl oxygen of
one subfragment and one amide nitrogen of the consecutive subfragment
are represented by curved dashed lines. (Reproduced from Ref. 56
with permission.)

TABLE 2

Valinomycin Complexes [56] Equilibrium Distances d (Å) to the
Ester Carbonyl Oxygens, Energies of Interaction ΔE_V of the
Cations Inside the K^+ Cavity, Experimental Dehydration
Enthalpies ΔE_S, Energy Balance δE, and Difference δ with
Respect to the Best Balance (energies in kcal/mol)

M^+	Na^+	K^+	Rb^+	Cs^+
d_1	3.05	2.80	2.74	2.70
d_2	2.95	2.68	2.73	2.72
d_3	2.81	2.73	2.77	2.75
d_4	2.42	2.62	2.69	2.72
d_5	2.47	2.72	2.68	2.69
d_6	2.67	2.77	2.75	2.78
ΔE_V	-108.5	-106.4	-102.7	-88.8
ΔE_S	106.0	85.8	79.8	72.0
δE	- 2.5	- 20.6	- 22.9	-16.8
δ	20.4	2.3	0	6.1

Source: Reprinted from Ref. 56 with permission.

(a) In the K^+ complex, the computed optimal distances fall in
the range 2.69-2.83 Å, in close agreement with the range of distances
observed in the crystal structure.

(b) K^+, Rb^+, and Cs^+ prefer an essentially central position in
the cavity. By contrast, Na^+ prefers a position shifted from the
center towards the vicinity of two carbonyls. It is interesting to
indicate in this connection that studies of infrared absorption
spectra of the sodium complex in a non-polar solvent [47,48], put
into evidence the existence of a heterogeneity in the carbonyl absorption bands: this heterogeneity (which was absent from the corresponding spectra of the K^+, Rb^+ and Cs^+ complex) was attributed to an
asymetric location of the sodium in the macrocycle cavity.

(c) The optimal values of the energies of interaction, ΔE_V,
are in the order: $Na^+ > K^+ > Rb^+ > Cs^+$. These values follow a trend

favoring the cations with the smallest ionic radii, and differ
greatly from the ordering observed experimentally.

(d) The order of the experimental desolvation enthalpies, ΔE_S,
is the same as the order of the interaction energies and thus does
not account either, by itself, for the experimental order of preferential binding.

(e) The energy balance for complexation of each cation, obtained
by subtracting the values of its ΔE_S from the values computed for its
ΔE_V, ranks in the order: $Rb^+ > K^+ > Cs^+ > Na^+$. Furthermore, Na^+ is
much more disfavored with respect to Rb^+ and K^+ than Cs^+, whereas K^+
is very near Rb^+ on the energy scale.

The ordering found as well as the relative spacing of the
cations are in striking agreement with the affinity sequence measured
in ethanol [45] or methanol [46] solvents. Hence *if one assumes that
the complexes all adopt a structure similar to that of the K^+ complex,
it is possible to account for the selectivity of association by taking
into consideration solely the complexation and the desolvation energies, the two terms operating in opposite directions, but with different differential effects.* In fact, the resulting energy balance for
complexation seems so unfavorable for Na^+ when compared to the other
cations that the very occurrence of the bracelet-like conformation
of valinomycin upon binding Na^+ in a polar solvent may be questioned.
Indeed, the circular dichroism spectrum of the Na^+ complex of valinomycin in methanol is characterized by a negative Cotton effect, whereas
the K^+, Rb^+, and Cs^+ complexes are characterized by a positive effect,
and this was taken as evidence for the involvement of a different Na^+-
binding conformation upon complexation in methanol [46]. A similar
conclusion was arrived at on the basis of ^{13}C and IR measurements in
methanol [49]. Such a situation is markedly different from the one
occurring in a nonpolar solvent in which the closed conformation of
valinomycin is involved in the binding of all four alkali cations
[47,48]. In fact, very recently the crystal structure of a valino-
mycin-picrate-sodium complex was established [59]. The configuration
of the crystal is very particular and is markedly different from the

one occurring in a corresponding valinomycin-picrate-potassium complex [60]: a water molecule is located in the internal region occupied by K^+ in the latter complex. Na^+ is external to the valinomycin cavity and binds to the three carbonyl oxygens adjacent to the lactyl residues, to the water, and to the picrate anion. Clearly, the occurrence of such a structure would deserve a special study.

4. SELECTIVE COMPLEXATION OF K^+, Na^+, AND NH_4^+ BY NONACTIN

Nonactin is a representative of another family of naturally occurring cyclic ionophores, the macrotetrolides antibiotics. It was isolated by Corbaz et al. in 1955 [61] from bacterial strains. Its structure was later determined [62] together with that of the related components dinactin, trinactin, and tetranactin [63]. It is composed of four tetrahydrofuran rings and four carbonylic esters linked by saturated hydrocarbon moieties, as shown in Fig. 2. Macrotetrolides were shown

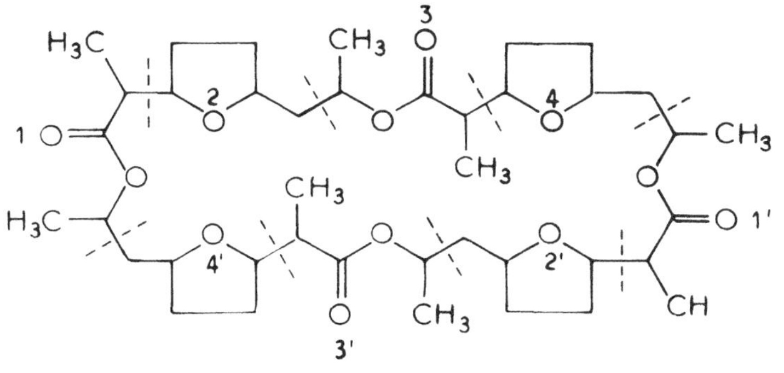

FIG. 2. The molecular structure of nonactin. Carbonyl oxygens have an odd subscript and ether oxygens have an even subscript. Oxygens belonging to the second, symmetry-related half of the molecule have a primed subscript. The dotted lines indicate the separations between fragments. (Reproduced from Ref. 74 with permission.)

to uncouple oxidative phosphorylation in mitochondria [64]. They enhance cation translocation across biological [64] or artificial [44] membranes, or cation extraction from an aqueous to an organic phase [65]. They display a high K^+/Na^+ selectivity ratio, amounting from 50 up to 100:1 [3,66,67]. The alkali, alkaline-earth, and ammonium complexes of nonactin were studied by several spectroscopic techniques, ^1H NMR [68,69], ^{13}C NMR [70], and IR [70]. We have investigated the preference of nonactin for three cations NH_4^+, K^+, and Na^+. The experimentally observed order is $NH_4^+ > K^+ > Na^+$ [1,65-67]. The same ordering prevails with respect to the ionophore-enhanced permeabilities across membranes [1,64].

The crystal structures of the K^+ [71,72] and the Na^+ [73] complexes are known. They both show the cation to be liganded by the four carbonyl oxygens and the four tetrahydrofuran ether oxygens in an internal cavity, but this cavity is more contracted in the Na^+ complex than in the K^+ complex. The knowledge of these differences in geometry has rendered possible the study in this case [74] of the respective roles of the *three* most conspicuous components of the energy balance: the cation-ionophore interaction energy, the desolvation energy of the cation, and the change in intramolecular energy of the ionophore upon going from the geometry of the K^+-specific cavity to that of the Na^+-specific cavity.

The subunits used to compute the multipole expansions for nonactin are given in Fig. 2. Each fragment contains one liganding oxygen, either carbonyl or ether. Nonactin is thus built out of eight subunits placed in the appropriate mutual disposition, the last four being equivalent to the first four, owing to the symmetry of the molecule. The necessary *ab initio* SCF wave functions of the subunits were computed using our usual basis set [24].

A comparison of the crystal structures of the two complexes indicates that the contraction of the size of the internal cavity, upon passing from the K^+-specific to the Na^+ complex, involves changes in torsional angles, which all remain below 10°, except for one (and its symmetric counterpart) O-C esteric bond of a carboxylic group, which is of 13°. Albeit small (and thus costing very little

in pure torsional energy), such concerted rotations bring about a
shortening of the separation of two liganding oxygens which may
reach up to 0.7 Å in two cases. This led us to attempt to evaluate
the increase of the repulsive interactions between the liganding
groups which is brought about by such a close approach. This was
done by computing the interfragment interaction energies involving
two distinct liganding groups, for both conformations, in the framework of our intermolecular procedure (see Ref. 74 for details).

4.1. Interaction of Nonactin with Na$^+$ and K$^+$

The first columns of Table 3 give the values of the cation-ionophore
interaction energies after energy minimization inside the appropriate
cavity, together with the corresponding oxygen-cation distances.
Table 4 summarizes the values of the components of the energy balance,
the indicated desolvation enthalpies being measured in methanol [57].

The comparison of the optimized cation-oxygen distances found
has shown a near-identity (within 0.01 Å) between the theoretical
values and those given by the crystallographic data. In particular,
Na$^+$ approaches the carbonyl oxygens more closely than the tetrahydrofuran oxygens by about 0.4 Å. This tendency is detectable also in
the K$^+$ complex, although in a much less pronounced fashion.

The numerical results of Table 4 indicate that the binding
energy of the Na$^+$-specific conformation is appreciably larger, by
32.7 kcal/mol, than the corresponding binding energy of K$^+$ to the
ionophore in its K$^+$-specific conformation. The desolvation enthalpy
operates in the right direction for reducing the difference between
the two ions, leaving them, however, in the reversed order with
respect to experimental observation. This outcome favoring the
smallest cation is only reversed when the ligand-ligand repulsive
interactions are incorporated in the energy balance, the repulsions
being stronger in the Na$^+$ cavity by 17 kcal/mol, making the overall
binding enthalpy about 4 kcal/mol more favorable for K$^+$ than for
Na$^+$ binding. These numbers come reasonably close to the range of

TABLE 3

Intermolecular Interaction Energies ΔE_n (kcal/mol) and
Distances (Å) to the Oxygens O_i[a] for the Optimized
Positions of the Cation in the Nonactin Cavity

	(A) Na^+	(B) K^+	(B) Na^+	(B) NH_4^+[b]
ΔE_n	-151.2	-117.5	-120.3	-131.8
O_1	2.45	2.75	2.79	2.63; 2.59
O_3	2.39	2.80	2.78	2.58; 2.83
$O_{1'}$	2.45	2.74	2.74	2.48; 2.76
$O_{3'}$	2.39	2.80	2.77	2.72; 2.60
O_2	2.75	2.81	2.82	1.90
O_4	2.79	2.90	2.92	1.99
$O_{2'}$	2.75	2.80	2.75	1.94
$O_{4'}$	2.79	2.90	2.92	2.00

[a]The four carbonyl oxygens are listed first.
[b]The distances indicated are HO_i distances; for the carbonyl oxygens,
distances to the two closest hydrogens are given (see text).
Note: (A) cavity of the Na^+ crystal; (B) cavity of the K^+ crystal.
Source: Reprinted from Ref. 74 with permission.

TABLE 4

Computed Contributions to the Binding of
K^+ and Na^+ by Nonactin[a]

	K^+	Na^+	δ
ΔE_n	-117.5	-151.2	+33.7
Desolvation[b]	+ 84.2	+104.9	-20.7
Repulsion	+ 9.3	+ 26.1	-16.8
Balance	- 24.0	- 20.2	- 3.8

[a]δ is the difference: K^+ - Na^+.
[b]In methanol.
Source: Reprinted from Ref. 74 with permission.

differences in complexation enthalpies which have been measured in
methanol (5.9-7.8 kcal/mol) [3].

It must be pointed out here that while the energy difference
is satisfactory and in the good direction, the numerical values of
the measured individual enthalpies are not quantitatively reproduced,
a feature which is due, on the one hand, to the approximate nature
of the calculation of the repulsions and, on the other hand, to the
fact that our ligand-cation binding energies tend to be somewhat
exaggerated by factors of approximately 1.15 for Na^+ and 1.17 for K^+
[17,23]. Accounting for this effect would bring the values closer
to the observed ones.

As a supplementary source of information, it is interesting to
consider what would be the characteristics of the binding if the Na^+
cation was constrained to utilize the K^+-specific cavity. The optimization of the energy in that case gives the results indicated in
the third column of Table 3. It is seen that in the computed optimal
position, the Na^+-oxygen distances are only slightly shifted with
respect to those obtained for K^+ with the small, already observed
tendency to come closer to the carbonyl oxygens than to the ether
oxygens. But on the whole, Na^+ in the K^+-specific cavity is not
strongly deported to one side, contrary to the situation found in
the case of valinomycin (vide supra) when we optimized the position
of Na^+ inside the cavity of the K^+ complex. Concerning the binding
energy, it is observed that its value for Na^+ is only 2.8 kcal/mol
larger than that for K^+. Comparison of columns 1 and 3 of Table 3
indicates that a considerable gain in binding energy is obtained
upon closing of the cavity onto Na^+ so as to allow a better approach
of the ion to the carbonyl oxygens.

4.2. Interaction of Nonactin with NH_4^+

The crystal structure of the NH_4^+ complex of nonactin is not available.
Taking into account the similarity of ionic radii of NH_4^+ and K^+, we

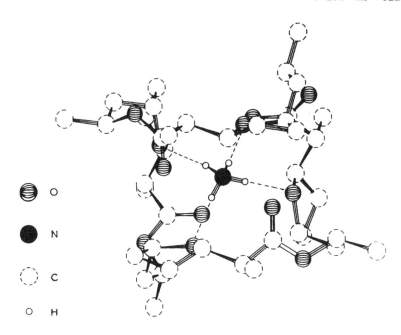

FIG. 3. The position of NH_4^+ after energy minimization inside the K^+-specific cavity. The hydrogens of NH_4^+ only are indicated for clarity. (Reproduced from Ref. 74 with permission.)

tentatively considered the possibility for NH_4^+ to utilize the K^+-specific cavity, the ammonium ion was given equivalent bond lengths of 0.90 Å and angles of 109.5°, and its interaction energy was optimized inside the cavity as for the other ions.

The results are reported in the last column of Table 3 and the corresponding structure is shown in Fig. 3.

A first observation is that, in its optimal position, each NH proton is directly hydrogen-bonded to an ether oxygen. The O-H distances are between 1.90 and 2.00 Å and the NH-O angles between 165 and 180°, the corresponding H-O bonds being nearly colinear with the direction of one lone pair centered on the *ether* oxygen, considered as having a standard sp^3 hybridization. Each carbonyl oxygen, on the other hand, is between the prolongation of two ammonium NH bonds, the closest interatomic distances ranging from 2.48 to 2.83 Å and the angles HN-O having values between 80 and 100°. These results confirm the conclusions arrived at on the basis of earlier infrared spectral determinations, which showed that the carbonyl stretching frequencies were not affected by the binding of NH_4^+, in contrast to the situation observed upon binding of K^+ [70].

Furthermore, the main characteristics of our computed equilibrium configuration are strongly similar to those observed in an x-ray crystallographic study of the complex of NH_4^+ with tetranactin [75], where the range of determined O (ether)-H distances is 1.95-2.15 Å, the angles O-HN having values in the range 168-176°, and the O (carbonyl)-H distances are between 2.72 and 2.87 Å, the angles O-HN ranging between 96 and 102°. It is remarkable that the special mode of binding of NH_4^+ to the ether oxygens of the nactins emerges from the theoretical calculations.

This example illustrates the fact that the two cations K^+ and NH_4^+ can display markedly different intrinsic binding preferences with respect to a polydentate ligand.*

Can our intermolecular procedure account for the preferential affinity of nonactin for NH_4^+ over K^+? The optimized interaction energy of NH_4^+ with nonactin is larger than that of K^+ by 14.3 kcal/mol, for the same conformation of the ionophore. On account of the absence of experimental determination of the solvation enthalpy of NH_4^+ in methanol, in which affinity measurements are generally performed (see, e.g., Ref. 46), one may tentatively compare the computed difference to the difference of hydration enthalpies of the ions: the experimental value of the hydration enthalpy of NH_4^+ amounts to 89 kcal/mol [78] (in Ref. 78; see especially the note added in proof of Ref. 28 of that paper), whereas that of K^+ ranges between 79.8 and 85.8 kcal/mol [3,57,79], yielding a difference in the range of 3.2 to 9.2 kcal/mol. Thus, when the balance is made with the desolvation enthalpies the overall preference, although decreased, remains largely in favor of NH_4^+. This is thus an example whereby

---

*Another example of two strikingly different behaviors of K^+ and NH_4^+ is provided by their differing roles in promoting self-aggregation of 5'-GMP dianions. See, e.g., Refs. 76 and 77 and references therein.

the ordering set by the cation-cavity interactions is so much in
favor of the preferred ion that it is not modified by the desolvation
enthalpies. From the quantitative point of view, an improvement in
the numerical values of the balance proper may still be expected
(see discussion in Ref. 74), without, however, affecting the conclusion brought by our analysis.

5. SELECTIVE BINDING OF Mg^{2+} AND Ca^{2+} BY IONOPHORE A23187

Ionophore A23187 (calcimycin) is an open-chain carboxylic polyether
which transports cations across membranes [80,81]. It belongs to a
class of compounds which encompasses molecules such as lasalocid,
monensin, nigericin, etc. (Ref. 82 and references therein). A dis-
tinctive property of A23187 is its marked selectivity for divalent
over monovalent ions [81]. Following the earlier demonstration of
its uncoupling effect on oxidative phosphorylation and inhibition of
ATPase in mitochondria [80], a considerable number of reports have
amply demonstrated its importance for the study of multifarious
physiological processes mediated by Ca^{2+}, such as exocytosis, contraction, and motility [83,84]. Its divalent cation complexes were
investigated by NMR spectroscopy [85,86], circular dichroism [87],
EPR [88], UV, and fluorescence spectroscopy [81].

X-ray crystal structures are available for the ionophore [89]
as well as for its 2:1 Ca^{2+} [90] and Mg^{2+} [91] complexes. The structural formula of A23187 is shown in Fig. 4. In both 2:1 crystal complexes, each deprotonated molecule of A23187 (symbol A^-) binds the
cation by one of its carboxylate oxygens, by the aromatic nitrogen of
its benzoxazole group, and by the carbonyl oxygen of its ketopyrrole
moiety. In both complexes also, an additional stabilization occurs
through the formation of two hydrogen bonds involving the NH group
of the ketopyrrole ring of one ionophore and the metal-bound anionic
oxygen of the benzoxazole ring of the other. The Ca^{2+} complex differs
from the Mg^{2+} complex by the inclusion of one water molecule in the

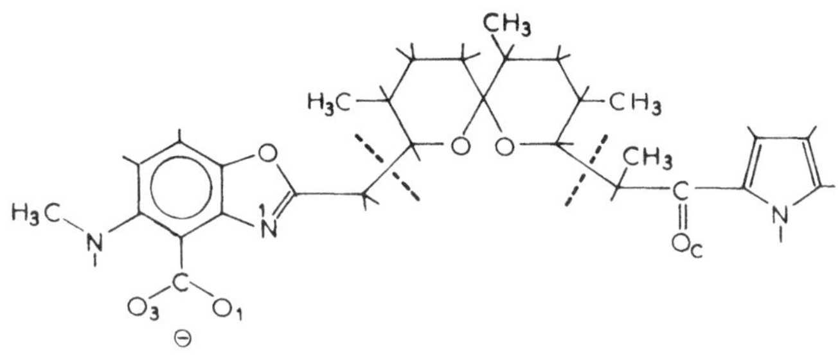

FIG. 4. The molecular structure of A23187. Notations adopted for the liganding atoms. The dotted lines indicate the separation between fragments. (Reproduced from Ref. 92 with permission.)

coordination polyhedron of the cation, which is thus sevenfold coordinated.

The availability of detailed x-ray coordinates for both the Ca^{2+} and the Mg^{2+} complex has led us to consider them in a comparative theoretical study [92]. We have been particularly interested in our investigation by the significance of the presence of the water molecule in the Ca^{2+} complex. The factors considered in the comparative energy balances were the cation-ionophore interaction, the desolvation energy of the cation, the intramolecular energy of each individual A^- ionophore in the 2:1 A^--M^{2+} complexes as a function of its detailed conformational angles, the ionophore-ionophore intermolecular interaction, and the additional stabilization brought to the Ca^{2+} complex by the presence of a water molecule.

The overall situation may be readily contrasted to the one encountered with nonactin, in which only the three first factors were encountered.

It should be pointed out that in contrast to the strong divalent/monovalent selectivity of A23187, its discrimination in favor of Ca^{2+} vs. Mg^{2+} is small [93]. Most reports indicate a selectivity ratio Ca^{2+}/Mg^{2+} close to 3 [88,94-98] (in Ref. 99 the ratio is in favor of Mg^{2+}). Such a ratio may be readily contrasted with the one displayed by nonactin, namely 50-100 [65], or valino-

mycin, where it amounts to 10^4 [47]. Furthermore, this ratio was
determined by means of a two-phase extraction technique [88,94,95,
98] or pertains to cation transport across membranes [96,97], in
which cation complexation is only the triggering step. In fact,
the determination of the affinity constants by potentiometric techniques indicates closely comparable values for Ca^{2+} and Mg^{2+} [100].

In the subdivision adopted to compute the multipolar expansion,
the ionophore is subdivided into three fragments, as shown in Fig. 4,
and containing the benzoxazole, the spiroketal, and the ketopyrrole
rings, respectively.

The molecular graphics of the 2:1 Mg^{2+} and Ca^{2+} complexes of
A23187 are given in Fig. 5, represented as viewed from a "window"
perpendicular to a plane passing through O_1, M^{2+}, and the middle of
a segment linking the two carbonyl atoms (see Sec. 6 for the general
use of such "windows").

The input data used for the internal geometries of the individual ionophores were taken from the crystal structures of their
2:1 complexes with Mg^{2+} [91] or Ca^{2+} [90]. Inspection of the respective values in the two complexes of the four most relevant torsion
angles in the hinge regions linking the spiroketal group to the
benzoxazole and to the ketopyrrole rings shows that whereas the two
A^- molecules do not differ markedly in terms of their torsion angles
in the Mg^{2+} complex, differences amounting up to 20° exist in their
Ca^{2+} complex. With respect to these angles, the first A^- molecule
in the latter complex is more closely related to either A^- in the
Mg^{2+} complex than to the other A^- moiety facing it. The variations
of the torsion angles occurring upon passing from the Mg^{2+} complex
to the Ca^{2+} complex result in the lengthening of the distances
between O_C and O_1, and, to a lesser extent, between O_C and N_1, most
notable in the first A^- molecule.

As done above for nonactin, we have attempted to evaluate the
variation of intramolecular energy within each ionophore resulting
from such conformational rearrangements, assuming that the principal
factor governing such variations is the repulsion between the liganding subunits of A^-, namely, benzoxazole and ketopyrrole. The inter-

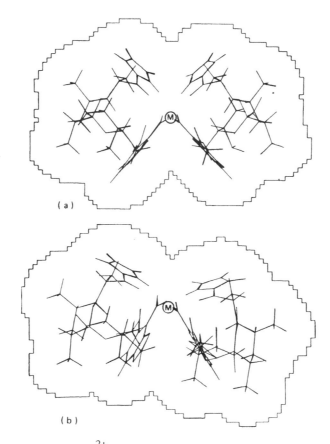

FIG. 5.   (a) The 2:1 Mg^{2+} complex of ionophore A23187 and the boundaries of the corresponding surface envelope.   (b) The 2:1 Ca^{2+} complex of ionophore A23187 and the boundaries of the corresponding surface envelope.   (Reproduced from Ref. 92 with permission.)

action energy explicitly computed between them in the respective conformations of the ionophore and reported in Table 5 under the label ΔE_i shows that the intramolecular repulsions are stronger in the Mg^{2+} complex and that their difference between the two structures favors the Ca^{2+} complex by 4.0 kcal/mol.

Let us now consider the intermolecular interaction energies in each cation-$(A^-)_2$ complex.   In view of analyzing the elements involved in the stabilities of the complexes, we have considered successively the complex $A^- - Mg^{2+} - A^-$, the complex $A^- - Ca^{2+} - A^-$ without

TABLE 5

Interaction Energies (kcal/mol) in the $A^- - M^{2+} - A^-$
Complexes of Calcimycin

	Mg^{2+}	Ca^{2+} (A)	Ca^{2+} (B)
M^{2+}		(A)	(B)
ΔE_i	+ 14.2	+10.2	
$\Delta E \ (A^-, \ M^{2+}, \ A^-)$	-682.5	-571.1	-573.1
$\Delta E_R \ (A^-, \ A^-)$	+ 30.8	+29.1	
$\Delta E_0 \ (M^{2+}, \ 2A^-)$	-769.7	-640.7	-643.1
ΔE_T	-668.3	-560.9	-562.9
$\Delta E \ (A^-, \ M^{2+}, \ W_a, \ A^-)$	-	-584.8	-585.6
$\Delta E \ (A^-, \ M^{2+}, \ W_a, \ W_b, \ W_c, \ A^-)$	-	-612.4	-612.8
ΔE_H	477.6	399.8	
δE	-190.7	-161.1	-163.1
$\delta E \ (W_a)$	-	-185.0	-185.8
$\delta E \ (W_a, \ W_b, \ W_c)$	-	-212.6	-213.0

(A) Calcium ligand distances of the crystal.
(B) Optimized calcium-ligand distances.
ΔE_i = intramolecular repulsions in the A^- moieties.
ΔE = intermolecular interaction energies between the entities in parenthesis treated together in the complex.
$\Delta E_R \ (A^-, \ A^-)$ = repulsion energy between the A^- moieties in the absence of M^{2+}.
$\Delta E_0 \ (M^{2+}, \ 2A^-)$ = interaction energy of M^{2+} with the two A^- entities without counting their interaction.
$\Delta E_T = \Delta E_i + \Delta E \ (A^-, \ M^{2+}, \ A^-)$.
ΔE_H = dehydration energies of the cations.
δE = energy balance for unhydrated complexes; $\delta E \ (W_a)$ = corresponding balance including W_a; same for W_a, W_b, W_c.
Source: Reprinted from Ref. 92 with permission.

including in its structure the water molecule observed in the crystal, then the effect of including this water molecule, and finally the possibility of including further molecules of hydration.

All the energies computed are assembled in Table 5. The equilibrium ion-ligand distances are given in Table 6.

TABLE 6

Cation-Ligand Distances ($\overset{\circ}{A}$) in Magnesium and
Calcium Cavities of Ionophore A23187

	(A)	(B)	(C)
O_1-M^{2+}	2.00	2.27	2.29
N_1-M^{2+}	2.23	2.69	2.58
O_c-M^{2+}	2.06	2.37	2.41
$O_{1'}$-M^{2+}	1.98	2.28	2.28
$N_{1'}$-M^{2+}	2.24	2.58	2.45
$O_{c'}$-M^{2+}	2.06	2.38	2.42

(A) = optimized distances in the Mg^{2+} complex (same as observed),
(B) = observed in the Ca^{2+} complex, (C) = optimized in the cavity
of the calcium complex.
Source: Reprinted from Ref. 92 with permission.

5.1. The Magnesium Complex

The mutual arrangement of the two ionophores in the Mg^{2+} complex is
defined by the parameters of the crystal structure [91], and the
location of the cation inside the so-defined cavity is optimized by
energy minimization. The value of ΔE at the outcome of the procedure is -682.5 kcal/mol. In this term are included the intermolecular interaction energy of Mg^{2+} with the two ionophores and the mutual
interactions of the ionophores with each other in the presence of the
cation. The first column of Table 6 lists the equilibrium distances
computed between Mg^{2+} and its six liganding atoms. These distances
are practically identical to the crystallographic values of Ref. 91.
Also reported in Table 5 are the values of the interaction energy
ΔE_R between the two A^- molecules in the absence of Mg^{2+} and of the
interaction energy ΔE_O of Mg^{2+} with the two A^- molecules when the
mutual interactions between the latter are *not* taken into account
in the calculation. It can be noted that the value of ΔE is much
smaller than that of the sum of ΔE_O and ΔE_R. This results from the

nonadditivity of the polarization contribution to the binding energy, which is proportional to the square of the electrostatic field exerted on the A^- molecules. When the interactions between the two A^- molecules are counted in the global computation, the polarizing field exerted by the dipositive charge of Mg^{2+} on one A^- molecule is opposed by the field exerted by the other A^- molecule. The resulting value of the total polarization contribution is much smaller than the value computed in the absence of interaction between the A^- molecules (-60.3 kcal/mol as opposed to -103.0 kcal/mol). This decrease of the polarization contribution is even larger than the repulsive value of ΔE_R, as computed between the two ionophores.

5.2. The Unhydrated Calcium Complex

The optimization of the position of Ca^{2+} inside the cavity defined by the crystal coordinates yields a set of calcium-ligand distances which compare quite favorably with those observed in the crystal of Ref. 90, two of them only showing a departure of 0.1 Å. To show that this does not seriously affect the value of the computed energy, we included two columns in Table 5, one corresponding to the Ca^{2+}-ligand distances of the crystal given in the corresponding column of Table 6, and another one corresponding to the optimized distances of our computation given in the last column of Table 6. It is seen that the energies differ only by 2 kcal/mol for a total of 563. As in the case of Mg^{2+}, the intermolecular energy comprises the interaction energy of the cation with the two ionophores as well as the mutual interactions between the two ionophores.

Also reported in Table 5 are the values of the interaction energy ΔE_R between the two A^- molecules in the absence of Ca^{2+} and of the interaction energy ΔE_O of Ca^{2+} with the two A^- molecules when the mutual interactions between the latter are not taken into account. The same remarks apply as for the Mg^{2+} complexes, concerning the non-additivity of the values of ΔE_O and ΔE_R with respect to that of ΔE. In the present case, the decrease of the value of the polarization

contribution is smaller than in the Mg^{2+} complex in ΔE with respect
to ΔE_O (-42.3 kcal/mol as compared with -72.2 kcal/mol).

If we now compare the values obtained in the complexes $A^- - Mg^{2+} -$
A^- and $A^- - Ca^{2+} - A^-$ for the total interaction energies ΔE_T (intra- plus
intercontributions), it appears at this level of treatment that the
difference between the interaction energies involving Ca^{2+} and Mg^{2+}
in their respective complexes is 105-107 kcal/mol in favor of the
smaller cation. This number is appreciably larger than the difference of 78 kcal/mol between the experimental dehydration enthalpies
ΔE_H of the cations [101] recalled in Table 5.

Therefore, an enthalpy balance δE_O in which the dehydration
enthalpy of the cation is subtracted from the interaction energy
ΔE_T in $A^- - M^{2+} - A^-$ would, at this stage, result in an appreciable
advantage (27-29 kcal/mol) in favor of Mg^{2+}.

5.3. The Hydrated Calcium Complex

Let us now study the influence on the energy balance of the incorporation of a water molecule (W_a) in the coordination polyhedron of Ca^{2+}.
We present here the results concerning the case when the cation is in
the position defined by the crystal coordinates. An identical conclusion is reached with the computed position of the cation inside
the cavity.

Starting from the data provided by the crystal structure (which
gives the position of the oxygen atom of W_a), energy minimization was
performed on the six variables (three translations and three rotations) which define the location of W_a with respect to the complex
$A^- - Ca^{2+} - A^-$. At the outcome of the procedure, a stable position of
the water molecule is found with an optimized distance $O_{Wa} - Ca^{2+}$ equal
to 2.35 Å, close to the value determined crystallographically of
2.38 Å. A limited lability was found for rotations of W_a around an
axis passing through Ca^{2+} and O_{Wa}. The optimized interaction energy
of W_a with the complex $A^- - Ca^{2+} - A^-$ is -27.4 kcal/mol. This value is
quite sizable. It is nevertheless appreciably weaker, on account of

the neutralization of the two charges of Ca^{2+}, than the one optimized for the interaction of a water molecule with a single Ca^{2+} ion, namely, -49.2 kcal/mol using our *ab initio* multipoles on W_a (with an optimized distance $O-Ca^{2+}$ equal to 2.28 Å).

The total stabilization energy ΔE of the resulting complex $W_a-A^--Ca^{2+}-A^-$ amounts to -584.3 kcal/mol. This value is smaller than the summed values of ΔE and $\Delta E(W_a)$, -588.3 kcal/mol. This feature is due again to the nonadditivity of the polarization contribution of the ionophores by the complexing entities, whether Ca^{2+} alone or $Ca^{2+} + W_a$. When the entity is $Ca^{2+} + W_a$, the polarizing field exerted by Ca^{2+} is opposed by the one, albeit much weaker, exerted by the water molecule, which approaches the complex through its oxygen atom; hence a weaker field and a reduced magnitude of the polarization energy of the ionophores.

Independent computations indicated, on the other hand, that there is no possibility of binding a water molecule as a seventh ligand to the cation in the crystal structure of the Mg^{2+} complex.

At this stage, the comparison of the stabilization energies of the Mg^{2+} and the Ca^{2+} complexes, including W_a in the latter, show that the difference in favor of Mg^{2+} has decreased to 5.7 kcal/mol. In fact one must furthermore take into account the energy expenditure necessary to extract the water molecule from the bulk of the solvent and transpose it to the Ca^{2+} complex. Assuming *tentatively* (see, e.g., Ref. 31) that this energy expenditure corresponds to the water-water dimerization energy (-5.3 kcal/mol as computed by the present procedure), the difference between the two complexes becomes 10 kcal/mol in favor of Mg^{2+}

Thus it appears at this point that although the presence of a water molecule as a seventh ligand to calcium in the complex brings about a large increment in the global energy, the balance between Ca^{2+} and Mg^{2+} remains nevertheless somewhat in favor of the Mg^{2+} complex.

It must be noted, however, that divalent cations are known to strongly bind and structure a second solvation shell [102,103]. This suggested an investigation of the possibility of binding two addi-

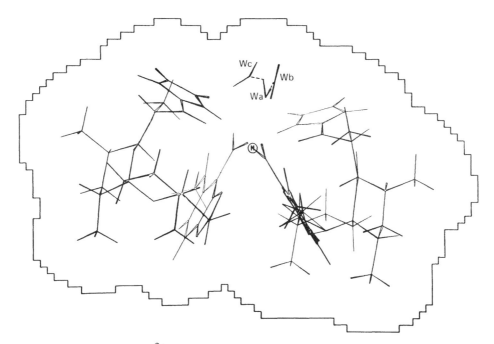

FIG. 6.   The 2:1 Ca^{2+} complex of ionophore A23187 bound to water molecules W_a, W_b, W_c.   (Reproduced from Ref. 92 with permission.)

tional water molecules, W_b and W_c, through the hydrogen atoms of W_a in a second solvation shell of Ca^{2+} by energy minimization of the set of the variables defining the orientation of W_b and W_c and simultaneous energy reminimization of the orientation of W_a in the presence of W_b and W_c.   The final configuration derived is schematized in Fig. 6.   In this configuration the binding energies of W_b and W_c amount to -17 kcal/mol for each and the stabilization energy of the whole complex W_a-W_b-W_c-A^--Ca^{2+}-A^- reaches then -612.4 kcal/mol.   The energy balance with the desolvation energy is then -212.6, i.e., 21.9 kcal/ mol larger than in the Mg^{2+} complex.   Even after subtraction of the energy admittedly necessary to detach three water molecules from the water bulk (15.9), the balance remains in favor of the calcium complex by 6 kcal/mol.

These results demonstrate therefore that the relative stabilization of the Ca^{2+} complex with respect to the Mg^{2+} complex is due to the fifth element in the energy balance outlined above, namely, the presence in the Ca^{2+} complex of water molecules coordinated with the cation. The computations suggest that this stabilization requires the intervention of not only the water molecule clearly distinguished in the crystal but of one or two supplementary such molecules bound to the former in a residual second hydration shell. That such a situation is a reasonable possibility in solution is supported by the fact that the crystal data of the Ca^{2+} complex [90] indicate the presence of an additional water molecule having a high thermal motion and probable partial occupancy, hydrogen-bonded to the well-defined calcium-bound water molecule and to the carboxylate oxygen of an adjacent complex.

6. STRUCTURAL PROPERTIES OF VALINOMYCIN AND A23187

Two important notions pertaining to the structure and reactivity of molecular systems have been developed recently which may have a bearing on the mechanism of functioning of ionophores. These are as follows:

1. The molecular electrostatic potential [104] (MEP) created in the surrounding space by the set of the nuclei and the electron density distribution of a molecule.
2. The atom accessibility, which can be defined [105] as the area of the portion of the van der Waals sphere of an atom in a molecule which can be touched by a reagent without the latter intersecting with the spheres of any other atoms.

These two notions have been shown to provide, when used simultaneously, a powerful tool for the exploration of the reactivity of biomolecules [34,106,107].

These two properties have been computed for the ionophores valinomycin [108] and A23187 [92]. The MEP was computed on the surface envelopes surrounding the investigated molecule or molecular complexes. These envelopes are formed by the intersection of spheres centered on each atom with radii proportional to their van der Waals radii with a proportionality factor of 1.7. They are looked at through planar windows carrying a fine grid of points from which perpendiculars are drawn to intersect the surface (see details in Ref. 109).

6.1. Valinomycin

The MEP and accessibilities of valinomycin have been computed [108] for two forms of the molecule, that of its K^+-complex and that of the uncomplexed form using the x-ray crystal data [110].

One of the most interesting results concerning the K^+ complexes was the demonstration that the positive potential of the K^+ ions is only moderately screened by the ionophore surrounding it so that the whole surface envelope of the K^+ complex presents appreciable positive potentials with a very strong maximum on the surface of the cation which is accessible in a pocket on the methyl-isopropyl side of the structure and another strong maximum in the narrower pocket on the other side of the molecule. The relevance of these findings to the ability of the complex to bind [60], and perhaps transport, anions is being explored further [111,112]. We will consider in more detail the results obtained for the uncomplexed form to illustrate the extent to which the notions of MEP and accessibility can be utilized to test the plausibility of some hypothesis concerning the mechanism of initiation of the ion capture, which have been advanced [113] on the basis of the observed crystal structure.

A diagrammatic drawing of the uncomplexed molecule is given in Fig. 7a. The corresponding MEP on the surface envelope is shown in FIG. 7b. The molecule is oval-shaped and has only four 1-4 H bonds of the type occurring in the K^+ complex involving four amide carbonyls; but it contains also two 1-5 H bonds involving ester carbonyls. This leaves two free amide carbonyls Q, Q', and four ester carbonyls P, P', M, M'. The mechanism envisioned implied that one of the pairs P, M, on the isopropyl-methyl face, or P', M', on the all-isopropyl face

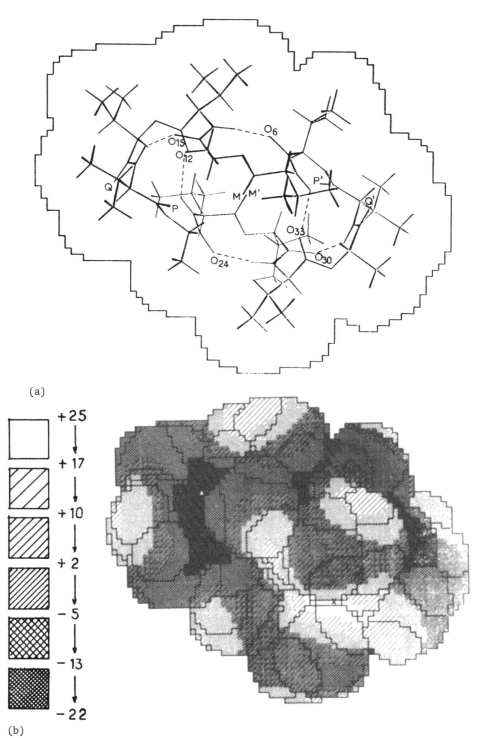

(a)

+25
↓
+17
↓
+10
↓
+2
↓
-5
↓
-13
↓
-22

(b)

FIG. 7. (a) Diagrammatic representation of the uncomplexed valinomycin molecule viewed toward face A. (b) Surface potentials of the
uncomplexed molecule (convention of shadings as indicated). (Reproduced from Ref. 108 with permission.)

could initiate the capture of K^+, leading to the consecutive rupture of 1-5 bonds and appropriate changes toward the bracelet-like conformation. Later considerations based on the similarity with the crystal structure of the analog isoleucinomycin led the authors [114] to propose the methyl-isopropyl face as more favorable for the initiation of the complex formation. Consideration of the potential and of the accessibilities can provide a clue as to the feasibility of the initiation, by allowing a distinction between the different carbonyl oxygens from the point of view of both their attractive character for a positive ion and their steric accessibilities. The essential results of the calculations are summarized as follows: among the free carbonyl oxygens, those of the amides Q and Q' have the largest negative potentials (-22 and -21 kcal/mol) and the largest accessibilities (4.9 and 3.8 Å, respectively). The four free ester carbonyl only P and P' have large potentials (-19 and -15 kcal/mol); but only P has a reasonable accessibility (of 1.4 Å). The two others M and M' are very unfavorable in potential and are poorly accessible. On the whole, the oxygens on the methyl-isopropyl face are more favored than those on the other face, a feature which would agree with Duax's contention, indicating, however, that the reasons for it are not essentially steric (accessibility) but rather electronic (potential). Furthermore the carbonyls M and M' appear rather disfavored, at least from the present static point of view. On the contrary, Q and Q' cannot a priori be eliminated although it is difficult to envision the succeeding steps if they start the complexation.

The study of the six other oxygens (involved in H bonding) from the same point of view has been instructive in showing that carbonyls involved in hydrogen bonding are not necessarily deprived either of their intrinsic attractive character for cations or of their accessibility: the two amide oxygens O_{12} and O_{24} are still accessible and are nearly as attractive as those of the free amides Q and Q', whereas the ester oxygens lose part or all of their attractivity but to a very different extent. This is a very clear demonstration of the importance of the overall molecular structure in

determining the reactivity properties of a site in a large complex
molecule [115]. Moreover only O_{30} is somewhat accessible.

It is obvious that such structural studies constitute only one
step toward the understanding of the mechanism of complexation and
that additional elements should be introduced, particularly a better
representation of conformational changes. Nevertheless it seems that
the image they give of the attractive character and accessibilities
of the carbonyl oxygens may help to put on a firmer ground some
hypothesis on the initiation of the complexation.

6.2. The 2:1 A23187-M^{2+} Complexes

In the complexes of calcimycin the use of the MEP enables to assess
the extent to which the distribution of the two negative charges
spread over the molecular skeleton is neutralized by the doubly
positive charge of the cation. Two windows have been used, A, perpendicular to a plane passing through O_1, M^{2+}, and the middle of a
segment linking the two carbonyl carbons, and B on the side of the
ketopyrrole group and perpendicular to the pseudosymmetry axis of
the complex. The profile of the limits of the surfaces for the two
unhydrated complexes viewed through the side window A can be seen
in Fig. 5. The contours and the molecular diagram of the $A^--Mg^{2+}-A^-$
complex viewed down from window B are given in Fig. 8a and a perspective view of the corresponding surface envelope is given in Fig. 8b.

The distribution of the MEP on the surface envelope of the
$A^--Mg^{2+}-A^-$ complex is illustrated on Fig. 9 for the two above-defined
views.

A wide region of negative potential can be seen in the center
of Fig. 9a. It corresponds to the surface envelope of the carboxylate
oxygen O_3. On Fig. 9b the two dominant negative regions correspond
to the free carboxylate oxygens O_3 and $O_{3'}$ viewed from above. The
potential values on the envelope extend from -20 to +20 kcal/mol.

Very similar features characterize the $A^--Ca^{2+}-A^-$ complex with
the difference, however, illustrated in Fig. 10, that the spread of

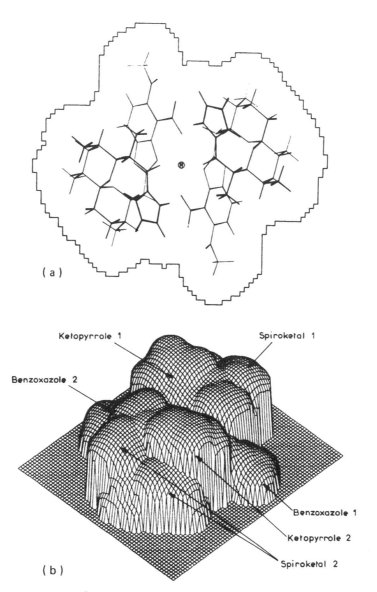

FIG. 8. The A^{-}-Mg^{2+}-A^{-} complex as viewed from window B. (a) Contours and molecular diagram. (b) Perspective view of the three-dimensional surface envelope. (Reproduced from Ref. 92 with permission.)

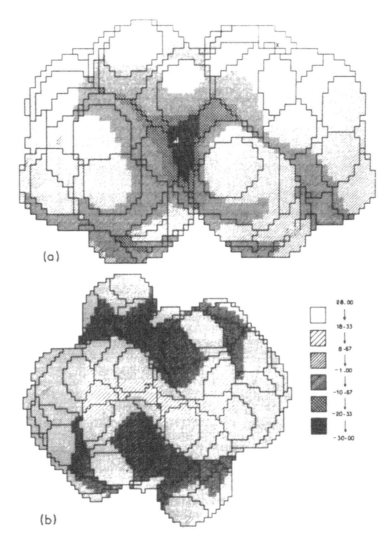

FIG. 9. The distribution of the MEP on the surface envelope of an
$A^- - Mg^{2+} - A^-$ complex. (a) Viewed from window A. (b) Viewed from
window B. (Reproduced from Ref. 92 with permission.)

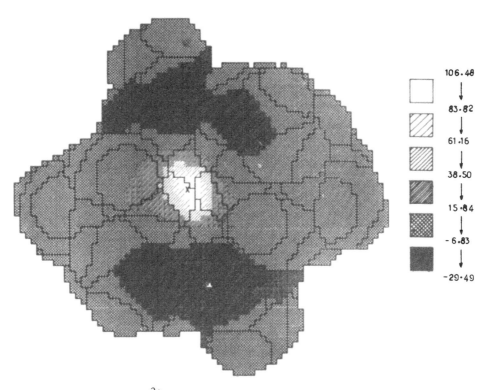

FIG. 10.   The $A^- \text{-} Ca^{2+} \text{-} A^-$ complex as viewed from window B.   Distribution of the MEP on the surface envelope.   (Reproduced from Ref. 92 with permission.)

the potential values over the surface envelope is much larger, going from -29 to -106 kcal/mol. The large positive value which is located in the center of the figure corresponds to the surface envelope of the ion, visible in a wide opening of the complex, as can be seen on the perspective view of the surface envelope given in Fig. 11a. On the other hand, two negative minima appear again on the surface envelopes of the carboxylate oxygens O_3 and O_3', about 5 kcal/mol deeper than in the magnesium complex.

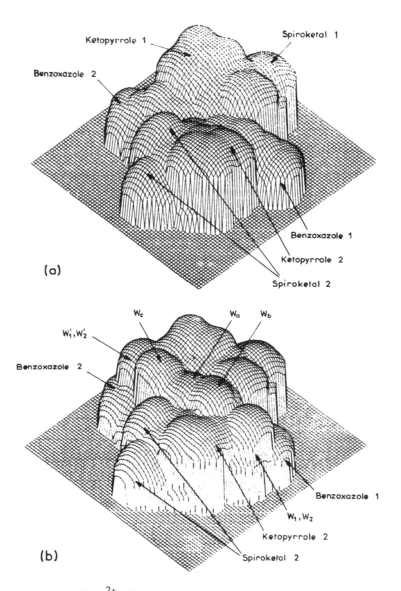

FIG. 11. The A^--Ca^{2+}-A^- complex as viewed from window B. (a) Perspective view of the surface envelope. (b) Perspective view of the surface envelope of the complex carrying W_a, W_b, W_c and two water molecules on each carboxylate. (Reproduced from Ref. 92 with permission.

The strong positive potential on the surface envelope of the
calcium ion indicates a large attraction toward a nucleophile, in
agreement with the observation that a water molecule is fixed in
this position in the crystal structure and our own computations,
summarized in Sec. 5, which indicated a significant stabilization
energy for the water molecule denoted W_a.

The location of two wide regions of negative potential on the
surface envelope of the free carboxylate oxygens in the two complexes
indicates the existence of a nonnegligible attraction for an electro-
phile in these regions. Its reality is confirmed by the observation
already mentioned that in the crystal a water molecule forms a bridge
between the molecule bound to the calcium cation and a carboxylate
oxygen on a neighbor complex. It suggests that in solution molecules
of water or solvent could hydrogen-bond to these oxygens. Such a
possibility has been explored [92] using our energy procedure, and
the results of the computations after energy minimization have shown
that binding of water is indeed possible and that despite the formal
"neutralization" of the anionic net charges of the two carboxylates
by the divalent ion, each of the two free carboxylate oxygens O_3 and
$O_{3'}$ can bind *two* water molecules with a sizable affinity, the individual stabilization energies ranging from 6.5 to 9.2 kcal/mol. This
is true in the unhydrated Mg^{2+} and Ca^{2+} complexes and remains true in
the Ca^{2+} complex carrying already either W_a alone or W_a, W_b, and W_c
in the positions investigated above in the intermolecular computations. Such results can be put in parallel to recent observations
on the apparent existence of a "residual hydrophilic character" in
some "external" zones of complexes of monensin (a monocarboxylic
ionophore) with Na^+ [116].

In relation to the possibility of binding water molecules to
the calcium complex, it is interesting to look at a perspective view
of the surface envelope of the Ca^{2+} complex carrying W_a, W_b, and W_c
plus two water molecules on each carboxylate, given in Fig. 11b.
Its comparison to Fig. 11a shows clearly that the "groove" of the

Ca^{2+} complex can easily accommodate not only the water molecule
bound directly to the ion but also two water molecules bound to the
first one, without this producing a local protrusion in the overall
surface envelope. Moreover the easy insertion of the water molecules
bound to each carboxylate is also visible in Fig. 11b where these
molecules are seen to fit into the cliff existing between the benzoxazole group of the first ionophore and the ketopyrrole of the
second.

These observations together with the appreciable interaction
energies obtained for water in the different sites raise the question
of the possibility that part of these molecules, if not all, could
remain bound to the respective complexes during the transport across
membranes [92]. The possibility of a cotransport of water molecules
was suggested [117] in the case of another ionophore, valinomycin.

7. ENERGY PROFILES FOR SINGLE AND DOUBLE OCCUPANCY BY Na^+ OF THE GRAMICIDIN A CHANNEL

The three ionophores investigated above are representatives of transmembrane cation *carriers*. We would like to describe briefly a recent
application [118] of the same theoretical methodology to the study of
a transmembrane *channel*: gramicidin A.

A head-to-head $\beta_{3.3}^{6.3}$ dimer was proposed [119,120] for the structure of the channel and rather precise helical parameters for this
model were deduced on this basis [120].

Although unanimity does not exist as to the dominant structure
of gramicidin A in the conducting state [121], the consideration of
the head-to-head dimer provides an interesting model for studying at
the microscopic level the possible mechanisms involved in channel
transport. The above-mentioned theoretical study aimed at a determination of the energy profile felt by the ion(s) inside the channel,
making use of a precise geometry (including all atoms) for the channel
backbone, generated from the structural data proposed in Ref. 120, and
applying the methodology described in Section 2 for the evaluation

of the energy of the ion(s) inside this backbone. Previous attempts
[122-129] at understanding the mechanism of ion-translocation in
gramicidin A have concentrated more on the possibilities of the
ion to jump over the successive barriers in the energy profile
than on an accurate determination of the energy profile itself,
often assuming it to be satisfactorily represented by the effect
of an array of dipolar ligand groups arranged so as to mimic the
disposition of the peptide carbonyls in the assumed structure.

It seems that a more realistic representation of the energy
profile can be obtained using the complete skeleton of the molecule
and introducing all the terms in the energy computations.

To derive the necessary multipolar expansions on gramicidin A,
each monomer was constructed from eight dipeptide subunits,
-CONHCH$_2$CONHCH$_2$- with alternate L and D conformations at the C$_\alpha$
carbons using the ϕ,ψ dihedral angles of reference [120], and
standard bond lengths and angles. The orientation of the second
monomer was obtained so as to form six appropriate CO-NH intermolecular hydrogen bonds and minimizing geometrically the repulsion
between the two formyl hydrogens. The first exploratory study has
dealt only with the oligopeptide backbone, the conformation of which
was kept rigid. Single and double occupancy for a sodium cation
were both considered.

7.1. Single Occupancy

The energy profile for a sodium ion in the channel was computed in
two steps: (1) constraining the ion to remain on the channel axis,
computing the ion-channel interaction energy at successive points
in steps of 0.5 Å; (2) allowing Na$^+$ to optimize its position in a
plane perpendicular to the axis at every previous point. The resulting energy profiles are given as curves A and B of Fig. 12.

Curve A of Fig. 12 indicates that the energy profile for one
ion constrained to remain on the axis presents two symmetric minima

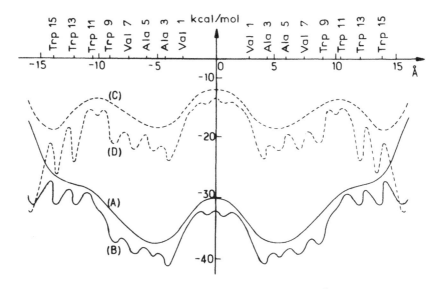

FIG. 12. Energy profile for single occupancy by Na$^+$ of the gramicidin
A channel. (A) = ion constrained to remain on the channel axis.
(B) = ion allowed to reach its preferred position at each point (see
text). (C) = pure electrostatic component corresponding to curve A.
(D) = pure electrostatic component corresponding to curve B (energies
in kcal/mol, distances in ångstroms). The location of the carbonyl
oxygens for the L residues is indicated for convenience. (Reproduced
from Ref. 118 with permission.)

with a central relatively small barrier, two higher barriers at the
exterior of the channel near its mouths, and two inflexion points
inside the channel close to the ends. When the ion is allowed to
find its preferred position at each step while progressing along the
channel, local minima and maxima appear in the energy profile, bringing about an appreciable overall lowering in energy near the extrem-
ities. The first minimum (-31.4 kcal/mol) is external to the channel,
with Na$^+$ bound to the carbonyl oxygen of Trp$_{15}$ at 2.02 Å from it,
slightly toward the axis. This value, which corresponds to single
binding of the ion to the most external carbonyl (in the presence of
the whole backbone), is 5 kcal/mol more negative than the corresponding energy of single binding of Na$^+$ to water computed by the same
method. The location of a minimum just outside the channel entrance

and of an inner deeper one agrees with the conclusions reached in Ref. 7 in the framework of a three-barriers, four-sites model.

It is notable that the profile obtained for the total cation-channel intermolecular interaction energy ΔE differs appreciably from the one deduced from the sole electrostatic contribution E_{MTP}, the separate evolution of which is shown in curves C and D of Fig. 12. The numerical values of ΔE are appreciably larger than those of E_{MTP}, the location of the absolute minimum of ΔE is closer toward the center of the channel, whereas that of E_{MTP} is toward the exterior of the channel, with a local minimum near the center. These differences are due to the polarization contribution E_{pol} which is proportional to the square of the electrostatic field exerted on the backbone by the positive charge of the cation; it increases regularly as the cation moves toward the center of the channel, translating the fact that an increasing number of polarizable sites are affected by its proximity.

7.2. Double Occupancy

Double occupancy by two Na^+ ions was considered in two ways: first the effect on the energy of the first ion entered, I_1, of a second ion I_2 placed at various spots along the axis was computed. The corresponding profiles for eight locations of I_2 are given in Fig. 13. Then, in view of the evaluation of the energy profile for double occupancy the *global* interaction energy ΔE was computed in the system gramicidin A plus two ions placed in symmetric positions on the axis. From this value the energy felt by each ion in the channel *in the presence of its partner* was obtained (see Ref. 118). This quantity plotted as a function of the position of the ions gives the energy profile of Fig. 14. In these computations the two ions have been constrained to remain on the axis.

The curves of Fig. 13 indicate the effect of the presence of a second ion on the energy profile of the first one: the most striking feature of these curves is the disappearance of the central

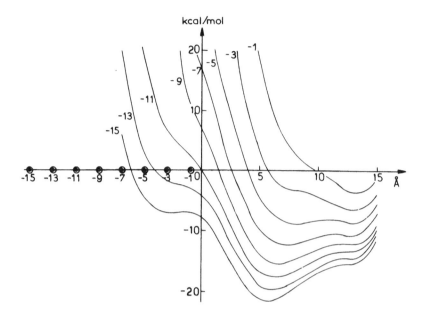

FIG. 13. The energy of I_1 moving along the axis under the influence
of a second ion I_2 at a fixed position. Each curve is labeled according to the position of I_2 marked by a correspondingly labeled dot.
Units as in Fig. 12.   (Reproduced from Ref. 118 with permission.)

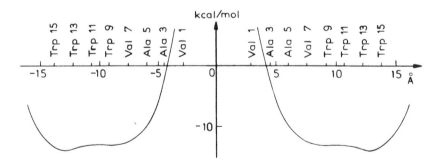

FIG. 14. The energy profile for symmetric double occupancy by Na^+
ions constrained to the channel axis. Units as in Fig. 12. (Reproduced from Ref. 118 with permission.)

energy barrier, a disappearance already practically achieved when I_2 is at -15 Å from the center, close to the entrance. For this location of I_2 an appreciable (although less deep than for I_1 alone) minimum remains at first on the right side of the center. When I_2 progresses, the minimum for I_1 becomes less and less negative and glides toward the exit while the barrier to exit becomes gradually smaller. Thus, upon progression of the second ion, the first one will have a tendency to move along the channel to reach its successive optimal positions and finally to leave. One notes that when I_2 reaches about -6 Å from the center, the position of I_1 becomes quasi-indifferent between +7 and +14 Å.

As concerns now the energy profile for double occupancy (Fig. 14) it is characterized by the presence of two symmetric energy wells separated by a high central barrier. The absolute minimum is at ±13 Å from the center and a quite flat region extends between ±13 and ±8 Å. It is satisfying that the location of these minima is in good agreement with the indications deduced from ion-induced ^{13}C chemical shifts of the carbonyl carbons in the gramicidin channel, which point to the involvement of the Trp 9, 11, and 13 carbonyls in the binding of Na^+ [130]. A more precise location will be obtained by allowing the ion to depart from the channel axis.

These results on the gramicidin A channel, although undoubtedly significant for the geometric model adopted here, must nevertheless be considered as preliminary. Refinement work is actively under way [131].

8. CONCLUSIONS

One of the most striking results pertains to the variability of the main factors responsible for the selective complexations in different types of ionophores. Thus in valinomycin the values of the interaction energies with the cavity and those of the desolvation energies of the cation rank in the same order from Na^+ to Cs^+, an order different from the experimental order of complexation; their differences, however, modify this order yielding the experimental one. In non-actin the same factors play similarly in the K^+/Na^+ selectivity but their interplay is not sufficient to account for the specificity observed and it is the different repulsions between the liganding groups in the two different cavities which are the decisive element. For the couple NH_4^+/K^+ the selectivity is already present in the cation-ionophore binding energies and desolvation does not change the order. In the A23187 ionophore, it is the water molecule(s) bound to the Ca^{2+} complex but absent in the Mg^{2+} complex which preferentially stabilizes, albeit slightly, the Ca^{2+} complex.

The examples presented here clearly show that the outcome of the energy balances imply generally relatively small differences between large numbers. Nevertheless these differences reproduce the order of selectivities in all the cases explored. This result can be attributed not only to the accuracy and care with which the estimation of the intermolecular interaction energies has been cali-brated on model *ab initio* computations, but also to the theoretical soundness of the expansion into analytic formulas of the different contributions of ΔE. At the present stage of development we believe that this reliability is ensured by the representation of the electro-static and the polarization contributions E_{MTP} and E_{pol} by means of a multicenter multipolar expansion of the *ab initio* SCF wave functions of the constituents subunits and by that of E_{rep} by a sum of bond-bond repulsions.

ABBREVIATIONS

A^-	deprotonated A23187 ionophore
A23187 ionophore	calcimycin
E_{MTP}	multipole-multipole interactions
MEP	molecular electrostatic potential
MTP	multipole expansion
OMTP	overlap MTP
SCF	self-consistent field

REFERENCES

1. G. Eisenman, S. Krasne, and S. Ciani, *Ann. N.Y. Acad. Sci. USA,*
    264, 34 (1975).

2. W. Lütz, P. Früh, and W. Simon, *Helv. Chim. Acta, 54,* 2767
    (1971).

3. C. Züst, P. Früh, and W. Simon, *Helv. Chim. Acta, 56,* 495
    (1973).

4. E. Kauffmann, J. M. Lehn, and J.-P. Sauvage, *Helv. Chim. Acta,*
    59, 1099 (1976).

5. B. Tümmler, G. Maas, F. Vogtle, H. Sieger, U. Heimann, and E.
    Weber, *J. Am. Chem. Soc., 101,* 2588 (1979).

6. Y. Pointud, J. Juillard, G. Jeminet, and L. David, *J. Chim.*
    Phys., Phys. Chim. Biol., 79, 67 (1982).

7. See for instance, G. Eisenman and R. Horn, *J. Membr. Biol., 76,*
    197 (1983) for a recent historical overview.

8. W. Simon, W. Morf, and C. Meier, *Struct. Bond., 16,* 113 (1973).

9. J.-M. Lehn, *Struct. Bond., 16,* 1 (1973).

10. Y. A. Ovchinnikov, V. T. Ivanov, and A. M. Shkrob, in *Membrane-*
    Active Complex, BBA Library, V. 12, Elsevier, Amsterdam, 1974,
    p. 229.

11. G. Eisenman, *Biophys. J., 2,* 259 (1962).

12. G. Eisenman and S. Krasne, *MTP Rev. Sci. Biochem.,* Ser. 2, 27
    (1975).

13. G. Eisenman, G. Szabo, S. McLaughlin, and S. Ciani, in *Molecu-*
    lar Mechanisms of Antibiotic Action on Protein Synthesis and
    Membranes (Proc. of a Symp. in Granada) (E. Muñoz, F. Garcia-
    Ferrandiz, and D. Vazquez, eds.), Elsevier, Amsterdam, 1972,
    p. 545.

14. W. Morf and W. Simon, *Helv. Chim. Acta, 54,* 794 (1971).

15. W. Morf and W. Simon, *Helv. Chim. Acta, 54,* 2683 (1971).

16. W. Morf, C. Züst, and W. Simon, in *Molecular Mechanisms of Anti-*
    biotic Action on Protein Synthesis and Membranes (Proc. of a
    Symp. in Granada) (E. Muñoz, F. Garcia-Ferrandiz, and D. Vazquez,
    eds.), Elsevier, Amsterdam, 1972, p. 523.

17. N. Gresh, P. Claverie, and A. Pullman, *Int. J. Quantum Chem.*
    Symp., 13, 243 (1979).

18. P. Ausloos (ed.), *Kinetics of Ion-Molecule Reactions,* Plenum
    Press, New York, 1979.

19. See, for instance, B. Pullman (ed.), *Intermolecular Interactions:*
    From Diatomics to Biopolymers, Wiley, New York, 1978.

20. H. Pericaudet and A. Pullman, *FEBS Lett.*, *34*, 222 (1973).

21. H. Berthod and A. Pullman, *Chem. Phys. Lett.*, *70*, 134 (1980) and references therein.

22. M. Dreyfus and A. Pullman, *Theor. Chim. Acta*, *19*, 20 (1970).

23. A. Pullman, H. Berthod, and N. Gresh, *Int. J. Quant. Chem. Symp.*, *10*, 59 (1976).

24. B. Pullman, N. Gresh, H. Berthod, and A. Pullman, *Theor. Chim. Acta*, *44*, 151 (1977).

25. See P. Claverie in Ref. 19, p. 69 ff.

26. M. Dreyfus, Thèse 3è Cycle, Paris, 1970.

27. A. Goldblum, D. Perahia, and A. Pullman, *J. Quant. Chem.*, *15*, 121 (1979).

28. A. Pullman and D. Perahia, *Theor. Chim. Acta*, *48*, 29 (1978).

29. N. Gresh and B. Pullman, *Theor. Chim. Acta*, *52*, 67 (1979).

30. N. Gresh, *Biochim. Biophys. Acta*, *597*, 345 (1980).

31. N. Gresh and B. Pullman, *Biochim. Biophys. Acta*, *608*, 47 (1980).

32. J. Langlet, P. Claverie, F. Caron, and J.-C. Boeuve, *Int. J. Quant. Chem.*, *20*, 299 (1981).

33. A. Pullman, K. Zakrzewska, and D. Perahia, *Int. J. Quant. Chem.*, *16*, 395 (1979).

34. A. Pullman and B. Pullman, *Quart. Rev. Biophys.*, *14*, 289 (1981).

35. H. Brockmann and G. Schmidt-Kastner, *Chem. Ber.*, *88*, 57 (1955).

36. M. Shemyakin, E. Vinogradova, M. Feigina, and N. Aldanova, *Tetrahedron Lett.*, *6*, 351 (1963).

37. H. Brickmann, M. Springorum, G. Traxler, and I. Hofer, *Naturwissenschaften*, *22*, 689 (1963).

38. W. McMurray and R. Begg, *Arch. Biochem. Biophys.*, *84*, 546 (1959).

39. C. Moore and B. C. Pressman, *Biochem. Biophys. Res. Commun.*, *15*, 562 (1964).

40. B. C. Pressman, *Proc. Natl. Acad. Sci. USA*, *53*, 1076 (1965).

41. B. C. Pressman, E. J. Harris, W. S. Jagger, and J. H. Johnson, *Proc. Natl. Acad. Sci. USA*, *58*, 1949 (1967).

42. D. Tosteson, P. Cook, T. Andreoli, and M. Tieffenberg, *J. Gen. Physiol.*, *50*, 2513 (1967).

43. T. Andreoli, M. Tieffenberg, and D. Tosteson, *J. Gen. Physiol.*, *50*, 2527 (1967).

44. P. Mueller and D. O. Rudin, *Biochem. Biophys. Res. Commun.*, *26*, 398 (1967).

45. M. Shemyakin, Y. A. Ovchinnikov, V. T. Ivanov, V. K. Antonov, E. I. Vinogradova, A. M. Shkrob, G. G. Malenkov, A. V. Evstratov, I. A. Laine, E. I. Melnik, and I. D. Ryabova, *J. Membr. Biol., 1,* 402 (1969).

46. E. Grell, T. Funck, and F. Eggers, in *Molecular Mechanisms of Antibiotic Action on Protein Synthesis and Membranes* (Proc. of Symp. in Granada) (E. Muñoz, F. Garcia-Ferrandiz, and D. Vazquez, eds.), Elsevier, Amsterdam, 1972, p. 656.

47. Y. A. Ovchinnikov, V. T. Ivanov, and A. M. Shkrob, in *Molecular Mechanisms of Antibiotic Action on Protein Synthesis and Membranes* (Proc. of Symp. in Granada) (E. Muñoz, F. Garcia-Ferrandiz, and D. Vazquez, eds.), Elsevier, Amsterdam, 1972, p. 459.

48. Y. A. Ovchinnikov and V. T. Ivanov, *Tetrahedron, 30,* 1871 (1974).

49. E. Grell, T. Funck, and H. Sauter, *Eur. J. Biochem., 34,* 415 (1973).

50. D. Haynes, A. Kowalsky, and B. C. Pressman, *J. Biol. Chem., 244,* 502 (1968).

51. P. Läuger, *Science, 178,* 24 (1972).

52. R. Benz and P. Läuger, *J. Membr. Biol., 27,* 171 (1976).

53. E. Grell, F. Eggers, and T. Funck, *Chimia, 26,* 632 (1972).

54. M. Pinkerton, L. K. Steinrauf, and P. Dawkins, *Biochem. Biophys. Res. Commun., 35,* 512 (1969).

55. K. Neupert-Laves and M. Dobler, *Helv. Chim. Acta, 58,* 432 (1975).

56. N. Gresh, C. Etchebest, O. de la Luz Rojas, and A. Pullman, *Int. J. Quant. Chem., Quant. Biol. Symp., 8,* 109 (1981).

57. G. Hedwig and A. Parker, *J. Am. Chem. Soc., 96,* 6589 (1974).

58. R. Staley and J. Beauchamp, *J. Am. Chem. Soc., 97,* 5920 (1975).

59. L. K. Steinrauf, J. Hamilton, and M. Sabesan, *J. Am. Chem. Soc., 104,* 4085 (1982).

60. J. A. Hamilton, M. N. Sabesan, and L. K. Steinrauf, *J. Am. Chem. Soc., 103,* 5880 (1981).

61. R. Corbaz, L. Ettlinger, E. Gaumann, W. Keller-Schierlein, F. Kradolfer, L. Neipp, V. Prelog, and H. Zähner, *Helv. Chim. Acta, 38,* 1445 (1955).

62. J. Dominguez, J. D. Dunitz, H. Gerlach, and V. Prelog, *Helv. Chim. Acta, 45,* 129 (1962).

63. J. Beck, H. Gerlach, V. Prelog, and W. Voser, *Helv. Chim. Acta, 45,* 620 (1962).

64. S. Graven, H. Lardy, D. Johnson, and A. Rutter, *Biochemistry, 5,* 1729 (1966).

65. D. Haynes and B. C. Pressman, *J. Membr. Biol., 18,* 1 (1974).

66. G. Eisenmann, S. Ciani, and G. Szabo, *J. Membr. Biol.*, *1*, 294 (1969).

67. M. Feinstein and H. Felsenfeld, *Proc. Natl. Acad. Sci. USA, 68*, 2037 (1971). '

68. J. Prestegard and S. I. Chan, *J. Am. Chem. Soc.*, *92*, 4440 (1970).

69. J. Prestegard and S. I. Chan, *Biochemistry, 8*, 3921 (1969).

70. E. Pretsch, M. Vašak, and W. Simon, *Helv. Chim. Acta, 55*, 1098 (1972).

71. B. Kilbourn, J. D. Dunitz, L. Pioda, and W. Simon, *J. Mol. Biol.*, *30*, 559 (1967).

72. M. Dobler, J. D. Dunitz, and B. Kilbourn, *Helv. Chim. Acta, 52*, 2573 (1969).

73. M. Dobler and R. Phizackerley, *Helv. Chim. Acta, 57*, 664 (1974).

74. N. Gresh and A. Pullman, *Int. J. Quant. Chem.*, *22*, 709 (1982).

75. Y. Nawata, T. Sakamaki, and Y. Iitaka, *Chem. Lett. Japan*, 151 (1975).

76. C. Detellier and P. Laszlo, *Helv. Chim. Acta, 162*, 1559 (1979).

77. C. Detellier and P. Laszlo, *J. Am. Chem. Soc.*, *102*, 1135 (1980).

78. D. Aue, H. Webb, and M. Bowers, *J. Am. Chem. Soc.*, *98*, 318 (1976).

79. R. Noyes, *J. Am. Chem. Soc.*, *84*, 513 (1962).

80. P. Reed and H. Lardy, *J. Biol. Chem.*, *247*, 6970 (1972).

81. D. Pfeiffer, P. Reed, and H. Lardy, *Biochemistry, 13*, 4007 (1974).

82. See, e.g., *The Polyether Ionophores* (J. Westley, ed.), Marcel Dekker, New York, 1982, and references therein.

83. H. Rasmussen and D. Goodman, *Physiol. Rev.*, *57*, 421 (1977).

84. R. Kretsinger, *Adv. Cyclic Nucleotide Res.*, *11*, 1 (1979).

85. M. Anteunis, *Bioorganic Chem.*, *6*, 1 (1977).

86. C. Deber and D. Pfeiffer, *Biochemistry, 15*, 132 (1976).

87. D. Pfeiffer and C. Deber, *FEBS Lett.*, *105*, 360 (1979).

88. J. Puskin and T. Gunter, *Biochemistry, 14*, 187 (1975).

89. M. Chaney, P. Demarco, N. Jones, and J. Occolowitz, *J. Am. Chem. Soc.*, *96*, 1932 (1974).

90. G. Smith and W. Duax, *J. Am. Chem. Soc.*, *98*, 1578 (1976).

91. A. Alleaume, G. Jeminet, and Y. Barrans, *Acta Crystallogr.*, section B, in press; private communication.

92. N. Gresh and A. Pullman, *Int. J. Quant. Chem., Quant. Biol. Symp.*, *10*, 215 (1983).

93. G. Krause, E. Grell, A. M. Albrecht-Gary, D. Boyd, and J. P. Schwing, in *Transmembrane Ion Motions. Proceedings of the 36th International Meeting of the Société de Chimie Physique*, Paris, 1982 (G. Spach, ed.), Elsevier, Amsterdam, p. 255 ff.

94. D. Pfeiffer and H. Lardy, *Biochemistry, 15*, 935 (1976).

95. D. Pfeiffer, R. Taylor, and H. Lardy, *Ann. N.Y. Acad. Sci. USA, 307*, 402 (1978).

96. R. Nakashima, R. Dordick, and K. Garlid, *J. Biol. Chem., 257*, 12540 (1982).

97. G. Pohl, R. Kreikenbohm, and K. Seuwen, *Z. Naturforsch. Biosci., 23C*, 562 (1980).

98. M. Debono, R. Molloy, D. Dorman, J. Paschal, D. Babcok, C. Deber, and D. Pfeiffer, *Biochemistry, 20*, 6865 (1981).

99. A. Caswell and B. C. Pressman, *Biochem. Biophys. Res. Commun., 49*, 292 (1972).

100. C. Teissier, J. Juillard, M. Dupin, and G. Jeminet, *J. Chim. Phys., Phys. Chim. Biol., 76*, 611 (1979).

101. S. Goldman and R. Bates, *J. Am. Chem. Soc., 94*, 1476 (1972).

102. H. Veillard, *J. Am. Chem. Soc., 99*, 7194 (1977).

103. H. Berthod, A. Pullman, and B. Pullman, *Int. J. Quant. Chem. Quant. Biol. Symp., 5*, 79 (1978).

104. R. Bonnacorsi, E. Scrocco, and J. Tomasi, *J. Chem. Phys., 52*, 5270 (1970).

105. R. Lavery, A. Pullman, and B. Pullman, *Int. J. Quant. Chem., 20*, 49 (1981).

106. R. Lavery and B. Pullman, *Nucl. Acids Res., 9*, 3765 (1981).

107. B. Pullman, R. Lavery, and A. Pullman, *Eur. J. Biochem., 124*, 229 (1981).

108. C. Etchebest, R. Lavery, and A. Pullman, *Studia Biophys., 90*, 7 (1982).

109. R. Lavery and B. Pullman, *Int. J. Quant. Chem., 20*, 259 (1981).

110. W. Duax, H. Hauptman, C. Weeks, and D. Norton, *Science, 176*, 911 (1972).

111. C. Etchebest, *Thèse de 3ème Cycle*, Paris (1982).

112. C. Etchebest and A. Pullman, to be published.

113. G. Smith, W. Duax, D. Lang, G. de Titta, J. Edmonds, D. Rohrer, and C. Weeks, *J. Am. Chem. Soc., 97*, 7242 (1975).

114. V. Pletnev, N. Galitskii, G. Smith, C. Weeks, and W. Duax, *Biopolymers, 19*, 1517 (1980).

115. B. Pullman, in *Catalysis in Chemistry and Biochemistry, Theory and Experiment*, 1979, p. 1.

116. Y. Barrans, M. Alleaume, and G. Jeminet, *Acta Crystallogr.*, *B38*, 1144 (1982).

117. D. G. Levitt, S. R. Elias, and J. M. Hautmann, *Biochim. Biophys. Acta*, *512*, 436 (1978).

118. A. Pullman and C. Etchebest, *FEBS Lett.*, *163*, 199 (1983).

119. D. Urry, *Proc. Natl. Acad. Sci. USA*, *68*, 672 (1972).

120. D. Urry, C. Venkatachalam, K. Prasad, R. Bradley, G. Parenti-Castelli, and G. Lenaz, *Int. J. Quantum Chem., Quantum Biol. Symp.*, *8*, 385 (1981).

121. Y. A. Ovchinnikov and V. T. Ivanov, in *Conformation in Biology* (R. Srinivasan and R. Sarma, eds.), Adenine Press, New York, 1982, p. 155.

122. A. Parsegian, *Nature*, *221*, 844 (1969).

123. D. G. Levitt, *Biophys. J.*, *22*, 209, 221 (1978).

124. P. Läuger, *Biophys. Chem.*, *15*, 89 (1982).

125. H. Monoï, *J. Theor. Biol.*, *102*, 69 (1983).

126. W. Fischer, J. Brickmann, and P. Läuger, *Biophys. Chem.*, *13*, 105 (1981).

127. J. Brickmann and W. Fischer, *Biophys. Chem.*, *17*, 245 (1983).

128. W. Fischer and J. Brickmann, *Biophys. Chem.*, *18*, 323 (1983).

129. H. Schröder, in *Physical Chemistry of Transmembrane Ion Motion. Proceedings of the 36th International Meeting of the Société de Chimie Physique*, Paris, 1982 (G. Spach, ed.), Elsevier, Amsterdam, p. 425.

130. D. Urry, K. Prasad, and T. Trapane, *Proc. Natl. Acad. Sci. USA*, *79*, 390 (1982).

131. (Added in proof.) See refined results in: C. Etchebest, S. Ranganathan, and A. Pullman, *FEBS Letters*, *163*, 199 (1984).

AUTHOR INDEX

Numbers in parentheses are reference numbers and indicate that an author's work is referred to although his name may not be cited in the text. Underlined numbers give the page on which the complete reference is listed.

387

G

SUBJECT INDEX

A

Absorption bands and spectra
(*see also* UV absorption
spectra), 59-61, 87, 88, 90,
104, 148, 305-307
π-Acceptor ligands, 87
Acetate (or acetic acid), 231
buffer, 60
α-hydroxyphenyl-, 167
Acetone, benzoyl-, 24
Acetophenone, 2-hydroxy-, 25
Acidity constants (*see also*
Microconstants), 23-34, 41-
43, 56, 58, 297
apparent, 261
Actin(s), 159, 164
macrolide, 7
non-, *see* Nonactin
Actinomyces, 231, 314
viridaris, 325
Actinomycin, 55
D, 93, 94
Actinomydura, 231
Actinonin, 329
Activation energy, 191
Adenine (and residues), 96
nucleotides (*see also* indi-
vidual names), 10
5'-ADP, 2, 4, 48
Adenosine 5'-diphosphate, *see*
5'-ADP
Adenosine 5'-triphosphate, *see*
5'-ATP
Adrenal, 287
Adriamycin, *see* Doxorubicin
Aerobactin, 317-319
Aflatoxin B$_1$, 94
Agar plate test, 316
Alafosfalin, 297
Alamethicin, 111, 141

α-Alanin (and residues)
amide, 84
β-amino, 82, 103, 104
β-chloro-D-, 297
D-, 297, 298
D-, analogs, 300
β-fluoro-D-, 297, 298, 309
L-, 176
ligase, 298
racemase, 298, 299
Albomycin, 314, 317, 319-323
antibiotic activity, 322
-resistant mutants, 322
structure, 321
Alborixin, 248, 286
transport selectivities, 258
Albumin
human serum, 86
serum, 319
Alkali ions (*see also* individual
names), 115-122, 149, 174,
208, 224, 253, 263, 268, 337,
341-344, 348
Alkaline earth ions (*see also*
individual names), 122-129,
238, 263, 268, 341, 348
Alkylating agents or drugs (*see
also* individual names), 298
Alopecia, 71
Aluminium(III), 37
Amines (*see also* individual names),
259
ethanol-, *see* Ethanolamine
steryl-, 258
Amino acids (*see also* individual
names)
D- (*see also* individual names),
175, 176, 295-309
residues, acylation of, 298
with antibiotic properties, 295-
309, 328

411

Printed and bound by CPI Group (UK) Ltd, Croydon, CR0 4YY

17/10/2024

01775696-0017